W9-DJK-576

MODERN GEOGRAPHY

AN ENCYCLOPEDIC SURVEY

Garland Reference Library of Humanities
(Vol. 1197)

Modern GEOGRAPHY

AN ENCYCLOPEDIC SURVEY

Edited by

Gary S. Dunbar
Professor of Geography (emeritus)
University of California, Los Angeles

Garland Publishing, Inc.

New York & London 1991

Library of Congress Cataloging-in-Publication Data

Modern geography : an encyclopedic survey / edited by Gary S. Dunbar.
p. cm. — (Garland reference library of the humanities ; vol. 1197)
Includes bibliographical references.
ISBN 0-8240-5343-5 (alk. paper)
1. Geography—Dictionaries. I. Dunbar, Gary S. II. Series.
G63.M57 1991
910′.3—dc20 90-3742
CIP

Design by Renata Gomes

Printed on acid-free, 250-year-life paper.
Manufactured in the United States of America.

Table of Contents

This book is dedicated to the five scholars to whom I owe the greatest intellectual debts:

Charles Julian Bishko

Henry Clifford Darby

Emyr Estyn Evans

Fred Bowerman Kniffen

John Kirtland Wright

Introduction

My intention has been to present an overview of developments in the field of geography from about 1890 to the present, with emphasis on personalities, institutions, major concepts, subfields, and the evolution of the discipline in various countries. Some excellent dictionaries of geography have been published in recent years, but no reference work has yet surveyed all the topics covered in this book. I have been extremely fortunate in having the cooperation of nearly 100 contributors. When I wrote to potential authors, I received almost unanimous acceptances, probably because it is difficult to turn down an assignment when the editor asks for an essay of only a few hundred words. Many of the contributors missed the deadlines that I originally set, but through reminding, cajoling, wheedling, and sometimes bribing, I eventually managed to extract most of the promised essays. I wrote the unsigned entries myself. If I had had more time and greater resources, I might have been able to produce a more comprehensive volume, but I think that the present work will be of considerable value to geographers and to all others who want to know how the field took shape. Almost two years will have passed between the submission of the first essay and the final publication of the book, and, although an effort has been made to update entries, the volume will inevitably include some data that are already obsolete.

I had originally hoped to get by with a minimum of editing, partly out of laziness and perhaps partly out of cowardice, but I was overruled by the publisher's concern for uniformity. May I be forgiven by the native speakers of those archaic dialects used in the islands on the north side of La Manche for Americanizing their spellings and punctuation and otherwise violating their sense of linguistic propriety.

I tried to restrain most entries to a few terse paragraphs in the hope of keeping down the size (and therefore the price) of the volume, of giving brief factual presentations that avoid polemic as much as possible, and of enticing contributors who might have rejected a longer assignment. Many of the writers were unable to rein in their muse, and their essays had to be pared down. Readers should therefore not be so narrow-minded as to think that these short entries represent complete introductions to the subject or that their lengths are an indicator of the editor's or authors' assessment of their relative importance.

The contributors were asked to include brief bibliographies—a maximum of six items for the longer essays or three citations for the shorter entries (those of 250 words or less). I sometimes had to add or subtract a few bibliographical items to fit my procrustean frame. I had hoped that the authors would add some statements about the historical evolution of their subject and also comments on international practice rather than confine their essays simply to the contemporary scene in their own country, but not all contributors attempted such broad coverage. Many said that they could not in good conscience comply with my request to include mention of some of the leading practitioners, for fear of stepping on tender toes.

Perhaps unfortunately, I did not exhibit a similar fear in choosing the subjects of biographical sketches, which might turn out to be the most controversial parts of the book. There will be many readers who will be disappointed in not finding themselves singled out for special mention and will therefore criticize the book for its incompleteness or lack of balance. I tried to make the biographical entries concise and uniform, containing the following data: full name, date of birth (and death, if deceased), highest academic qualification, institutional affiliations, major publications (maximum of four titles), and some biographical references. Obviously the book does not include all of the important geographers of the past 100 years, but it certainly presents a generous sample. Among living geographers, emphasis has been on the older generation, those who have been largely responsible for making the field of geography what it is today. The outstanding younger geographers who were not included still have a great number of productive years ahead of them, and they will undoubtedly receive mention in future encyclopedias.

I might have included more terms or concepts, but these are already well covered in recent dictionaries, such as *The Dictionary of Human Geography*, edited by R. J. Johnston, D. Gregory, and D. M. Smith (2d ed. 1986), and *The Encyclopaedic Dictionary of Physical Geography*, edited by A. S. Goudie *et al.* (1985), and so I retained only those concepts that would be considered essential to any book such as this—*e.g.*, Diffusion, Environment, Environmental Determinism, Place, Region, and Space. In any case, no one should expect this modest volume to cover the entire field perfectly. It does, however, present ample factual coverage of an interesting array of people, institutions, subfields, and national developments that some future editor might use as a base for a much grander edifice.

G. S. Dunbar
Cooperstown, New York

Acknowledgments

I should like to thank the contributors, who are listed in the next section, and also the following individuals who rendered various forms of assistance: Professor David H. K. Amiran of Hebrew University of Jerusalem; Professor Jacqueline Beaujeu-Garnier of the University of Paris; Professor James Bird of the University of Southampton; Professor John Borchert of the University of Minnesota; Professor Harold Brookfield of the Australian National University; Professor George Carter of Texas A&M University; Professor J. T. Coppock of the University of Edinburgh; Professor Leslie Curry of the University of Toronto; Professor H. Clifford Darby of the University of Cambridge; Professor Lucio Gambi of the University of Bologna; Professor William Garrison of the University of California, Berkeley; Professor R. Louis Gentilcore of McMaster University; Professor Jean Gottmann of the University of Oxford; Professor Harold Haefner of the University of Zurich; Professor Torsten Hägerstrand of the University of Lund; Professor Louis-Edmond Hamelin of Laval University; Professor John Fraser Hart of the University of Minnesota; Professor Richard Hartshorne of the University of Wisconsin; Professor Leslie Hewes of the University of Nebraska; Professor Robert Holz of the University of Texas; Dr. Brian Hoyle of the University of Southampton; Professor Johannes Humlum of the University of Aarhus; Professor Stig Jaatinen of the University of Helsinki; Mr. J. B. Jackson of Santa Fé, New Mexico; Professor Ronald Johnston of the University of Sheffield; Professor Emrys Jones of the London School of Economics; Professor Robert Kates of Brown University; Professor Fred Kniffen of Louisiana State University; Professor Walter Kollmorgen of the University of Kansas; Professor Peirce Lewis of Pennsylvania State University; Professor David Lowenthal of University College London; Professor J. Ross Mackay of the University of British Columbia; Dr. Maria Claudia Mancini of the University of Rome; Professor Rhoads Murphey of the University of Michigan; Professor Gunnar Olsson of Nordplan; Professor Christiaan van Paassen of the University of Amsterdam; Professor James Parsons of the University of California, Berkeley; Professor Philippe Pinchemel of the University of Paris; Dr. Robert H. T. Smith of the Australian National Board of Employment, Education and Training; Professor O. H. K. Spate of the Australian National University; Professor David Stoddart of the University of California, Berkeley; Professor Arthur Strahler of Columbia University; Professor Waldo Tobler of the University of California, Santa Barbara; Professor Jean Tricart of the University of Strasbourg; Professor Gilbert White of the University of Colorado; Professor Herbert Wilhelmy of the University of Tübingen; Professor Michael Wise of the London School of Economics; Professor M. Gordon Wolman of the Johns Hopkins University; Professor Julian Wolpert of Princeton University; and Professor William Wonders of the University of Alberta.

Special thanks go to the translators, Professor C. Julian Bishko of the University of Virginia (Spanish and Portuguese) and Mrs. Ursula Willenbrock Martin of Pacific Palisades, California (German). The initial suggestion for this book came from Marie Ellen Larcada of Garland Publishing, Inc., and so I should thank her for providing me with an excuse for not taking up golf in the first years of my retirement.

Of inestimable value were the libraries of Syracuse University, Cornell University, and the State University of New York College at Oneonta. I am grateful to Professors José Betancourt and C. W. Woolever of SUNY Oneonta for arranging for the copying of the first draft of the book.

Contributors

KLAUS AERNI
Geographical Institute
University of Bern
Bern, Switzerland

JOHN AGNEW
Department of Geography
Syracuse University
Syracuse, New York

R. WARWICK ARMSTRONG
Department of Health and Safety Studies
University of Illinois at Urbana-Champaign
Champaign, Illinois

JÓSEF BABICZ
Institute of the History of Science, Education and Technology
Polish Academy of Sciences
Warsaw, Poland

TUNKU SHAMSUL BAHRIN
Faculty of Arts and Social Sciences
University of Malaya
Kuala Lumpur, Malaysia

HANNO BECK
Geographical Institute
University of Bonn
Bonn, Germany (FRG)

YEHOSHUA BEN-ARIEH
Department of Geography
Hebrew University of Jerusalem
Jerusalem, Israel

GEORGE BENNEH
Department of Geography
University of Ghana
Legon, Accra, Ghana

NILO BERNARDES
Department of Geography
Catholic University of Rio de Janeiro
Rio de Janeiro, Brazil

BRIAN J. L. BERRY
Professor of Political Economy
University of Texas at Dallas
Richardson, Texas

JOSÉ BETANCOURT
Department of Geography
State University of New York
Oneonta, New York

JOHN N. H. BRITTON
Department of Geography
University of Toronto
Toronto, Canada

BRIAN T. BUNTING
Department of Geography
McMaster University
Hamilton, Canada

KARL W. BUTZER
Departments of Geography and Anthropology
University of Texas
Austin, Texas

TONY CAMPBELL
Map Library
The British Library
London, England

HORACIO CAPEL SÁEZ
Department of Human Geography
Faculty of Geography and History
University of Barcelona
Barcelona, Spain

ILARIA LUZZANA CARACI
Institute of Geographical Sciences
University of Genoa
Genoa, Italy

A. J. CHRISTOPHER
Department of Geography
University of Port Elizabeth
Port Elizabeth, Republic of South Africa

WILLIAM A. V. CLARK
Department of Geography
University of California
Los Angeles, California

PAUL CLAVAL
Department of Geography
University of Paris IV
Paris, France

PEDRO CUNILL GRAU
School of Geography
Central University of Venezuela
Caracas, Venezuela

STEPHEN J. DANIELS
Department of Geography
University of Nottingham
Nottingham, England

GALAL ELDIN ELTAYEB
Department of Geography
University of Khartoum
Khartoum, Sudan

J. NICHOLAS ENTRIKIN
Department of Geography
University of California
Los Angeles, California

JEROME D. FELLMANN
Department of Geography
University of Illinois
Urbana, Illinois

GRAZIELLA GALLIANO
Institute of Geographical Sciences
University of Genoa
Genoa, Italy

JOÃO CARLOS GARCIA
Center of Geographical Studies
University of Lisbon
Lisbon, Portugal

MOHAMED EL-SAYED GHALLAB
Institute of African Studies
University of Cairo
Cairo, Egypt

J. A. VAN GINKEL
Geographical Institute
University of Utrecht
Utrecht, The Netherlands

REGINALD G. GOLLEDGE
Department of Geography
University of California
Santa Barbara, California

A. S. GOUDIE
School of Geography
University of Oxford
Oxford, England

PETER GOULD
Department of Geography
Pennsylvania State University
University Park, Pennsylvania

OLAVI GRANÖ
Chancellor
University of Turku
Turku, Finland

HOWARD F. GREGOR
Department of Geography
University of California
Davis, California

LOUIS GRIVETTI
Department of Geography
and Graduate Group in Nutrition
University of California
Davis, California

MARÍA TERESA GUTIÉRREZ DE MACGREGOR
Institute of Geography
National Autonomous University of Mexico
Mexico City, Mexico

CHAUNCY D. HARRIS
Committee on Geographical Studies
University of Chicago
Chicago, Illinois

R. L. HEATHCOTE
School of Social Sciences
The Flinders University
Bedford Park, South Australia

GORDON L. HERRIES DAVIES
Department of Geography
Trinity College
University of Dublin
Dublin, Ireland

ARILD HOLT-JENSEN
Department of Geography
University of Bergen
Bergen, Norway

DAVID J. M. HOOSON
Department of Geography
University of California
Berkeley, California

RONALD JONES
Department of Geography
Queen Mary and Westfield College
University of London
London, England

TERRY G. JORDAN
Department of Geography
University of Texas
Austin, Texas

MARTIN S. KENZER
Department of Geography
University of Southern California
Los Angeles, California

GEORGE KISH
William Herbert Hobbs Professor of Geography
University of Michigan
Ann Arbor, Michigan

WILLIAM A. KOELSCH
Graduate School of Geography
Clark University
Worcester, Massachusetts

JAMES S. KUS
Department of Geography
California State University
Fresno, California

CHAN LEE
Department of Geography
Seoul National University
Seoul, Korea

WALTER LEIMGRUBER
Geographical Institute
University of Fribourg
Fribourg, Switzerland

GORDON R. LEWTHWAITE
Department of Geography
California State University
Northridge, California

ELISABETH LICHTENBERGER
Institute of Geography
University of Vienna
Vienna, Austria

MAX LINKE
Geography Section
University of Halle-Wittenberg
Halle, East Germany

HUNG-HSI LIU
Department of Geography
Chinese Culture University
Taipei, Taiwan

DAVID N. LIVINGSTONE
School of Geography, Archaeology and Palaeoecology
The Queen's University of Belfast
Belfast, Northern Ireland

AKIN L. MABOGUNJE
Department of Geography
University of Ibadan
Ibadan, Nigeria

GEOFFREY MARTIN
Department of Geography
Southern Connecticut State University
New Haven, Connecticut

ROBERT B. MCMASTER
Department of Geography
University of Minnesota
Minneapolis, Minnesota

W. R. MEAD
Department of Geography
University College London
London, England

MARVIN W. MIKESELL
Committee on Geographical Studies
University of Chicago
Chicago, Illinois

JAMES K. MITCHELL
Department of Geography
Rutgers University
New Brunswick, New Jersey

ROBERT D. MITCHELL
Department of Geography
University of Maryland
College Park, Maryland

JANICE MONK
Southwest Institute for Research on Women
University of Arizona
Tucson, Arizona

MARK MONMONIER
Department of Geography
Syracuse University
Syracuse, New York

STANLEY A. MORAIN
Technology Application Center and Department of Geography
University of New Mexico
Albuquerque, New Mexico

WARREN MORAN
Department of Geography
University of Auckland
Auckland, New Zealand

W. T. W. MORGAN
Department of Geography
University of Durham
Durham, England

RICHARD MORRILL
Department of Geography
University of Washington
Seattle, Washington

ROBERT A. MULLER
Department of Geography and Anthropology
Louisiana State University
Baton Rouge, Louisiana

GLADSTONE OLIVA GUTIÉRREZ
Institute of Geography
Academy of Sciences of Cuba
Havana, Cuba

JOHN O'LOUGHLIN
Department of Geography and Institute of Behavioral Science
University of Colorado
Boulder, Colorado

KENNETH OLWIG
Geographical Institute
Royal Danish School of Educational Studies
Copenhagen, Denmark

OOI JIN BEE
Department of Geography
National University of Singapore
Singapore

JOSEFINA OSTUNI
Institute of Geography
National University of Cuyo
Mendoza, Argentina

PATRICK O'SULLIVAN
Department of Geography
Florida State University
Tallahassee, Florida

PHILIP W. PORTER
Department of Geography
University of Minnesota
Minneapolis, Minnesota

THOMAS M. POULSEN
Department of Geography
Portland State University
Portland, Oregon

THOMAS A. REINER
Regional Science Department
University of Pennsylvania
Philadelphia, Pennsylvania

J. LEWIS ROBINSON
Department of Geography
University of British Columbia
Vancouver, Canada

HÉCTOR F. RUCINQUE
Program of Postgraduate Studies in Geography
Pedagogical and Technological University of Colombia
Bogotá, Colombia

JOSEPH E. SCHWARTZBERG
Department of Geography
University of Minnesota
Minneapolis, Minnesota

DAVID SEAMON
Department of Architecture
Kansas State University
Manhattan, Kansas

SUSAN SMITH
Department of Social and Economic Research
University of Glasgow
Glasgow, Scotland

KEITH SUTTON
School of Geography
University of Manchester
Manchester, England

KEIICHI TAKEUCHI
Laboratory of Social Geography
Hitotsubashi University
Tokyo, Japan

KEITH TINKLER
Department of Geography
Brock University
St. Catharines, Canada

EROL TÜMERTEKIN
Institute of Geography
Istanbul University
Istanbul, Turkey

DAVID TURNOCK
Department of Geography
University of Leicester
Leicester, England

JAMES E. VANCE
Department of Geography
University of California
Berkeley, California

CHRISTIAN VANDERMOTTEN
Laboratory of Human Geography
Free University of Brussels
Brussels, Belgium

THOMAS T. VEBLEN
Department of Geography
University of Colorado
Boulder, Colorado

PHILIP L. WAGNER
Department of Geography
Simon Fraser University
Burnaby, Canada

HELEN WALLIS
Map Library
The British Library
London, England

MICHAEL WOLDENBERG
Department of Geography
State University of New York
Buffalo, New York

JOHN A. WOLTER
Geography and Map Division
Library of Congress
Washington, D.C.

WILBUR ZELINSKY
Department of Geography
Pennsylvania State University
University Park, Pennsylvania

Abbreviations

AAAG	*Annals of the Association of American Geographers*
AAG	Association of American Geographers
AG	*Annales de géographie*
anon.	anonymous
b.	born
bibliog.	bibliography
B.Litt.	Bachelor of Letters (a postgraduate degree in Britain)
B.S., B.Sc.	Bachelor of Science
BSGI	*Bollettino della Società Geografica Italiana*
CG	*Canadian Geographer*
comp.	compiler
CWW	*Canadian Who's Who*
DAB	*Dictionary of American Biography*
DBF	*Dictionnaire de biographie française*
D. de l'Univ.	Docteur de l'Université (France)
D.-ès-L.	Docteur ès Lettres (France)
D.-ès-Sc.	Docteur ès Sciences (France)
D.Let.	Doutor de Letras (Portugal)
D.Litt.	Doctor of Letters
DNB	*Dictionary of National Biography* (Great Britain)
D.Phil.	Doctor of Philosophy (Great Britain)
Dr.phil.	Doktor der Philosophie (Germany)
DSB	*Dictionary of Scientific Biography*
ed., eds.	edited, edition, editor(s)
EG	*Economic Geography*
Fil.Dr.	Filosofie doktor (Sweden)

GBS	*Geographers: Biobibliographical Studies*
GJ	*Geographical Journal*
GOF	*Geographers on Film*
GR	*Geographical Review*
grad.	graduate
GSE	*Great Soviet Encyclopedia*
GZ	*Geographische Zeitschrift*
hab., habil.	habilitation (qualification for teaching in German universities)
IBG	Institute of British Geographers
IESS	*International Encyclopedia of the Social Sciences*
IGU	International Geographical Union
JG	*Journal of Geography*
LGF	*Les géographes français*
Lic.Phil.	Licentiate in Philosophy (Sweden and Finland)
M.A.	Master of Arts
M.Eng.	Master of Engineering
MOGG	*Mitteilungen der Österreichischen Geographischen Gesellschaft*
M.S., M.Sc.	Master of Science
NDB	*Neue Deutsche Biographie*
obit., obits.	obituary, obituaries
PG	*Professional Geographer*
PGM	*Petermanns Geographische Mitteilungen*
Ph.D.	Doctor of Philosophy
PHG	*Progress in Human Geography*
p., pp.	page(s)
rev.	revised, revision
RGI	*Rivista Geografica Italiana*
SGM	*Scottish Geographical Magazine*
TESG	*Tijdschrift voor economische en sociale geografie*
TIBG	*Transactions of the Institute of British Geographers*
trans.	translated, translation, translator
UK	United Kingdom
vol.	volume
WW	*Who's Who* (Great Britain)
WWA	*Who's Who in America*
WWF	*Who's Who in France*
WWWA	*Who Was Who in America*

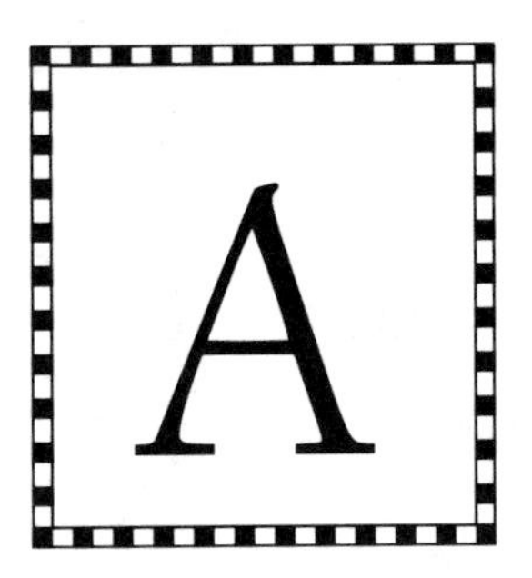

Ackerman, Edward Augustus (1911-1973)

Ph.D., Harvard University, 1939. Harvard University, 1940-48; University of Chicago, 1948-55 *passim*; Resources for the Future, 1954-58; and Carnegie Institution of Washington, 1958-73. *New England's Fishing Industry* (1941); *American Resources* (with J. R. Whitaker) (1951); *Japan's Natural Resources* (1953); *Technology in American Water Development* (with G. Löf and C. Seipp) (1959). Obits. by G. F. White in *AAAG*, 64 (1974), 297-309; and by D. J. Patton in *GR*, 64 (1974), 150-153.

Agricultural Geography

Modern agricultural geography seeks to identify, measure, and explain spatial differentiation in agriculture. Although its existence as an organized field in its own right barely goes back to the end of the last century, it had not gone unrepresented in earlier periods. Descriptions of agricultural landscapes were prominent at least as far back as antiquity, as in the accounts by the Romans and Greeks. But most such efforts went little beyond simple description and failed to analyze the combinations and relative strengths of the agents responsible for these landscape variations.

Only with the arrival of the nineteenth century were the beginnings of a more scientific agricultural geography apparent, particularly as represented by the theories of economist D. Ricardo and agricultural scientist and farmer J. H. von Thünen. Ricardo emphasized the regionally differentiating role of soil fertility, whereas Thünen stressed the influence of market distance and transport costs on the spatial differentiation of production and its intensity. Finally, beginning with the twentieth century, historian V. Hehn and geographer E. Hahn became major precursors of the time perspective in the agricultural-geographic viewpoint in their outlining, respectively, of the broad world patterns of plant and animal migrations and the developmental sequences of economic forms.

It is these and other findings that agricultural geographers have combined with still other phenomena in pursuit of their primary goal, the comprehension of spatial variation in the agricultural area. Their efforts at first were mostly descriptions of farm commodity distributions, with an emphasis on relationships to the physical environment. Today, that viewpoint has broadened to include all items of spatial interest, and with much more emphasis on the human side. Associated with this more comprehensive viewpoint was a greater effort to see how these factors interrelated in the formation of agricultural regions. Methodological foundations for the regional perspective were established by L. Waibel in his "agricultural formation," in which he viewed the structure of an agricultural region as not just a functional, or production, unit, but a combination of ecological and socioeconomic factors as well. Although still not without debate as a concept, agricultural regionalization in one form or another is actively pursued in a variety of ways, not only for its offering of a deeper understanding of the agricultural landscape but because of its value for regional inventory and planning. Recent instances of efforts in this direction are the works of the International Geographical Union's commissions on Agricultural Typology (Chair: J. Kostrowicki) and Comparative Research on Food Systems of the World (Chair: M. Shafi).

Concern for resource problems has always been an important theme in agricultural-geographic research, but never so much as now when many resource pressures are at their greatest and sophisticated research tools like computers, mathematical-statistical techniques, and satellite photography are becoming widely available. Loss of agricultural land to urban expansion has attracted considerable attention, although the traditional gloomy prognosis for periurban farming is now beginning to be called into question in the face of evidence indicating certain zones of persistence and even improving efficiency of farming operations. Agricultural industrialization, with its widening and deepening impacts on natural resources and rural society, is now recognized as having at least three significant spatial components: widespread convergence of systems, regardless of ideological boundaries; increasing areal concentration, representing operational specialization; and increasing concentration on urban margins, reflecting market and land-cost stimulations.

The effects of agricultural industrialization on farm populations and communities, as well as the growing penetration of rural areas by urban settlement and functions, has also evoked an increasing number of more sociologically oriented studies, which in part may explain the current resurgence of interest in the broader field of rural geography. Under this rubric, resource considerations have magnified, the subject being limited only by the requirement of location in a rural area. The impact of governmental policies on agricultural patterns and particularly the differing regional reactions to these actions is also attracting the efforts of a growing number of researchers, especially in North America and Western Europe. Meanwhile, agricultural geographers continue to pursue the more traditional research themes, such as intensive mapping of land use, but now with broader perspectives and using more advanced mapping techniques. Impressive examples are the national and international atlases of farming systems.

Bibliography

Bowler, I. R., "Agricultural Geography," *PHG*, 8 (1984), 255–262; 9 (1985), 255–263; 10 (1986), 249–257.

Gregor, H. F., *Geography of Agriculture: Themes in Research* (1970).

Grigg, D., *An Introduction to Agricultural Geography* (1984).

Ilbery, B. W., *Agricultural Geography—A Social and Economic Analysis* (1985).

Pacione, M., ed., *Progress in Agricultural Geography* (1986).

Sick, W.-D., *Agrargeographie* (1986).

• *Howard F. Gregor*

Ahlmann, Hans W:son (1889–1974)

Fil. Dr., University of Stockholm, 1915. University of Stockholm, 1929–50; Swedish ambassador to Norway, 1950–56; President of IGU, 1956–60. *The Economical Geography of Swedish Norrland* (1921); *Land of Ice and Fire* (1938); *Glaciological Research on the North Atlantic Coasts* (1948); *Glacier Variations and Climatic Fluctuations* (1953). L. P. Kirwan, C. M. Mannerfelt, C. G. Rossby, and V. Schytt, "Glaciers and Climatology: Hans W:son Ahlmann's Contribution," pp. 11–13 in *Glaciers and Climate: Geophysical and Geomorphological Essays Dedicated to Hans W:son Ahlmann* (1949); and obit. by V. Schytt in *Svensk Geografisk Årsbok*, 50 (1975), 279–282.

Algeria. *See* North Africa.

Almagià, Roberto (1884–1962)

Laurea, University of Rome, 1905. University of Padua, 1911–15; University of Rome,

1915–58. *La Geografia* (1919, 2nd ed. 1922); *Monumenta Italiae Cartographica* (1929); *Monumenta Cartographica Vaticana* (4 vols., 1944–55); *Fondamenti di geografia generale* (2 vols., 1945–46, 3rd ed. 1954–55). A. Sestini, "Roberto Almagià stimolatore ed organizzatore di studi geografici," *RGI*, 92 (1985), 3–10 (plus other articles in the same issue); obits. by O. Baldacci in *BSGI*, ser. 9, 3 (1962), 257–273; and E. Migliorini in *RGI*, 70 (1963), 2–25.

American Geographical Society

The American Geographical and Statistical Society of New York was founded in 1851. The words "and Statistical" were deleted twenty years later. The Society assumed national and international prominence early in the twentieth century, after it was moved to new quarters at Broadway and 156th Street with the financial support of Archer Huntington (1911), Isaiah Bowman was appointed Director (1915), and the Society's *Bulletin* was transformed into the *Geographical Review* (1916). The Society prospered in the interwar period but declined in the postwar era. Its magnificent library and map collections (600,000 items) were moved to Milwaukee, Wisconsin, in 1978. The AGS building was sold, but a small staff remains in New York, where the *Geographical Review* continues to be published. The Society's important bibliographical periodical, *Current Geographical Publications*, which has been issued ten times a year since 1938, has been published by the American Geographical Society Collection of the University of Wisconsin-Milwaukee Library since 1978.

BIBLIOGRAPHY

Roselle, W. C., ed., "The Move to Milwaukee" (Commemorative Issue of *Current Geographical Publications*, December 1978).

Wright, J. K., *Geography in the Making: The American Geographical Society, 1851–1951* (1952).

Amiran, David Horace Kallner (b. 1910)

Dr. phil., University of Bern, 1935. Hebrew University of Jerusalem, 1949–79. (ed., with others) *Atlas of Israel* (1964, 3rd ed. 1985); (ed., with others) *Atlas of Jerusalem* (1973); *Coastal Deserts: Their Natural and Human Environments* (with A. W. Wilson) (1973); "Land Transformation in Israel," in *Land Transformation in Agriculture*, ed. by M. G. Wolman and F. G. A. Fournier (1987). A. Shachar, "David Amiran—A Tribute to His Scholarly Work," *Studies in the Geography of Israel*, 11 (1980), 3–14 (in Hebrew, includes bibliography).

Anuchin, Dmitriy Nikolayevich (1843–1923)

D.Sc., University of Moscow, 1889. University of Moscow, 1880–1923. *Konspekt lektsiy po fizicheskoy geografii* (1904); *Izbrannyye geograficheskiye raboty* (1947); *Geograficheskiye raboty* (1954). V. A. Esakov in *GBS*, 2 (1978), 1–8; V. A. Esakov in *DSB*, 1 (1970), 173–175; *GSE*, 2 (1973), 186–187; and *Who Was Who in the USSR* (1972), 25.

Applied Geography

The traditional definition of applied geography has been the *application* of geographic and other perspectives to the so-called real world, *i.e.*, life outside academia. This has generally meant imposing on a given problem a geographer's view of the world. In short, from a geographer's vantage point problem-solving constituted an applied geographical endeavor. This view is probably as old as the discipline itself, and was practiced by both human and physical geographers alike. Most applied work occurred during times of crisis, *e.g.*, during and after wars, or in preparation for future military advantage.

More recently, however, and most notably in the English-speaking world, applied geography has become synonymous with business or marketing geography. It now denotes an emphasis on nonteaching job opportunities for geographers, mostly in the areas of private consulting and government employment. Hand-in-hand with its original concern for problem-solving (in both the private sector and government) has been a dramatic increase in its emphasis on geographic and quasi-geographic techniques, particularly geographic information systems (GIS), especially among human geographers.

A recent trend has been a growing insistence that applied geography be regarded as a subdiscipline—an end in and of itself—which has resulted in the creation of applied geography and GIS conferences, applied and GIS newsletters, annual GIS workshops, two international journals, association specialty groups, and concrete curricula changes in geography programs. In turn, this has spawned antagonism between those who regard themselves as applied geographers and those who profess greater allegiance to "pure" or "theoretical" geography; neither side, however, can exist effectively without the other.

A significant new development is the rise of academic departments specializing in applied geography. A few have even instituted a program in applied geography (all emphasizing technique courses), and at least one (Southwest Texas State University) now offers an M.A.G. (Masters in applied geography), while Ryerson Polytechnical Institute (Toronto) has a School of Applied Geography.

Bibliography

Frazier, J. W., "On the Emergence of an Applied Geography," *PG*, 30 (1978), 233-237.

——— , ed., *Applied Geography: Selected Perspectives* (1982).

Harrison, J. D., "What *Is* Applied Geography?," *PG*, 29 (1977), 297-300.

Kenzer, M. S., ed., *Applied Geography: Issues, Questions, and Concerns* (1989).

Oliver, J. E., *Perspectives on Applied Physical Geography* (1977).

Sant, M., *Applied Geography: Practice, Problems and Prospects* (1982).

• *Martin S. Kenzer*

Arbos, Philippe (1882-1956)

D.-ès-L., University of Grenoble, 1922. University of Clermont-Ferrand, 1919-52. *La vie pastorale dans les Alpes françaises* (1922); *L'Auvergne* (1932); *Les populations rurales du Puy-de-Dôme* (with others) (1933). M. Derruau in *GBS*, 3 (1979), 7-12; obits. by M. Sorre in *AG*, 66 (1957), 182-183; L. Gachon in *Revue de géographie alpine*, 45 (1957), 5-7; and R. Blanchard in *Revue de géographie de Lyon*, 23 (1957), 57-58.

Areal Differentiation

The study of the areal variation of complexes of heterogeneous phenomena that are causally linked and spatially proximate. Areal differentiation has been most closely associated with the study of place and region, and derives its significance in part from our naive awareness and interest in the fact that places vary. The intellectual division of the earth's surface into regional units based on similarities and differences among areas is one means of studying such variations.

Critics of this approach have attacked its relatively atheoretical and descriptive character. For example, it has been described as an areal inventory, or as a synthesis that posits no mechanisms for explaining differences. Other critics have noted the seeming incompatibility between an approach that emphasizes areal differences and a modern world that appears to be growing more homogeneous and interconnected.

Modern proponents have attempted to counter these criticisms by theorizing mechanisms that create areal variation and that sustain a *process* of areal differentiation. For example, Marxist geographers have argued that differences among places and regions are the necessary consequences of conflicting tendencies in modern capitalism that cause patterns of uneven development. Also, post-modern geographers have argued for the continuing significance of areal differences as an indication of the failure of theories of modernization to account for the diversity of the contemporary world.

Bibliography

Agnew, J., *Place and Politics: The Geographical Mediation of State and Society* (1987).

Entrikin, J. N., and S. D. Brunn, eds., *Reflections on Richard Hartshorne's* The Nature of Geography (1989).

Gregory, D., "Areal Differentiation and Post-Modern Human Geography," pp. 1-27 in *New Horizons in Human Geography*, ed. by D. Gregory and R. Walford (1989).

• *J. Nicholas Entrikin*

Argentina

Two circumstances—the foundation in 1922 of the Argentine Society for Geographical Studies, and academic institutionalization in the period 1940–55, with the organization of a professional program in several universities—contributed to structuring a geography more in conformity with that found in countries where the science was well consolidated. At the same time, the physiographic approach, so characteristic of the territorial descriptions carried out by foreign and Argentine explorers in the period before 1940, fell into decline.

Buenos Aires centralizes geographical studies around the Sociedad Argentina de Estudios Geográficos GAEA (Argentine Society for Geographical Studies) and the institutes of geography of the University of Buenos Aires and the Secondary Teachers Normal School. Such figures as, on human geographical topics, Romualdo Ardissone, Federico Daus, and Horacio Difrieri, and, in physical geography, Mario Grondona, have been leaders in Argentine geography, fortifying its humanistic character and regional outlook under the revitalizing influence of French, German, and U.S. geographers.

In the period 1940–55, the Institute of Tucumán, profiting from the presence of the German geographers Wilhelm Rohmeder, Fritz Machatschek, Gustav Fochler-Hauke, and Willi Czajka, promoted studies emphasizing a chorographic focus and treatment of landscapes. The subsequent departure of these Germans, even though some of their students continued their teachings, resulted in a falling-off of this research drive.

The Institute of Geography of the National University of Cuyo also occupies a prominent position. Although molded in fruitful and constant interchange with French geography, its academic and institutional policies have always been controlled by its own graduates, who—like the specialists Mariano Zamorano in human geography and Ricardo Capitanelli in physical geography—have discharged important offices in such international organizations as the International Geographical Union and the Panamerican Institute of Geography and History. Succeeding generations, ever alert to new perspectives, are participating through seminars and the *Boletín de Estudios Geográficos* in the diffusion of the advances made in teaching and research.

Since the 1970s there has been a renewal of themes and focuses of interest in Argentine geography. The number of institutes and centers of geographical research is growing, because of the presence of new universities and of the National Council of Scientific and Technical Research (CONICET). The Argentine Society for Geographical Studies, with its publications and its "Geography Weeks" (held annually since 1931), maintains its vigorous role, bringing together geographers on different teaching levels. The Academia Nacional de Geografía (National Academy of Geography), of more recent foundation, has had as yet only a limited impact.

The same changes can be seen in teaching as in research. While the universities have made important advances, their programs of study display no homogeneity; some reflect a trend towards analysis, others towards a more broadly comprehensive approach. At the secondary level, the natural science view remains strongly rooted.

In Argentina, as in other countries of Latin America, geography is subject to the influence of movements originating in Europe and the United States. Nevertheless, the distance from these centers of diffusion of new perspectives, and the different realities they confront, allow Argentine geographers to reshape these concepts into a distinct pattern of their own.

BIBLIOGRAPHY

Bolsi, A. S., "Presencia europea en la investigación y enseñanza de la Geografía argentina" (unpublished manuscript).

Capitanelli, R., "La investigación geográfica en Argentina," pp. 269–289 in *La Geografía y la Historia en la identidad nacional*, ed. by P. H. Randle (Vol. 1, 1981).

Jorge, C., "Geografía," in *Evolución de las ciencias en la República Argentina, 1923–1972* (Vol. 10, 1988).

Ostuni, J., *et al.*, *Treinta años de labor en el Instituto de Geografía* (Cuyo) (1977).

Vilá-Valentí, J., "The Iberian Peninsula and Latin America," pp. 264–281 in *Geography Since the Second World War*, ed. by R. J. Johnston and P. Claval (1984).

Zamorano, M., "Los cuarenta años de existencia del Instituto de Geografía," Universidad Nacional de Cuyo, Instituto de Geografía, *Boletín de Estudios Geográficos*, no. 85 (1987).

• *Josefina Ostuni*
(Translated by C. Julian Bishko)

Association of American Geographers

Professor William Morris Davis of Harvard University founded the Association of American Geographers in 1904 with 48 charter members. Davis envisioned a "society of mature geographical experts," in contradistinction to the National Geographic Society, which was moving in other directions. The Association remained a rather small, elitist organization until 1948, when it merged with the American Society of Professional Geographers, which had been established in 1943 by a group of young professionals (mostly in Washington, D.C.) who had chafed at the exclusivity of the older association. At the time of the merger the membership of the ASPG outnumbered that of the AAG by more than three to one (1,094 to 306). The journal of the younger organization, *The Professional Geographer* (begun in 1946), was continued as a forum for discussion. It contains short articles, news notes, and book reviews, and complements the *Annals of the Association of American Geographers*, which was established in 1911 to publish scholarly articles. Today the AAG has a membership of about 6,000, a permanent central office staff in Washington, and a lively publication program centered around the two journals. The annual meetings, normally held in the spring in hotels or convention centers in large North American cities, are attended by ever-increasing numbers (more than 3,000 in 1989).

BIBLIOGRAPHY

James, P. E., and G. J. Martin, *The Association of American Geographers: The First Seventy-Five Years, 1904-1979* (1978).

Linke, M., "Die Association of American Geographers—ihr Beitrag zur Entwicklung der Geographie in den USA," *PGM*, 133 (1989), 115-120.

Association of French Geographers (Association de géographes français)

The *Association de géographes français* was founded in 1920 by Emmanuel de Martonne. At that time academics held no position of responsibility in the Paris Geographical Society, whose resources had been considerably diminished by World War I. Thus the decision was made to create a place where university geographers could meet. The Vidalian orientation was stressed by the choice of the first president, Lucien Gallois. There was also a strong emphasis on geomorphology; Emmanuel de Margerie, Emmanuel de Martonne, André Cholley, and Pierre Birot presided over the destiny of the association. Georges Chabot was the only human geographer to become president. The association played a major role in the selection of geographers for university positions, but it lost a good deal of its former significance when new procedures for university appointments were instituted in the 1950s. The monthly meetings of the association provided—and still provide—young scholars with a public for their findings, which are then quickly published in the *Bulletin de l'Association de géographes français*.

• *Paul Claval*

Atwood, Wallace Walter (1872-1949)

Ph.D., University of Chicago, 1903. University of Chicago, 1901-13; Harvard University, 1913-20; Clark University (President), 1920-46. *The Physiographic Provinces of North America* (1940); *The Rocky Mountains* (1945). W. A. Koelsch in *GBS*, 3 (1979), 13-18; W. A. Koelsch, "Wallace Atwood's 'Great Geographical Institute,'" *AAAG*, 70 (1980), 567-582; obits. by G. B. Cressey in *AAAG*, 39 (1949), 296-306; and S. Van Valkenburg in *GR*, 39 (1949), 675-677.

Australia

Geographical activities in Australia have long reflected the historical context of the past 200 years of European settlement, namely, a concern for the exploration of "new" lands, the appraisal of their material resources and the links between people and the places they modified and were, in turn, influenced by. Thus the legacy of European scientific rationalism and romantic idealism, alongside British military and commercial imperialism and utilitarianism, can be seen to have initially guided and subsequently left their mark on Australian geography.

Beginning with the first permanent European settlement, in 1788, exploration and resource inventory were major foci of activity by both private and official parties. They were high on the agendas of the geographical societies, formed in the 1880s as branches of what was to become in 1886 the Royal Geographical Society of Australasia. The branches were founded in 1883 in New South Wales and Victoria, and in 1885 in Queensland and South Australia. Formed on the mold of the Royal Geographical Society of London, their membership included merchants, entrepreneurs, government employees, scientists and interested layfolk. The South Australian and Queensland branches have survived and continue to publish their annual proceedings; the Victorian branch merged with the local history society and the New South Wales branch's activities lapsed until revived in 1927 as the Geographical Society of New South Wales.

Paralleling this activity, in 1888 came the foundation of the Australasian Association for the Advancement of Science (again in the British mold) with Section E: Geography having as its first president John Forrest, the West Australian explorer and politician. This connection has been maintained as the association—now the Australian and New Zealand Association for the Advancement of Science (ANZAAS)—grew, so that "Geographical Sciences" is now included as Section 21.

This institutionalization of geography was extended in 1960 with the formation of the National Committee for Geography, chaired by the recently appointed professor of geography at the new Australian National University, O. H. K. Spate. This was one of the committees of the Australian Academy of Science (itself established only in 1954) to liaise in this case with the International Geographical Union and to further encourage the development of geography in Australia. In the same year an Institute of Australian Geographers—initially, exclusively tertiary academics—was founded. Its annual publication (*Australian Geographical Studies*) first appeared in 1963. An Australian Geography Teachers Association was formed in 1967 and its journal (*Geographical Education*) first appeared in 1969.

As late as the 1880s, the teaching of geography in schools relied mainly upon British text books. Despite the subsequent formation of the geographical societies and the foundation of universities at Sydney in 1850 and Melbourne in 1853, geographical instruction at tertiary level was not available nationally until tertiary education was generally expanded by the creation of seven new universities in the 1960s—all but one of which contained a geography department. A one-year course in geography had been available at Adelaide University since 1904, but geography was not recognized as a university discipline until the "Father of Australian Geography"—T. Griffith Taylor (1880-1963)—was appointed as associate professor at Sydney University in 1921, and Dr. Charles Fenner (1884-1955), the Superintendent of Technical Education in South Australia, was appointed part-time lecturer in geography at Adelaide University. In 1951 these were still the only two geography departments at the tertiary level in Australia; by 1961 there were seven; by 1971 seventeen (including the University of Papua New Guinea, which had links with the Australian National University in Canberra). Currently there are seventeen university departments excluding Papua New Guinea, and many of the colleges of advanced education have geographers on their staff.

Geographical research, however, preceded the creation of geography departments. Official and private explorers contributed new knowledge, and officials in the departments of lands and surveys (concerned with the appraisal and allocation of land and the human imprint on the landscape), in the departments of agriculture (concerned with new plant and animal varieties, soil erosion and the control of diseases), and in the departments of woods and forests (concerned with conserving native, and trying out exotic, timbers), all directly or indirectly were contributing to geographical knowledge, and in their annual and special reports in the parliamentary papers were providing analyses of environmental problems that often had a strong geographical flavor.

The archetype official organization carrying out research that was often highly geographical was the Commonwealth Scientific and Industrial Research Organisation (CSIRO), which had been set up in 1927 to apply science to the increasingly complex environmental problems facing land settlement. Its virtual monopoly of government research funding until the 1960s was the result of the success in increasing agricultural yields through impressive solutions to some of those problems (control of the prickly pear infestations and of bovine pleuro-pneumonia, and the recognition of trace element deficiencies in Australian soils). Its technique for rapid resource inventory (the land unit approach), developed in 1960, was to remain a basic geographical tool until satellite imagery provided a superior reconnaissance technology. Resource inventory received a further boost in 1960 with the publication by the Commonwealth Department of National Mapping of the first edition of the *Atlas of Australian Resources*. Now in its third edition, the atlas remains a fundamental geographical contribution to knowledge of the continent, and has been complemented recently by state atlases in commemoration of anniversaries of European settlement (e.g., *Atlas of South Australia* [1986] and *Atlas of New South Wales* [1987]). A further national atlas (*Australians: A Historical Atlas*) has also been provided by geographers for the Australian Bicentennial Historical Project.

In 1965 the Commonwealth Government created the Australian Research Grants Scheme, by which for the first time the universities were offered national funds on a competitive basis. The resultant surge of research among the newly formed geography departments covered most avenues of geographical enquiry—in human geography from quantification and Marxist ideological explanations to behavioral studies, concern for natural hazards and, most recently, for gender questions and the aging population; while in physical geography the continent's ecology, and the terrains and processes affecting the arid inland and the coast, were the main foci.

Since the 1920s, geography has often been the center of controversy, with friction between optimistic resource boosters and the more circumspect geographical analyses taking various forms. Initially it was Griffith Taylor's media battles with the boosters of the arid inland's settlement potential and A. Grenfell Price's more sanguine hopes for white settlement in the tropics; in the 1940s it featured Taylor's successor J. Macdonald Holmes (1896–1966) and a concern for soil erosion and the need for regional planning for post-World War II development; more recently has come further concern for land degradation, coastal erosion and wilderness destruction. Significantly, the first Director of the Australian Conservation Foundation (set up in 1965) was a geographer (Dr. J. G. Mosley).

In August 1988 the 26th International Geographical Congress was held in Sydney, the first in Australia and only the second in the Southern Hemisphere. The delegates received copies of the two major research journals and a volume of essays, designed collectively to provide an overview of geography in Australia. The April 1988 number of *Australian Geographical Studies* provided reviews of various branches of the discipline, from "geomorphology" to "geographical education and its Australian heritage," while the May number of the *Australian Geographer* offered articles grouped under the theme of "Environment and Development in Australia"—a theme echoed in the volume of essays entitled *The Australian Experience: essays in Australian land settlement and resource management*. A further review of geographical research, published at the same time by the Academy of the Social Sciences in Australia, also reflected this environmental focus: *Land Water and People: geographical essays in Australian resource management*.

This continuing professional focus upon environmental issues has been complemented recently by renewed popular interest in the environment for its own sake, as well as its conservation. The success of two new popular magazines illustrates this: *Geo (Australasia's Geographical Magazine)*, which appeared in 1979, and electronic entrepreneur Dick Smith's *Australian Geographic* (a "journal of discovery and adventure"), which appeared in 1985. Both seemed to be trying—with evident success—to emulate *National Geographic Magazine*. Geography, in its many forms, appears to be thriving in Australia. *See also* Geographical Society of New South Wales; Institute of Australian Geographers.

Bibliography

Heathcote, R. L., ed., *The Australian Experience: essays in Australian land settlement and resource management* (1988).

Heathcote, R. L., and J. A. Mabbutt, eds., *Land Water and People: geographical essays in Australian resource management* (1988).

Home, R. W., ed., *Australian Science in the Making* (1988).

Jeans, D. N., and J. L. Davies, "Australian Geography 1972-1980," *Australian Geographical Studies*, 22 (1984), 3-35.

Powell, J. M., "Geographical education and its Australian heritage," *Australian Geographical Studies*, 26 (1988), 214-230.

Spate, O. H. K., and J. N. Jennings, "Australian Geography 1951-1971," *Australian Geographical Studies*, 10 (1972), 113-140.

• *R. L. Heathcote*

Austria

Scientific geography came into being in a large empire, the Austro-Hungarian Monarchy. The Austrian Geographical Society, founded in 1856, numbered members of the royal household among its patrons and financed expeditions to places such as East Africa and the Near East. Historical geography developed in connection with the export trade to the Near East and in close cooperation with archeological and philological research carried out by the Austrian Academy of Sciences.

After the collapse of the Austro-Hungarian Empire scholarly relations between Vienna and Berlin remained intact. Despite the existential crisis of the small Austrian state in the interwar period, famous geographers were teaching there, *e.g.*, Otto Maull in Graz, Fritz Machatschek in Vienna, Johann Sölch in Innsbruck and Vienna, and Hugo Hassinger in Vienna, the latter laying the foundations for the Vienna School of Urban Geography.

Even in the postwar era the traditional historical spatial connections were still effective and resulted in the publication of pertinent books and atlases. Two of the latter, being of very high scholarly standard, deserve special attention. The atlas of the Tyrol, published by the Institute of Geography in Innsbruck, includes the South Tyrol as well. An atlas of Southeastern Europe was edited by J. Breu for the Institute of Southeast European Studies in Vienna.

Present-day Austria is an Alpine country, so research into glaciers and the pleistocene has always been of a high standard. In the postwar period interdisciplinary cooperation was further emphasized in physical geography. Within quaternary research, systematic investigation into loess became an independent branch of study in Vienna, together with specialized research into dendrochronology, palynology, and isotope measuring techniques, in close cooperation with pedology both in the sense of a historical geoscience and of a practically oriented discipline evaluating the economic potential of soils.

An Institute for High Alpine Research is affiliated with the Department of Geography of Innsbruck University. Its staff has made photogrammetric measurements of 925 glaciers and entered the data into a glacier information system. Synoptic climatology became a new field of research, cultivated by geographers such as Franz Fliri of Innsbruck.

Investigations in human geography have two foci: research into high mountainous areas and urban studies. Thus, farming and tourism in the Austrian Alps are research themes of long standing.

Urban research in Vienna has perennially taken up new problems, in close contact with the city's planning authorities, *e.g.*, the trend towards a dichotomization in retailing; the formation of the CBD; urban blight and urban renewal; the formation of a new lower class made up of guestworkers and foreign refugees; a dissolution of traditional lebensraum units and, thus, of specific urban quarters; the segmentation of the housing, labor and leisure markets in the social welfare state, etc. Demographic research is also an expanding discipline. As urban research has done for quite a long time, population geography now includes studies in foreign countries.

International communication focuses on the Federal Republic of Germany, in whose biennial "Geographers' Meetings" Austrian geographers participate. Books and journals published in Germany are widely read. There is, however, some diffusion of ideas from the English-speaking countries, mainly taken up by Viennese geographers.

A wide spectrum of institutions is engaged in geography in one way or another, and includes the Austrian Geographical Society and the Institute for Urban and Regional Research of the

Austrian Academy of Sciences. Besides, there is practically oriented research conducted by semi-official institutes with some affinity to geography, such as the Austrian Institute for Regional Planning. Geography graduates occupy many of the top positions in this institute, as they also do in the Central Office of Statistics, in the statistical offices and the urban and regional planning offices of the various Bundesländer, and in the Austrian Regional Planning Conference.

BIBLIOGRAPHY

Bernleithner, E., "Sechshundert Jahre Geographie an der Wiener Universität," pp. 55-125 in *Studien zur Geschichte der Universität Wien*, 3 (1965).

Hassinger, H., *Österreichs Anteil an der Erforschung der Erde* (1949).

Lichtenberger, E., "Forschungsrichtungen der Geographie. Das osterreichische Beispiel 1945-1975," *MOGG*, 117 (1975), 1-115.

——— , "Geographical Research in Austria at the Universities and the Commission for Regional Research of the Austrian Academy of Sciences," pp. 11-48 in *Contemporary Essays in Austrian and Hungarian Geography*, ed. by E. Lichtenberger and M. Pécsi (1988).

——— , "The German-Speaking Countries," pp. 156-184 in *Geography Since the Second World War*, ed. by R. J. Johnston and P. Claval (1984).

• *Elisabeth Lichtenberger*

Austrian Academy of Sciences, Institute for Urban and Regional Research

The Institute was established in 1989, replacing the Commission for Regional Research (and Reconstruction), which had been founded in 1946 by Hugo Hassinger to carry out basic research. From 1954 to 1984, Hans Bobek was head of the commission. During his tenure, the commission prepared the national atlas of Austria, which will remain a document of the research status of geosciences, cultural history, and geographical disciplines at that time. The commission also issued numerous publications on climate, vegetation, industry, transport, and the regional structure of Austrian society. Within the institutional framework of the commission Bobek and his collaborators carried out important work on central places. In 1984, Elisabeth Lichtenberger became head of the commission. A geographical information system on Austria and Vienna was set up. Basic research into theoretical problems of the interrelationships between space and society became the focus of the commission's work.

BIBLIOGRAPHY

Beiträge zur Stadt- und Regionalforschung, vols. 1-9 (1975 ff.).

Bobek, H., ed., *Atlas der Republik Österreich* (1964-80).

• *Elisabeth Lichtenberger*

Austrian Geographical Society (Österreichische Geographische Gesellschaft)

The Austrian Geographical Society was founded in 1856. According to geography's scholarly prestige at that time, it numbered among its founding members representatives of both the political and the economic elites: high-ranking officers and civil servants, noblemen, and magnates of industry. It participated in the exploration of Africa and the Near East, and its publications (*Mitteilungen* and *Abhandlungen*) were internationally renowned. Native and foreign explorers, geographers, geologists, botanists, geophysicists, anthropologists, ethnologists, etc., gave lectures, and boards of experts met to discuss specific problems.

Due to the effects of the two world wars the number of members and their composition have undergone marked changes. There are no patrons any longer, and the number of members has declined. The society had to adapt to the conditions prevalent in a small country and to the competition of other media marketing geographical information. All the university geography departments now have their own publication series, and the same is true of various research institutions outside the universities.

BIBLIOGRAPHY

Bernleithner, E., "Die Geographische Gesellschaft in Wien und ihr Anteil an der Entwicklung der Landeskunde von Deutschland und Österreich," *Berichte zur deutschen Landeskunde*, 21 (1958), 294-324.

• *Elisabeth Lichtenberger*

de Béthune. Physical geography has become increasingly more specialized, and today the split with human geography is complete. Geomorphology is the dominant specialization: erosivity in a temperate climate (J. De Ploey), karst study (C. Ek), planation surfaces and periglacial phenomena (A. Pissart and E. Juvigné), morphology of Recent terrains and of the North Sea basin (G. De Moor), tropical morphology (L. Peeters and J. Alexandre), and glaciology (R. Souchez).

Geography occupies a minor place in the secondary education system. Regional geography has dominated for a long time, as, for example, in the textbooks by J. Tilmont. Since the 1970s, topical or thematic studies have grown in importance, and physical geography has nearly disappeared. The secondary teachers belong to two dynamic societies: the Fegepro (French-speaking) and the Vla (Flemish). The employment of geographers outside the education system has recently developed but remains far less important than in, for example, the United Kingdom.

BIBLIOGRAPHY

Christians, C., and L. Daels, "Belgium, A Geographical Introduction to Its Regional Diversity and Its Human Richness," *Bulletin de la Société géographique de Liège* (1988).

De Moor, G., ed., "Belgian Physical Geographers at Home and in the World between 1950 and 1985," *Bulletin de la Société belge d'études géographiques*, 57 (1988).

Denis, J., ed., "Geography in Belgium," *Ibid.*, 53 (1984).

Pissart, A., ed., *Géomorphologie de la Belgique: Hommage au Professeur Macar* (1976).

• *C. Vandermotten*

Bennett, Hugh Hammond (1881-1960)

B.S., University of North Carolina, 1903. U.S. Department of the Interior, 1933-35; U.S. Department of Agriculture, 1903-33, 1935-52. *The Soils and Agriculture of the Southern States* (1921); *Soil Erosion, A National Menace* (with W. R. Chapline) (1928); *The Cost of Soil Erosion* (1934); *Soil Conservation* (1939). W. Brink, *Big Hugh, The Father of Soil Conservation* (1951); L. Filler in *DAB*, Supplement 6 (1980), 52-53; obits. by C. P. Barnes in *AAAG*, 50 (1960), 506-507; and A. L. Patrick in *GR*, 51 (1961), 121-124.

Berg, Lev Semyonovich (1876-1950)

Doctorate, University of Moscow, 1909. University of Leningrad, 1916-50. *Klimat i zhizn'* (1922); *Landshaftno-geograficheskiye zony SSSR* (1931); *Ocherki po istorii russkikh geograficheskikh otkrytiy* (1946); *Geograficheskiye zony SSSR* (2 vols., 1948-49). *Who Was Who in the USSR*, p. 64; I. P. Gerasimov *et al.*, "The Centennial of L. S. Berg," *Soviet Geography*, 18 (1977), 1-32; E. M. Murzayev in *GBS*, 5 (1981), 1-7.

Berlin Geographical Society (Gesellschaft für Erdkunde zu Berlin)

The Berlin Geographical Society was founded in April 1828. Capitalizing on the great interest shown in Alexander von Humboldt's public lectures in 1827-1828, his friend and collaborator, Heinrich Berghaus, called for the establishment of a geographical society. Carl Ritter was elected the first president, and he remained the leading figure in the society until his death in 1859. The society supported German exploration in Africa in the latter half of the nineteenth century and in the polar regions during the presidency of Ferdinand von Richthofen around the turn of the century. The society benefitted greatly from the sustained interest of Richthofen and Albrecht Penck. It suffered during the Nazi period, especially during World War II, when its building was destroyed. Although most of the society's books and maps had been evacuated to what is now East Germany, they have never been returned. In 1948 the society was able to resume its activities in West Berlin and rebuild its collections. A new building was constructed for the society in 1967 with financial aid from the Volkswagen corporation. The society's journal, the *Zeitschrift*, which began publication in 1853 (succeeding earlier publications going back to 1833), ceased in

1944 but was reborn as *Die Erde* five years later. In 1957 the new journal renumbered its volumes to continue the numbering of the *Zeitschrift*.

BIBLIOGRAPHY

Lenz, K., "The Berlin Geographical Society (1828-1978)," *GJ*, 144 (1978), 218-223. See also Lenz' longer paper, "150 Jahre Gesellschaft für Erdkunde zu Berlin," *Die Erde*, 109 (1978), 15-35.

Bader, F. J. W., "Rückblick nach 100 Bänden der Zeitschrift der Gesellschaft für Erdkunde zu Berlin," *Die Erde*, 100 (1969), 93-117.

Bernard, Augustin (1865-1947)

D.-ès-L., University of Paris, 1895. Ecole supérieure des lettres (Algiers), 1894-1902; University of Paris, 1902-33. *L'archipel de la Nouvelle-Calédonie* (1895); *Le Maroc* (1913); *L'Algérie* (1929); *L'Afrique septentrionale et occidentale* (1937, 1939). M. Prévost in *DBF*, 6 (1954), cols. 48-49; M. Larnaude in *LGF*, 107-118; K. Sutton in *GBS*, 3 (1979), 19-27; obit. by M. Larnaude in *AG*, 57 (1948), 56-59.

Berry, Brian Joe Lobley (b. 1934)

Ph.D., University of Washington, 1958. University of Chicago, 1958-76; Harvard University, 1976-81; Carnegie-Mellon University, 1981-86; University of Texas at Dallas, 1986 to the present. *The Geography of Market Centers and Retail Distribution* (1967); *The Human Consequences of Urbanisation* (1973); *Contemporary Urban Ecology* (with J. Kasarda) (1977); (ed.) *The Nature of Change in Geographical Ideas* (1978). *WWA*; *GOF* (1971); P. Halvorson and B. M. Stave, "A Conversation with Brian J. L. Berry," *Journal of Urban History*, 4 (1978), 209-238.

Biasutti, Renato (1878-1965)

Studied at the University of Florence in late 1890s but did not complete Laurea degree. University of Naples, 1913-27; University of Florence, 1907-13, 1927-53. *La casa rurale della Toscana* (1938); *Le razze e i popoli della terra* (3 vols., 1941; 3rd ed., 4 vols., 1959); *Il paesaggio terrestre* (1947, 1962). E. Cerulli, "Il contributo di Renato Biasutti agli studi etnologici," *RGI*, 87 (1980), 305-312; A. Sestini, "Renato Biasutti e gli inizi degli studi antropogeografici in Italia," *RGI*, 87 (1980), 313-323; obits. by B. Nice in *RGI*, 72 (1965), 313-337; and R. Riccardi in *BSGI*, 6 (1965), 169-180.

Biogeography

Biogeographers study the constantly changing distributions of plants and animals over a range of temporal and spatial scales. Geographic practitioners of biogeography are also concerned with the natural and anthropogenic processes that shape plant and animal communities. The antecedents of modern biogeography in both biology and geography lie in the eighteenth- and nineteenth-century period of scientific exploration by naturalists who were concerned with describing and classifying the world's flora and fauna. Early approaches to biogeography included both historical explanations of the distributions of individual taxa (species, genera, families) and attempts to relate present vegetation patterns to the modern environment, particularly climate.

Biogeography as practiced by contemporary geographers shares these nineteenth-century antecedents with biological biogeography and also has its roots in some of the major themes of twentieth-century physical and human geography. Thus, biogeography reflects the systems perspective of integrating physical and biological processes that has been popular in ecosystem ecology and physical geography since the 1960s. It also reflects the concern with human agency in modifying landscapes associated with the "man-land" tradition of geography exemplified by the work of Carl O. Sauer and his students during the mid-twentieth century. Much contemporary biogeographic research by geographers combines the traditions of physical geography and the man-land theme using methodologies common to modern ecology and Quaternary studies. The major research themes of geographic biogeography during the 1970s and 1980s have been human/biota interactions, vegetation/environment relations, paleoecology, and vegetation dynamics.

In the area of human/biota interactions, biogeographers have examined the use of wild biota by traditional societies and processes of plant and animal domestication. An area of much research activity has been the investigation of the roles of humans in modifying plant and animal communities through the introduction of alien wild plants and wild animals, and also through the combined effects of grazing, logging, burning, and fuelwood gathering.

Biogeographic work on vegetation/environment relations has been conducted mainly at a landscape scale. This includes the study of variation in vegetation composition in relation to a spatially varying physical environment based on the use of ordination and gradient analyses. It also includes physiographic plant geography, which is the study of plant distributions at the scale of landforms and relates vegetation patterns to physiographic processes. Broad-scale vegetation mapping is now being revolutionized by the application of remote sensing techniques. Also at regional to global scales, geographers have been active in modeling climate influences on ecosystem properties and plant life forms.

In addition to the study of spatial variation in vegetation, many biogeographers focus on changing patterns over time. The study of the processes and patterns of successional change and regeneration dynamics by geographers grew out of the physical geographer's interest in the frequency and magnitude of physical events and the cultural-historical geographer's concern with the ecological consequences of anthropogenic disturbances. A major research theme has been the role of humans in altering vegetation by changing fire regimes. Linkages of biogeography to geomorphology are reflected by research on vegetation disturbances such as mass movements, snow avalanches, and flooding. Based on the analysis of fossil pollen records, biogeographers have been active in the study of vegetation history over longer time spans in relation to both human-caused disturbance and climatic fluctuation.

BIBLIOGRAPHY

Kellman, M. C., *Plant Geography* (2nd ed., 1980).

Sauer, C. O., "The Agency of Man on the Earth," pp. 49-69 in *Man's Role in Changing the Face of the Earth*, ed. by W. L. Thomas (1956).

Sauer, J. D., *Plant Migration* (1988).

Taylor, J. A., ed., *Biogeography: Recent Advances and Future Directions* (1984).

Vale, T. R., *Plants and People: Vegetation Change in North America* (1982).

Veblen, T. T., "Biogeography," pp. 28-46 in *Geography in America*, ed. by G. L. Gaile and C. J. Willmott (1989).

• *Thomas T. Veblen*

Bird, James Harold (b. 1923)

Ph.D., University of London, 1953. King's College London, 1954-63; University College London, 1964-67; University of Southampton, 1953-54, 1967-88. *The Major Seaports of the United Kingdom* (1963); *Seaports and Seaport Terminals* (1971); *Cities and Centrality* (1977); *The Changing Worlds of Geography* (1989).

Blanchard, Raoul (1877-1965)

D.-ès-L., University of Lille, 1906. University of Grenoble, 1906-48. *La Flandre* (1906); *Grenoble, étude de géographie urbaine* (1911); *Les Alpes françaises* (1925); *Les Alpes occidentales* (7 vols. in 13 parts, 1938-56). J. Taton in *DSB*, 2 (1970), 190-191; *Mélanges géographiques canadiens offerts à Raoul Blanchard* (1959); *In Memoriam: Raoul Blanchard* (1966); P. Guichonnet and J. Masseport in *LGF*, 133-144; obit. by J. Dresch and P. George in *AG*, 75 (1966), 1-5 (followed by bibliog., 5-25); and two autobiographical works: *Ma jeunesse sous l'aile de Péguy* (1961) and *Je découvre l'université* (1963).

Bobek, Hans (b. 1903)

Dr.phil., University of Innsbruck, 1926; (hab.) University of Berlin, 1935. University of Berlin, 1931-40; University of Freiburg, 1946-48; Vienna School of Economics, 1949-51; University of Vienna, 1951-71. *Innsbruck* (1928); *Iran* (1962); (ed.) *Österreich-Atlas* (1964-80); *Das System der Zentralen Orte Österreichs* (with M. Fesl) (1978). H. Bobek, "Some Comments toward a Better Understanding of My Scholarly Life-Path," pp. 167-185 in *The Practice of Geography* (1983), ed. by Anne Buttimer; "Festschrift Hans Bobek," ed. by K. Wiche, *MOGG*, 105

(1963) (see especially Heft I/II, pp. 5-22, "Der Weg zur Sozialgeographie: Der wissenschaftliche Lebensweg von Professor Dr. Hans Bobek," by W. Hartke).

Boesch, Hans (1911-1978)

Dr.phil., University of Zurich, 1937 (hab., 1939). University of Zurich, 1939-78. *Amerikanische Landschaft* (1955); *USA—Die Erschliessung eines Kontinentes* (1956); *A Geography of World Economy* (1964); *Japan* (1978). H. Kishimoto, "Biographie Hans Boesch 1911-1978," pp. 127-129 in *Geography and Its Boundaries: In Memory of Hans Boesch* (1980), followed by "Bibliographie," pp. 131-135; obit. by E. Spiess in *Geographica Helvetica*, 33 (1978), 169-172.

Borchert, John Robert (b. 1918)

Ph.D., University of Wisconsin, 1949. University of Minnesota, 1949-89. "The Climate of the Central North American Grassland," *AAAG*, 40 (1950), 1-39; "American Metropolitan Evolution," *GR*, 57 (1967), 301-332; "Instability in American Metropolitan Growth," *GR*, 73 (1983), 127-149; *America's Northern Heartland: An Economic and Historical Geography of the Upper Midwest* (1987). *GOF* (1975), *WWA*.

Bose, Nirmal Kumar (1901-1972)

M.Sc., University of Calcutta, 1925. Editor of *Man in India* (1951-72), taught geography in the University of Calcutta from 1945; *Cultural Anthropology* (1925); (ed.) *Peasant Life in India* (1961); *Calcutta* (1964); *Calcutta: A Social Survey* (1968). S. Mookerjee in *GBS*, 2 (1978), 9-11; obits. in *Geographical Review of India*, 34 (1972), 404-406; and *Indian Geographical Journal*, 47 (1972), 36-39.

Bowman, Isaiah (1878-1950)

Ph.D., Yale University, 1909. Yale University, 1905-15; American Geographical Society, 1915-35; Johns Hopkins University (president), 1935-48. *Forest Physiography* (1911); *The New World* (1921); *The Pioneer Fringe* (1931); *Limits of Land Settlement* (1937). G. J. Martin, *The Life and Thought of Isaiah Bowman* (1980); obits. by G. F. Carter in *AAAG*, 40 (1950), 335-350; and G. M. Wrigley in *GR*, 41 (1951), 7-65.

Brand, Donald Dilworth (1905-1984)

Ph.D., University of California, 1933. University of New Mexico, 1934-47; University of Michigan, 1947-49; University of Texas, 1949-75. *Prehistoric Settlements of Sonora* (with C. Sauer) (1931); *Aztatlán* (with Sauer) (1932); *Quiroga* (1951); *Mexico: Land of Sunshine and Shadow* (1966). *60 Years of Berkeley Geography* (1983), 15-16.

Brazil

The efflorescence of scientific geography in Brazil took place in the 1940s as the result of the activities of pioneer institutions created in the preceding decade. These were: the first university chairs of geography (1934) at São Paulo and Rio de Janeiro; the setting up of the National Council of Geography (1937), incorporated into the Brazilian Institute of Geography (IBGE); and the Association of Brazilian Geographers (AGB), founded in São Paulo (1934). Previously, the production of geographical knowledge by Brazilians had been, with rare exceptions, nonscientific or the result of contributions by specialists in related sciences like geology or botany.

The initial motivation for university courses in geography and history was the training of teachers for secondary education; but, due above all to the influence of French instructors, research was associated with this objective from the beginning and, thanks to the work of the AGB in diffusion of geographical thought, was taken up by the country's other centers, particularly in the 1950s.

From the beginning, the models for the geography practiced in Brazil reflected and, with some time-lag, went along with what was being produced at academic centers in the northern hemisphere. There must also be taken into account, however, the relationship between the paradigms adopted and the topics given priority in the face of changes in Brazil's economic, political and social makeup since World War II. As an exporter of primary goods that had in the early 1950s a social structure that was still essentially agrarian and pre-capitalistic, Brazil underwent an intensive industrialization, accompanied by the formation of an internal market, explosive urbanization, and a technological modernization that was fostered by the expansion of capitalism in city and countryside. Along with this, growing political and administrative demands began to create the first challenges to geographers. The federal government, chiefly through the IBGE, promoted research aimed at speedier and more extensive information gathering, the preparation of studies and surveys, and, above all, the regionalization of the national territory. At the same time, the job market for geographers gradually expanded at all administrative levels in local and regional planning.

Since World War II, there have been four major phases in the development of geographical research in Brazil. The first followed the Vidalian possibilist model of the French regional school, with the production of monographs under the influence of the French professors Pierre Deffontaines (1894–1978), Pierre Monbeig (1908–1987), and François Ruellan (1894–1975). The preoccupation with man-environment relations fitted well into the concept of an essentially agrarian Brazil and placed strong emphasis on the role of physical geography. Following this, there was some influence of the Hettnerian school of areal differentiation, under the direct influence of Leo Waibel (1888–1951) and the works of Preston James (1899–1986), which promoted concentration on thematic geography and stressed the role of the economy.

A new period began in the aftermath of the 18th International Geographical Congress (Rio de Janeiro, 1956). Major influences included those of the French geographer Pierre George (b. 1909), the new regional theories stemming from the United States, and the renewal of studies in urban geography. The vision of an urban-industrial country then in the process of formation stirred strong interest in spatial interaction, but research then was still largely idiographic. At the same time, applied research increased throughout the 1960s, with the penetration of government planning at all levels.

The next period was characterized by the introduction of quantitative methods and theoretical geography, and intense controversies that pit a group of geographers of the IBGE (where the new geography flourished most) against those of other centers such as the University of São Paulo. The spread of quantitative and theoretical geography owed much to John P. Cole's sojourn in Rio (in 1969 and 1970) and the several visits of Brian J. L. Berry. The peak of this new paradigm came in the 1970s. At that time the pattern of domestic manufacturing to reduce imports became dominant, and important questions began to be raised regarding the effects of the explosive urbanization and the rapid transformations in agriculture that came from the intensified borrowing of capital. These facts dictated ways of applying geography and the themes given preference, such as new orientations in agrarian geography that went well beyond mere profession of faith in a geography with economic base as a social science.

These things also came to dictate a major involvement of anti-quantitative investigators in social geography and therefore the spread of a radical geography in the form of "critical geography." Although not winning unanimity of adherence like theoretical geography, this new wave became characteristic of Brazilian geography in the 1980s, and its propagation owes much to the return of Milton Santos to Brazil in 1978. Among other themes, it placed heavy emphasis on the social production of space.

The development of a social geography and the responses to theoretical locational geography made more acute the problems of practice divorced from physical and human geography. However, environmental questions, combined with contrasts in occupation of territorial zones and the controversy over their potentialities, conferred a social meaning on physical geography, provoking a confluence of the two traditionally dichotomous fields. Among other factors, obligations imposed by the federal government in the working out of previous studies and in turning out environmental impact reports for various projects that required engineering forced geographers to develop a special awareness of the need for interdisciplinary dialogue.

On the other hand, spatial consequences of every kind arising out of governmental planning and current politico-administrative activity aroused geographers to the theme of territorial management, and thus to the recent development of a new political geography. Also contributing to professional diversification are advances in new techniques, particularly those of "geoprocessing." Brazil has centers where geographers apply themselves to interpretation of satellite imagery and employ computers to produce, among other things, sophisticated thematic maps.

At the present time, the situation of the profession is virtually the reverse of what it was fifty years ago. University courses aimed at training secondary school teachers are threatened by a crisis due to low salaries for teachers, while post-graduate courses are gaining in strength and number. Scientific research in the universities flourishes and expands according to available resources, and opportunities multiply for the professional exercise of geography as a technical tool, whether in governmental and planning agencies or in enterprises of an engineering or consulting nature. With all this, the days when research was monopolized by Rio de Janeiro and São Paulo are long gone.

BIBLIOGRAPHY

Becker, B. K., "Geography in Brazil in the 1980s: Background and Recent Advances," *PHG*, 10 (1986), 157-183.

Bernardes, N., "Address Delivered by Professor Nilo Bernardes on the Occasion of the Closing Session of the Second Latin-American Regional Conference at Rio de Janeiro on 20 August 1982," *IGU Bulletin*, 33 (1983), 12-22.

Geiger, P., "Industrialização e Urbanização no Brasil, conhecimento e atuação da geografia," *Revista Brasileira de Geografia*, 50 (1988), 59-84.

James, P. E., and G. J. Martin, *All Possible Worlds* (2nd ed., 1981).

Monteiro, C. A., *A Geografia no Brasil (1934-1977), avaliação e tendências* (1980).

• *Nilo Bernardes*
(translated by C. Julian Bishko)

Brazilian Institute of Geography and Statistics (IBGE) (Fundação Instituto Brasileiro de Geografia e Estatística)

The IBGE came about through the union in 1937 of the National Council of Statistics (CNE), founded in 1935, and the National Council of Geography (CNG). The former's primary task had been to carry out the decennial censuses and maintain regular annual statistics, while to the latter initially fell the task of preparing the 1:1,000,000 world map, doing larger-scale maps, and serving as a member of the International Geographical Union (IGU). In pursuit of these objectives, the CNG from the start promoted geographical research and recruited professional geographers, creating in 1949 a Division of Geography especially devoted to systematic and regional studies. When the Ministry of Planning and General Coordination was established in 1964, the IBGE became one of its bureaus. Today, following several reorganizations, the IBGE is a foundation that consists of three directorates: the Directorate of Research and Inquiries, in charge of the censuses, current statistics, and national indicators; the Directorate of Geosciences, including the Departments of Geodesy and Cartography and of Natural Resources and Environment Studies, which have both specialists in physical geography and other professionals; and the Directorate of Geography, with specialists in topical human geography and regional studies. The last is the computing directorate. The IBGE produces an enormous quantity of maps and charts at various scales, censuses and other statistical compilations, topical atlases (outstanding among which is the National Atlas of Brazil), and various sorts of geographical syntheses, publications, and reports. Since January 1939 the IBGE has also published the *Revista Brasileira de Geografia*, the most important journal of its kind in the country.

• *Nilo Bernardes*
(translated by C. Julian Bishko)

Brigham, Albert Perry (1855-1932)

M.A., Harvard University, 1892. Colgate University, 1892-1925. *Textbook of Geology*

(1901); *Introduction to Physical Geography* (with G. K. Gilbert) (1902); *Geographic Influences in American History* (1903); *The United States of America* (1927). *AAAG*, 20 (1930), 51-104; P. James in *GBS*, 2 (1978), 13-19; obits. by L. Martin in *GR*, 22 (1932), 499-500; and by O. E. Baker in *AAAG*, 23 (1933), 27-32.

British Library, Map Library

The library departments of the British Museum (founded in 1753) were transferred in 1973 to the newly founded British Library. With this administrative change the Map Room of the British Museum became the Map Library of the British Library, which ranks as the national library of the United Kingdom. (Scotland and Wales also have their own national libraries, in Edinburgh and Aberystwyth, respectively.)

The map collections of the British Museum became a separate administrative unit in 1844, with Richard Henry Major in charge as Keeper of the Department of Maps and Charts. In 1892 the Map Room became part of the Department of Printed Books. A reorganization of the British Library in 1985 established the Map Library as part of Special Collections, in company with Manuscripts, Music and Stamps. Officers in charge of the Map Room, later Map Library, succeeding R. H. Major, were R. K. Douglas, Basil Soulsby, John A. J. de Villiers, Frederick Sprent, Edward Lynam, R. A. Skelton, and Helen Wallis. The present Map Librarian, Tony Campbell, who succeeded Helen Wallis in 1986, has indirect responsibility for the whole of the map collections of the British Library.

Total holdings of printed and manuscript maps, atlases, globes and other cartographic items in the Map Library number some 1,750,000. The most requested materials are from King George III's Topographical Collection and from the comprehensive holding of Ordnance Survey production. The Map Library, under the Copyright Act of 1911, is the national repository of British map production, past and present. It is also the retrospective archive of remotely sensed imagery for the United Kingdom.

The Map Library is situated in Bloomsbury, London, in the King Edward Building of the British Museum, its home for more than seventy years. In 1996 it will move to the new British Library building now under construction at St. Pancras, about a mile to the north. There it will be possible for cartographic materials to be brought together from the whole British Library for combined consultation, and pollution-free storage will be provided.

BIBLIOGRAPHY

Wallis, H., "The Map Collections of the British Museum Library," pp. 2-20 in *My Head is a Map*, ed. by H. Wallis and S. Tyacke (1973).

Wallis, H., Y. Hodson, P. Barber, A. S. Cook, and S. Tyacke, "British Library Map Collection," a special issue of *The Map Collector*, no. 28 (September 1984).

• *Helen Wallis and Tony Campbell*

Broek, Jan Otto Marius (1904-1974)

Ph.D., University of Utrecht, 1932. University of California, Berkeley, 1936-46; University of Utrecht, 1946-48; University of Minnesota, 1948-70. *The Santa Clara Valley, California* (1932); *The Economic Development of the Netherlands Indies* (1942); *Geography, Its Scope and Spirit* (1965); *A Geography of Mankind* (with J. Webb) (1968). *GOF* (1970); *WWWA*.

Brookfield, Harold Chillingworth (b. 1926)

Ph.D., London School of Economics, 1950. University of Natal, 1952-55; University of New England, 1955-57; McGill University, 1970-75; University of Melbourne, 1976-82; Australian National University, 1957-69, 1982 to the present. *Melanesia* (with D. Hart) (1970); (ed.) *The Pacific in Transition* (1973); *Interdependent Development* (1975); *Land Degradation and Society* (with P. Blaikie) (1987). *Who's Who in Australia.*

Brooks, Charles Franklin (1891-1958)

Ph.D., Harvard University, 1914. Yale University, 1915-18; Clark University, 1921-32; Harvard University (Blue Hill Observatory), 1931-57. *Why the Weather?* (1924); *Climatology of the West Indies* (with R. DeC. Ward) (1934);

Climatic Maps of North America (with A. J. Connor *et al.*) (1936); *The Climates of North America* (with R. Ward) (1936). *WWWA*; obits. by S. Van Valkenburg in *AAAG*, 49 (1959), 461-465; and R. G. Stone in *GR*, 48 (1958), 443-444.

Brown, Ralph Hall (1898-1948)

Ph.D., University of Wisconsin, 1925. University of Colorado, 1925-29; University of Minnesota, 1929-48. "Materials Bearing upon the Geography of the Atlantic Seaboard," *AAAG*, 28 (1938), 201-231; "The American Geographies of Jedidiah Morse," *Ibid.*, 31 (1941), 145-217; *Mirror for Americans* (1944); *Historical Geography of the United States* (1948). L. M. Coppens in *GBS*, 9 (1985), 15-20; H. C. Darby in *IESS*, 2 (1968), 155-156; and obits. by S. Dodge in *AAAG*, 38 (1948), 305-309; and anon. in *GR*, 38 (1948), 505-506.

Brückner, Eduard (1862-1927)

Dr.phil., University of Munich, 1885. University of Bern, 1888-1904; University of Halle, 1904-06; University of Vienna, 1906-27. *Klimaschwankungen seit 1700* (1890); *Die feste Erdrinde und ihre Formen* (1897); (with A. Penck) *Die Alpen im Eiszeitalter* (3 vols., 1901-09); *Klimaschwankungen und Völkerwanderungen* (1912). *Österreichisches Biographisches Lexikon 1815-1950*, 1 (1957), 120; H. Kinzl, "Eduard Brückner: Ein führender Gletscher- und Eiszeitforscher," *Geographisches Taschenbuch 1970/72* (1970), 262-265.

Brunhes, Jean (1869-1930)

D.-ès-L., University of Paris, 1902. University of Fribourg, 1896-1912; University of Lausanne, 1907-12; Collège de France, 1912-30. *L'irrigation* (1902), *La géographie humaine* (1910), *Géographie humaine de la France* (2 vols., 1920-1926), *La géographie de l'histoire* (with C. Vallaux) (1921). Y. Chatelain in *DBF*, 7 (1956), cols. 554-555; J. Taton in *DSB*, 2 (1970), 538-539; M. J.-B. Delamarre in *LGF* (1975), 49-80; P. Deffontaines, "La aportación geografica de Jean Brunhes," *Revista de geografía* (Barcelona), 2 (1968), 161-167; obits. by E. de Martonne in *AG*, 39 (1930), 549-553; and C. Vallaux in *La Géographie*, 34 (1930), 237-239.

Bryan, Kirk (1888-1950)

Ph.D., Yale University, 1920. U.S. Geological Survey, 1912-27; Yale University, 1914-1917; Harvard University, 1927-50. *Ground Water for Irrigation in the Sacramento Valley, California* (1916); *The Papago Country, Arizona* (1925); *Geological Antiquity of the Lindenmeier Site in Colorado* (with L. Ray) (1940). S. Judson in *DSB*, 2 (1970), 548-549; R. DeFord in *DAB*, supplement 4 (1974), 118-119; obits. by L. Ray in *GR*, 41 (1951), 165-166; and D. Whittlesey in *AAAG*, 41 (1951), 88-94.

Butzer, Karl Wilhelm (b. 1934)

Dr.phil., University of Bonn, 1957. University of Bonn, 1957-59; University of Wisconsin, 1959-66; University of Chicago, 1966-84; University of Texas, 1984 to the present. *Environment and Archeology* (1964, rev. 1971); *Early Hydraulic Civilization in Egypt* (1976); *Geomorphology from the Earth* (1976); *Archaeology as Human Ecology* (1982). *GOF* (1980); *WWA*; *Dictionary of International Biography*.

Camena d'Almeida, Pierre (1865-1943)

D.-ès-L., University of Paris, 1893. University of Caen, 1888-99; University of Bordeaux, 1899-1935; *Les Pyrénées, développement de la chaîne* (1893); *La France* (with P. Vidal de la Blache) (1902, rev. 1909); *Géographie universelle*, vol. 5; *Etats de la Baltique, Russie* (1932). L. Papy in *GBS*, 7 (1983), 1-4; obit. by H. Cavaillès in *AG*, 43-44 (1945), 67-68.

Canada

The origins of geography teaching and geography as a university discipline in Canada are difficult to define. In most Canadian provinces, geography courses modeled after British traditions of education were given in the late nineteenth century in elementary and secondary schools. Geographers from France, and Canadians educated in France, gave a few geography courses in some of the Quebec universities in the 1920s. The first university to create a partial department of geography was the University of British Columbia, in 1923, and the first full department was established by Griffith Taylor at the University of Toronto in 1935.

Although mapmakers were called "geographers" in the federal government during the 1930s, the first professional geographer to be appointed was J. Lewis Robinson in the Division of Northern Affairs in 1943. This was followed by the establishment of a federal Geographical Branch, headed by J. Wreford Watson, in 1947. All geographical work in the federal government was meant to be centralized into this research and publication branch. In 1967, however, the branch was reduced to deal only with the Atlas of Canada work; the number of government geographers was greatly increased and they were dispersed through many government departments and branches. Geographical research units were started in most provincial governments during the 1950s and 1960s and geographers now handle a variety of tasks related to provincial natural resource and regional planning matters.

About fifty academic geographers came together in Montreal in 1951 to found the Canadian Association of Geographers. By 1990 it had a membership of more than 1,000 academic and professional geographers, and the association published two periodicals, *The Canadian Geographer* and *The Operational Geographer*. Because education is under provincial jurisdiction in Canada there has been no national organization or publication for geography school teachers.

The development of geography as a discipline was led by geography departments in Canadian universities. All of the large universities established geography departments during the 1950s, and by the 1960s most of the small universities and colleges had departments. This was twenty to forty years later than the founding of geography departments in many U.S. universities. By 1990 there were geography departments in more than forty Canadian universities.

The first graduate work was started at the University of Toronto in the 1940s and at McGill University and the University of British Columbia in the 1950s. In these early years most Canadian geography students received their graduate training in universities in the United States or, occasionally, in France or Britain. By the 1970s, however, most Canadian graduate students could remain in Canada for graduate-level education and, in addition, many foreign students now come to Canada for graduate training in geography.

The discipline of geography in Canada has continued to subdivide into numerous subfields and specializations, as in other countries. In the larger geography departments more than fifty undergraduate courses are now available for geography majors and undergraduate electives; these departments have twenty to thirty geography faculty and are generally larger than most U.S. geography departments.

Physical geography courses and research were significant in the early years of Canadian academic and government geography, partly because of the need for knowledge and understanding of the physical environment of the vast unoccupied parts of the country. Economic geography work expanded later as Canada and its provinces turned to natural resource developments and industrialization during the 1960s. Eighty percent of Canada's population now live in urban areas and therefore urban geography—often leading to urban planning—is the most popular course in the universities and colleges. Much geographical research and publication is now focused on social and economic concerns in urban areas. Regional geography, particularly for areas outside Canada, has declined in terms of faculty research and student enrollments.

Because education curricula differ slightly in each Canadian province, geography has different significance across Canada at elementary and secondary school levels. In the four western provinces geography is usually combined with history, and perhaps sociology or civics, into "social studies." Depending on the previous training and interests of the individual teacher, geography concepts may be either lost or important in such courses. On the other hand, geography is well-established and strong in schools in Ontario, where there is an active geography teachers organization with its own geography periodical. In Quebec also, as in France, geography is well-known in schools. A general lack of school geography in the Atlantic Provinces may be related to the few and smaller geography departments in the universities of that region.

Professional geography is becoming known in Canadian business and industry. Many large firms employ geographers to advise on environmental matters, marketing strategy, or store locations. Many undergraduate geographers receive further training in planning schools or faculties and seek employment in environmental consulting firms or in urban and regional planning organizations. Geography in Canada is both an academic subject and discipline in the schools and universities and also a professional and applied field of employment.

Bibliography

Dean, W. G., *et al.*, "Canadian Geography, 1967," *CG*, 11 (1967), 195-371.

Hamelin, L.-E., "Petite histoire de la géographie dans le Québec et à l'Université Laval," *Cahiers de géographie de Québec*, no. 13 (1963), 137-152.

Hamelin, L.-E., and L. Beauregard, eds., *Retrospective 1951-1976* (25th anniversary of the Canadian Association of Geographers) (1979).

Robinson, J. L., "The Development and Status of Geography in the Universities and Governments of Canada," *Yearbook of the Association of Pacific Coast Geographers*, 13 (1951), 3-13.

——, "The Production and Employment of Geographers in Canada," *PG*, 29 (1977), 208-214.

——, "Geography in Canada," *Ibid.*, 38 (1986), 411-417.

• *J. Lewis Robinson*

Canadian Association of Geographers

Professor Griffith Taylor of the University of Toronto tried without success to persuade the Royal Society of Canada, of which he was a fellow, to start a geographical section, and so he had to join the section on geological sciences.

He also tried unsuccessfully to get the Royal Canadian Geographical Society, a popular society analogous to the National Geographic Society, interested in doing more with professional geography. As geography expanded in Canadian universities after World War II, geographers saw a need for some sort of professional association. Initial discussions took place in eastern Canada in 1949, and an informal group began in the West in 1950. On 29 August 1950 a number of university and government geographers met, drew up a provisional constitution, and chose Wreford Watson as interim president. The Meeting of Establishment of the Canadian Association of Geographers took place 30 May 1951, and Donald Putnam was elected president. The association's journal, *The Canadian Geographer*, was issued irregularly in the first decade but has been a quarterly since 1961. Responding to interest in applied geography, the CAG has published a new journal, *The Operational Geographer*, since 1983. The association's annual meetings often coincide with those of the other Learned Societies ("The Learneds"). A genuine effort has been made to integrate francophone Canadian geographers in the activities and publications of the association.

BIBLIOGRAPHY

Fraser, J. K., "The Development of the Canadian Association of Geographers, 1951-1976," pp. 5-17 in *Retrospective 1951-1976*, ed. by L.-E. Hamelin and L. Beauregard (1979).

Watson, J. W., "The Canadian Association of Geographers: A Sketch of the Preliminaries," *CG*, no. 1 (1951), 1-3.

——— , "The Development of Canadian Geography: The First Twenty-Five Volumes of the *Canadian Geographer*," *Ibid.*, 25 (1981), 391-398.

Capel Sáez, Horacio (b. 1941)

Doctorate, University of Barcelona, 1972. University of Murcia, 1963-66; University of Barcelona, 1966 to the present. *Capitalismo y morfología urbana en España* (1975); *Filosofía y ciencia en la geografía contemporánea* (1981); *Ciencia para la burguesía* (with others) (1983); *Geografía humana y ciencias sociales* (1987).

Capot-Rey, Robert (1897-1977)

D.-ès-L., University of Paris, 1934. University of Algiers, 1935-66. *La région industrielle sarroise* (1934); *Géographie de la circulation sur les continents* (1946); *Le Sahara français* (1953); *Borkou et Ounianga* (1961). J. Bisson in *GBS*, 5 (1981), 13-19; obit. by J. Bisson and P. Rognon in *AG*, 87 (1978), 59-73 (plus bibliog. on pp. 74-77 comp. by F. Grivot).

Caraci, Giuseppe (1893-1971)

Laurea, University of Florence, 1917. University of Milan, 1928-30; University of Messina, 1932-36; University of Pisa, 1936-46; University of Rome, 1946-64. *Tabulae Geographicae Vetustiores in Italia adservatae* (3 vols., 1926-32); *Tre fiorentini del Rinascimento: Indagini sulla storia della geografia all'epoca delle grandi scoperte geografiche* (6 vols., 1952-57); "Italiani e Catalani nella primitiva cartografia nautica medievale," Rome, Istituto Scienze Geografiche e Cartografiche della Facoltà di Magistero, *Memorie Geografiche*, 5 (1959); "La Vinland Map," *Studi Medievali* (Rome), 1966, 509-615. Obits. by R. Riccardi in *BSGI*, ser. 10, 1 (1972), 1-28; and O. Baldacci in *RGI*, 79 (1972), 71-79.

Carter, George Francis (b. 1912)

Ph.D., University of California, Berkeley, 1942. Johns Hopkins University, 1943-67; Texas A&M University, 1967-78. *Plant Geography and Culture History in the American Southwest* (1945); *Pleistocene Man at San Diego* (1957); *Man and the Land* (1964); *Earlier than You Think* (1980). *GOF* (1970), *60 Years of Berkeley Geography* (1983).

Cartography

Defined concisely as "the art and science of map making," cartography may have a broad scope or a narrow focus. The United States government, for example, uses the job titles *Cartographer* and *Cartographic Technician* across a wide range of expertise and duties, including inspection and evaluation of maps and aerial photographs, photogrammetric compilation from stereomodels, manual digitizing from printed and manuscript maps, map drafting, production planning and project management, quality inspection of work by outside contractors, map analysis, and the development and testing of mapping software. Yet in the minds of many persons within and outside government the term cartography is a synonym for *map finishing*, a production phase traditionally confined to graphic design, line tracing, and type placement. In the 1980s rapid development of computer-assisted cartography added further to the scope of cartography's broad definition by more fully integrating map design and map analysis, and by making the narrow map-finishing function a convenient and useful tool widely available to artists and illustrators, geographers, journalists, marketing and demographic analysts, and physical and social scientists. Ironically, the demise of the paper map as the principal device for storing geographic data accompanies an explosion in mapmaking in general and customized maps in particular; in addition to countless ephemeral maps on video screens, paper maps are now more abundant and more widely used than before computer-assisted cartography.

The heart if not the mind of geography, cartography plays two roles in academic geography: communication and analysis. Long used to describe locations, networks, boundaries, terrain, and spatial patterns, maps also communicate the results of geographic analysis and synthesis. Small-scale generalized maps collected in atlases or at the end of a report summarize salient points in a spatial format. Maps also present for evaluation and implementation the results of location-allocation modeling, regionalization and redistricting, and other forms of spatial modeling. The geographer-cartographer chooses a combination of map scale, projection, and symbolization to organize and represent relationships among selected geographic features. Geographers and others who use maps to communicate need to know the uses and limitations of map projections, the variety and appropriate uses of map symbols, and the relevant principles of graphic design, aesthetics, and visual perception. Cartographic design is an indispensable part of a solid academic program in geography.

But the map is also a tool for analysis. In exploratory geographic analysis, maps are useful for refining the problem, understanding the data, suggesting hypotheses, and designing the research. Interactive computer graphics supports efficient visual inspection for spatial trends, case clusters, interesting anomalies, and relationships among various distributions. High-interaction graphics workstations foster the integration of map analysis with statistical graphics and numerical analysis. As a recognized subfield within cartography, *analytical cartography* promotes geographic analysis through trend-surface analysis, spatial interpolation and error analysis, bi-dimensional regression, automated point-pattern analysis, overlay analysis and buffering in a geographic information system (GIS), and other techniques for enhancing, manipulating, and interrelating cartographic images.

Always a part of academic geography, cartography did not emerge as a distinct specialization within geography until the early 1950s. Although public administrators, navigators, and soldiers have used maps for several thousand years to record boundaries, plan journeys, and organize military intelligence, use of the map as a research tool made little progress until the rise of the earth and social sciences in the mid-nineteenth century, when mapping and map analysis aided the development of geology, meteorology, and epidemiology. Despite their obvious value, analytical cartography and map design tended to be informal and intuitive, and during the first half of the twentieth century the geographic literature said little about cartography per se. Early exceptions include the writings of J. Paul Goode, of the University of Chicago, who founded a popular classroom atlas and developed improved projections for small-scale world maps, and Arthur R. Hinks, of Cambridge University, who between 1913 and 1944 published five editions of *Maps and Surveys*, a concise and comprehensive examination of the roles of geodesy and surveying in systematic large-scale mapping. Erwin Raisz, at Harvard University, perfected the physiographic diagram as a tool for studying the geography of landforms and wrote *General Cartography* (1938), a comprehensive text on the design and production of both large- and small-scale maps.

Richard Edes Harrison, who worked as a cartographic illustrator for *Fortune*, *Life*, and *Time* magazines from the late 1930s through the post-World War II period and also taught at Syracuse University, produced many dramatic maps with innovative, earth-from-space projections.

In the 1950s cartography attained the critical mass of a geographic subdiscipline. Undisputed leader in the field's development was Arthur Robinson, of the University of Wisconsin. Robinson's influential textbook, *Elements of Cartography* (1953), stimulated formal undergraduate and graduate coursework on map design, and his scholarly research monograph *The Look of Maps* (1952) served for two decades as an exemplar of the application of experimental scientific methods to the evaluation and improvement of map symbols. Robinson sponsored numerous doctoral dissertations on numerical cartography, the perception of cartographic symbols, and the history of cartography. During the late 1950s, and the 1960s and 1970s, Robinson's doctoral program at Wisconsin and doctoral programs led by George Jenks at the University of Kansas and John Sherman at the University of Washington trained nearly fifty Ph.D.s, most of whom became academic geographers identified principally as cartographers.

Although few of these new academics distinguished themselves as geographic illustrators or as editors of atlases, many focused their research on how people acquire information from maps. During the 1960s and 1970s, some doctoral students and other researchers embraced the *psychophysical paradigm* and, in an often-vain attempt to fine-tune the scaling of symbols on statistical maps, tried to adapt to cartographic design the subject-testing procedures of experimental perceptual psychology. Although psychophysical testing generally confirmed effects observed years earlier for graphics far simpler than maps, attempts to "rescale" map symbols to trick the "average map reader" into seeing what the map author wanted him to see often failed. Not only do map viewers vary widely in their information-seeking goals, geographic expertise, and spatial ability, but printing often thwarts fine-tuning by distorting graytones and colors unpredictably.

While some cartographers attempted to follow the psychologist's footsteps further, from perception into cognition, other investigators abandoned psychophysics but pursued subject-testing for a serious and comprehensive evaluation of particular symbolization schemes. For example, Judy Olson, at Boston University and later at Michigan State University, explored the appropriate use of color in two-variable choropleth maps. With the support of the US Geological Survey, John Sherman, Joseph Wiedel, and others evaluated the effectiveness of various types of tactile maps for improving the mobility of the blind. Despite the inconclusiveness of psychophysical and cognitive studies, subject-testing remains an important, useful tool for improving cartographic communication.

Cartographic researchers have been adept at using the computer to create new, more effective map symbols and better means of preprocessing and filtering map data. George Jenks devised a method for selecting an optimum, minimum-error set of categories for choropleth maps. Waldo Tobler, at the University of Michigan and later at the University of California, Santa Barbara, developed a means of avoiding classification altogether by plotting a continuum of graytone area symbols. Mark Monmonier, at Syracuse University, designed an algorithm for choosing optimal viewing azimuths for oblique views of three-dimensional statistical surfaces. Tobler devised an algorithm for computing contiguous area cartograms, and John Snyder, at the US Geological Survey, made a number of important contributions to the theory and computation of map projections. David Rhind and his associates at the University of Durham and later at Birkbeck College, University of London, produced a number of innovative experimental computer-generated atlases, including an interactive electronic atlas with massive amounts of geographic, pictorial, and text information stored on videodisk.

Cartography in the 1990s confronts a number of important unsolved problems, including automated map generalization, automated label placement, automated generation of sequences of views for multivariate spatial-temporal datasets, automated blunder protection for interactive graphics systems, and the measurement and graphic representation of error and uncertainty. Because graphic display and efficient human visualization are key elements of operational GIS technology, cartographic research is at once both basic and applied. Effective and efficient solutions to practical problems in cartographic display depend upon the development of better theory and new algorithms, and a fuller understanding of the map viewer.

Bibliography

Monmonier, M., *Technological Transition in Cartography* (1985).

Monmonier, M., and G. A. Schnell, *Map Appreciation* (1988).

Muehrcke, P. C., "Whatever Happened to Geographic Cartography?," *PG*, 33 (1981), 397-405.

Olson, J. M., "Future Research Directions in Cartographic Communication and Design," pp. 257-284 in *Graphic Communication and Design in Contemporary Cartography*, ed. by D. R. F. Taylor (1983).

Petchenik, B. B., "A Map Maker's Perspective on Map Design Research," pp. 37-68 in *Ibid.*

Robinson, A. H., R. D. Sale, J. L. Morrison, and P. C. Muehrcke, *Elements of Cartography* (5th ed., 1984).

• *Mark Monmonier*

Central-Place Theory

Although there were antecedents, the formulation of central-place theory is properly attributed to the German geographer Walter Christaller in his 1933 book *Die zentralen Orte in Süddeutschland.* What Christaller sought was a "general deductive theory" designed "to explain the size, number and distribution of towns" in the belief that "some ordering principles govern the distribution." This theory, he thought, "could be designated as the theory of urban trades and institutions," to be set alongside Thünen's theory of agricultural location and Weber's theory of industrial location. He argued that:

(1) The basic function of a city is to be a central place providing goods and services for a surrounding tributary area. The term "central place" is used because to perform such a function efficiently, a city locates at the center of minimum aggregate travel of its tributary area, *i.e.*, central to the maximum profit area it can command.

(2) The centrality of a city is a summary measure of the degree to which it is such a service center; the greater the centrality of a place, the higher is its "order."

(3) Higher order places offer more goods, have more establishments and business types, larger populations, larger tributary areas and tributary populations, do greater volumes of business, and are more widely spaced than lower order places.

(4) Low order places provide only low order goods to low order tributary areas; these low order goods are generally necessities requiring frequent purchasing with little consumer travel. High order places provide not only low order goods but also high order goods sold by high order establishments with greater conditions of entry. These high order goods are generally "shopping goods" for which the consumer is willing to travel longer distances, although less frequently.

(5) The higher the order of goods provided, the fewer are the establishments providing them, the greater the conditions of entry and trade areas of the establishments, and the fewer and more widely spaced are the towns in which the establishments are located. Because higher order places offer more shopping opportunities, their trade areas for low order goods are likely to be larger than those of low order places, since consumers have the opportunity to combine purposes on a single trip, and this acts like a price-reduction.

(6) Central places fall into a hierarchy comprising discrete groups of centers. Centers of each higher order group perform all the functions of lower order centers plus a group of central functions that differentiates them from and sets them above the lower order. A consequence is a "nesting" pattern of lower order trade areas within the trade areas of higher order centers, plus a hierarchy of routes joining the centers.

(7) The hierarchy may be organized according to (a) a market principle, according to which nesting follows a rule of threes. Deviations from the market principle may be explained by (b) the principles of traffic, which give rise to linear patterns and nesting according to fours, or (c) the socio-political or administrative "separation" principles, with the hierarchy organized according to a rule of sevens, and with low order twin cities where one could predict a high order place to be located if only the marketing principles were operative.

Christaller's theory was ignored by his contemporaries but played a key role in geography's misnamed "quantitative revolution," which unfolded from beginnings in the late 1950s as economic and urban geographers explored the theoretical foundations of their craft. As the science of geography has evolved, so has central-place theory. Numerous alternatives to Christaller's ideas have been advanced and subsequently integrated into a common mathematical frame. The theory also has become a key component of the regional planner's growth center strategies of economic development and the applied marketing geographer's techniques of retail trade area evaluation. A contemporary assessment and review of the theory is provided by Brian J. L. Berry and John B. Parr, with Bart J. Epstein, Avijit Ghosh and Robert H. T. Smith, in *Market Centers and Retail Location: Theory and Applications* (1988). *See also* Location Theory.

BIBLIOGRAPHY

Berry, B. J. L., *Geography of Market Centers and Retail Distribution* (1967).

——— , *et al.*, *Market Centers and Retail Location: Theory and Applications* (1988).

Christaller, W., *Die zentralen Orte in Süddeutschland* (1933) (trans. by C. W. Baskin, *Central Places in Southern Germany*, 1966).

Hottes, R., "Walter Christaller: Ein Uberblick über Leben und Werk," *Geographisches Taschenbuch 1981/1982* (1981), 59–70 (abridged trans. by G. Weigend in *AAAG*, 73 [1983], 51–54).

• *Brian J. L. Berry*

Chabot, Georges (1890–1975)

D.-ès-L., University of Paris, 1927. University of Dijon, 1928–45; University of Paris, 1945–60. *Les plateaux du Jura central* (1927); *Les villes* (1948); *Traité de géographie urbaine* (with J. Beaujeu-Garnier) (1964); *Géographie régionale de la France* (1966). Obit. by J. Beaujeu-Garnier in *AG*, 85 (1976), 98–100; bibliog. by F. Grivot in *Ibid.*, 341–347.

Chatterjee, Shiba Prasad (1903–1989)

Ph.D., University of London, 1936. University of Rangoon, 1926–32, 1948–49; University of Calcutta, 1936–73; President of IGU, 1964–68. (ed.) *Bengal in Maps* (1949); (ed.) *The National Atlas of India* (preliminary Hindi edition 1957, English edition issued serially 1959 onward); (ed.) *Planning Atlas of the Damodar Valley Region* (1969). Obits. by P. Nag in *GJ*, 155 (1989), 449; and S. C. Chakraborty in *Annals of the National Association of Geographers, India*, 9 (1989), 93–94.

Chile

Geography is both old and new in Chile. As a school subject it has had an active presence but a low profile since the establishment of public schools early in the nineteenth century. Most of the time it has been taught in tandem with history, and until recently aspiring teachers would have had a double major in history and geography.

As an academic subject in higher education, geography has a decidedly shorter history. In the mid-1850s the University of Chile started granting degrees to "Geographer-Engineers" (within the same faculty granting degrees to Mining Engineers and Bridge Engineers). These engineers devoted their energies to satisfying the government's need for survey and cartographic data and the delimitation of the national territory. The creation of the Instituto Pedagógico in the Faculty of Philosophy and Humanities of the University of Chile in 1889 initiated the permanent presence of geography in the university curriculum. For more than half a century the Pedagógico was the only higher education institution in the country training geographers (or, rather, geography teachers).

German geographers had a marked influence on the development of geography in Chile. Perhaps the single most important person was Hans Steffen (1865–1936), who arrived in Chile as a very young Doctor of Geography in 1889 and taught in the Pedagógico until 1913, while performing important research in Western Patagonia. Well into the twentieth century, the physical landscape was *the* object of study in Chilean geography.

A change in the purpose of academic geography and in the content of the subject was brought about by Humberto Fuenzalida V. (1904–1966), who, in the early 1940s, succeeded in creating the Instituto de Geografía at the University of Chile. The purpose of the Instituto was to increase the output of geographic research pursued by professional geographers in conjunction with academics interested in related fields of the natural and social sciences. The courses dictated at the Instituto would, ideally, lead to the formation of professional geographers not linked to school teaching. The actual implementation of this ideal did not occur until decades later.

Although a physical geographer and geologist by training and interest, Fuenzalida's postgraduate work in France (1926–1930) helped to reorient geography in Chile. Since the 1950s a number of French geographers have been invited to teach and conduct research at the Instituto; a few U.S. graduate students did the same, and slowly the deep imprint of German physical geography was complemented by an increasing interest in human geography.

Concurrently, the Catholic University of Chile started its own Escuela de Pedagogía (1942), which incorporated from the beginning a teacher preparation track in history, geography and civics education. In the next decade other universities followed suit: the Universidad Católica de Valparaíso, Universidad de Concepción, and the Universidad Técnica del Estado (now the Universidad de Santiago). The Universidad Austral de Valdivia created an Institute of Geosciences.

The national reform of the university system that occurred in the early 1980s has reinforced the role of geography in all universities and professional institutes that offer teacher education. On the other hand, few universities have made a commitment to keeping viable research institutes in geography. Thus, the importance of those institutes can hardly be overemphasized, especially now that a fairly significant number of Chilean geographers have pursued graduate studies and some have completed doctoral degrees in universities abroad (Spain, France, Germany, USA, etc.).

Perhaps the most important group today is the Instituto de Geografía of the Catholic University of Chile (IGUC). Started in the mid-1960s as a Center for Geographic Research within the Department of History and Geography, it became a full-fledged Institute of Geography in 1970, charged with conducting research and educating professional geographers. The preparation of geography teachers is done jointly by IGUC and the Faculty of Education. IGUC is currently initiating a Master's program in geography that will crown a remarkable blossoming of geography in one of the senior universities in the country.

Relatively few non-textbooks on geography have been published in Chile. Perhaps the most remarkable work was CORFO's *Geografía Económica de Chile* (4 vols., 1950–1962), which owed its inspiration and a number of chapters to H. Fuenzalida. There are several journals. The Sociedad Chilena de Historia y Geografía, founded in 1911, continues to publish its *Revista Chilena de Historia y Geografía*, but this journal is devoted predominantly to history. The Instituto Geográfico Militar, established in 1922 and the main mapping agency in the country, has published since 1948 the *Revista Geográfica de Chile "Terra Australis."* In 1986 *Terra Australis* became the official organ of the National Section of the Pan American Institute of Geography and History. The Instituto de Geografía (University of Chile) started its *Informaciones Geográficas* in 1951. It is now published by the "new" Department of Geography, which incorporated the Instituto (1968) and has been part of the Faculty of Architecture and Urbanism since 1985. The Catholic University of Valparaíso started its *Revista Geográfica de Valparaíso* in 1967. In 1974 the Instituto de Geografía of the Catholic University of Chile (IGUC) initiated publication of the *Revista de Geografía Norte Grande*, originally a project of the Taller del Norte Grande (working group on the "Great North," *i.e.*, the desert provinces of Chile). The newest addition to the list is *Revista Chilena de Geopolítica* (1984), published by the Instituto Geopolítico de Chile, directed by Dr. Hernan Santis A. of the IGUC.

Although the professional identity of geographers is not so clear in Chile as it is in other Latin American countries, *e.g.*, in Brazil, geography seems to have a solid future, mainly because of the commitment made by some of the leading universities to the education of professional geographers and the pursuit of graduate programs in the discipline.

BIBLIOGRAPHY

Gangas Geisse, M., *La evolución de la geografía chilena durante el siglo XX. Contextos, tendencias y autores* (1985).

Santis, A. H., and M. Gangas Geisse, "Notas para la historia de la geografía contemporánea en Chile (1950-80)," *Revista de Geografía* (Barcelona), 16-17 (1982-1983), 5-21.

Santis, A. H., and M. Gangas Geisse, "La geografía de Chile en cinco obras publicadas entre 1890 y 1962. Estudio y análisis crítico," *Revista Geográfica de Chile Terra Australis*, 28 (1984-1985), 143-172.

• *José F. Betancourt*

China (People's Republic)

The history of geography in China is at least as long as that of any other part of the world, but it has also suffered from very long periods of isolation from the intellectual currents and discoveries of the rest of the world. The first works in geography—mainly regional surveys of human and physical phenomena—appeared about the fifth century BC, and the first maps followed about the time of Christ. Although the studies were almost entirely concerned with China itself, there were notable expeditions along the land routes to the Mediterranean and also around the Indian Ocean, culminating with the fleets led by Cheng Ho in the early fifteenth century.

However, the institutions of Confucian and feudal China then sank into a stagnant parochialism, unaffected by the European discoveries, Renaissance and succession of revolutions—industrial, political, and scientific. Therefore, when the Western world (including the Russians) forced its way into China in the nineteenth century, it found a crumbling, still-feudal empire that had forgotten its ancient civilization, including geographical knowledge. Apart from a few individual scholars, such as Ferdinand von Richthofen, it was not until the 1920s and 1930s that Western geographers worked and taught in China and Chinese students began to study in Europe and America. The first university geography department dates from 1928. Good scientific work was done on physical geography, *e.g.*, by Zhu Kezhen (1890-1974) in climatology, and in human and regional geography by Weng Wenhao (1889-1971), among others.

But over the past half-century conditions in China have conspired to plunge geography, along with most other civilized endeavors, into darkness for years at a time and to distort the development of the subject. During the dozen years of the Sino-Japanese war and the following civil war and "liberation," chaotic conditions in education and reseach perforce prevailed, although remarkably some good work continued in the south-west interior. Then in the 1950s the Chinese system of education and research was completely reorganized on the Soviet model. This meant an unprecedented concentration of research power and resources in the Academy of Sciences, at the expense of the universities, which had just before been organized on Western lines. For geography in particular this involved a very heavy emphasis on physical geography, especially geomorphology, and the rigid separation of physical and human geography, as decreed by the Soviet leaders. Moreover, by the time of the "Anuchin revolution" in the Soviet Union around 1960, aimed at reforming this system and restoring the balance, China had cut off relations with the Soviet Union. Thus, Chinese geography around 1980 was, in its basic structure and emphasis, probably more frozen into the Stalinist model than was Soviet geography at that time. Human geography had been neglected, criticized and practically forbidden, and, moreover, during the "Cultural Revolution" (1966-76) most academic activities came to a standstill.

In the 1980s there was a wholehearted movement toward the abolition of the physical-human split in geography (which flew in the face of the long-standing Chinese philosophical conviction about the unity and harmony of man and nature, as did the Western and Soviet notions of the "conquest" or "transformation" of nature). It was also then decided to shift the center of gravity from physical towards human geography and also to promote integrated and regional studies. This reversal of theory and practice—the re-establishment of respect for native Chinese traditions as well as those of foreign countries—seems to have gone deep enough to survive the severe convulsions and setback of May-June 1989. The first International Geographical Conference ever held in China, scheduled for August 1990, may provide some further light on these crucial questions. China became a full participating member of the IGU only in 1984, and with a fifth of the world's population, an enormous range of problems of great interest to geographers, a very long tradition in geography, and a rich store of geographical talent—young and old—neither the world nor China can afford yet another period of isolation.

BIBLIOGRAPHY

Hou Ren Zhi, ed., *Historical Atlas of Beijing* (in Chinese) (1988).

Mirsky, J., ed., *The Great Chinese Travelers* (1964).

Needham, J., and Wang Ling, *Science and Civilisation in China*, vol. 3, *Mathematics and the Sciences of the Heavens and the Earth* (1959).

Wu, Chuanjun, *et al.*, *Geography in China* (1984).

• *David J. M. Hooson*

China (Taiwan)

There were no geographical institutions in Taiwan until 1946. In that year, the first geographical institution, the Department of Geography and History at the Taiwan Provincial Teachers College, was established in Taipei. The college expanded and became the National Taiwan Normal University in 1957, and the original department was divided into separate geography and history departments.

Following the withdrawal from the mainland in the late 1940s, some of the leading Chinese geographers resettled in Taiwan (*e.g.*, Professor Hsueh-chuen Sha, Dr. Chi-yun Chang, and Dr. Tang-yueh Sun). Prof. Sha was the chairman of the geography department in Taiwan Normal University from 1949 to 1969. More than a thousand young geographers were trained in his department and served in colleges and middle schools as geography teachers. During the period from 1954 to 1958, Dr. Chi-yun Chang was the Minister of Education in the central government of the Republic of China, and he founded the geography department in National Taiwan University. In 1961, Dr. Chi-yun Chang founded a private university, the Chinese Culture University, located in Yang-min-shan, Taipei, where a complete geography department consisting of both undergraduate and graduate studies was established. All three universities now offer doctorates in geography.

The Geographical Society of the Republic of China consists of around a thousand members. The society has been an active member of the International Geographical Union since 1946. The major geographical journals published in Taiwan are *Geographical Studies*, an annual publication of the department of geography of the National Taiwan Normal University; *Geographical Research*, published annually by the Graduate Institute of Geography of the Normal University; *Science Reports*, issued once a year by the department of geography of the National Taiwan University; and *Science Reports*, an occasional publication of the Institute of Geography of the Chinese Culture University.

Brief mention of some of the leading contributors to the field of geography in Taiwan follows. The late Dr. Chi-yun Chang was the founder of two geography departments and the Chinese Culture University. His specialties were historical geography and the geography of China. Dr. Tang-yueh Sun was the former deputy minister of education and chairman and director of the department of geography in the Chinese Culture University. He specializes in economic geography, cultural geography, and the geography of Europe. Professor Hsueh-chuen Sha was the chairman of the geography department of the Normal University for twenty years. He specializes in political geography and the geography of China. Dr. Jen-hu Chang, son of Chi-yun Chang, is a well known climatologist. After retiring from the University of Hawaii, where he was a professor for more than twenty years, he succeeded his father as chairman of the board of directors at the Chinese Culture University. Professor Hung-hsi Liu specializes in physical geography and the geography of China. He retired from the Normal University after serving as a professor for thirty years (1958–88) and is now the chairman of the geography department at the Chinese Culture University. Professor C. Y. Wang, a specialist in quantitative geography, hydrology, cultural geography, and regional planning, was chairman of the geography department of the National Taiwan University in the 1970s. Dr. Tsai-tien Shih, a geomorphologist, teaches at the Normal University and has served as president of the Geographical Society of the Republic of China since 1987.

BIBLIOGRAPHY

Fuchs, R. J., and J. M. Street, eds., *Geography in Asian Universities* (1976).

Jacobs, J. B., J. Hagger, and A. Sedgley, compilers, *Taiwan: A Comprehensive Bibliography of English-Language Publications* (1984).

Williams, J. F., and C. Chang, "Geography in Taiwan," *PG*, 37 (1985), 219-220.

• *Hung-hsi Liu*

Chisholm, George Goudie (1850-1930)

B.Sc., University of Edinburgh, 1883. University of Edinburgh, 1908-23. *The Two Hemispheres* (1882); *The World As It Is* (1883-84); *Handbook of Commercial Geography* (1889, last ed. 1962); (ed.) *Longmans Gazetteer of the World* (1895). K. Maclean in *GBS*, 12 (1988), 21-33; K. Maclean, "George G. Chisholm: His Influence on University and School Geography," *SGM*, 91 (1975), 70-78; M. J. Wise, "A University Teacher of Geography," *TIBG*, 66 (1975), 1-16; obit. by A. G. Ogilvie and D. C. T. Mekie in *SGM*, 46 (1930), 101-104.

Cholley, André (1886-1968)

D.-ès-L., University of Paris, 1925. University of Lyon, 1923-27; University of Paris, 1927-56. *Les préalpes de Savoie* (1925); *Guide de l'étudiant en géographie* (1942); *Recherches morphologiques* (1957). J. Gras in *LGF* (1975), 153-171; obit. by P. Birot in *AG*, 78 (1969), 129-130.

Chorley, Richard John (b. 1927)

D.Sc., University of Cambridge, 1974. Columbia University, 1952-54; Brown University, 1954-57; University of Cambridge, 1958 to the present. *The History of the Study of Landforms* (with A. Dunn and R. Beckinsale) (2 vols., 1964, 1973); *Frontiers in Geographical Teaching* (with P. Haggett) (1965); *Network Analysis in Geography* (with Haggett) (1969); *Geomorphology* (1984). *GOF* (1982); *WW*.

Chorography/Chorology

As synonyms for regional geography and areal differentiation, the old terms chorography and chorology were revived by nineteenth-century German geographers, Friedrich Marthe and Ferdinand von Richthofen in particular. As Richard Hartshorne noted in 1939, "Richthofen distinguished between a first step, chorography, which is non-explanatory description, providing material for systematic geography, and chorology, a final step, the explanatory study of regions, based on systematic geography; but this separation has not been followed." The modern use of the terms tends to emphasize the logic associated with the idea of the objective, or scientific, study of specific place and region. The progression topography-chorography-geography for studies of areas of increasing size (or decreasing map scale) seems to have been consigned to the dustbin of history. *See also* Areal Differentiation; Regional Geography.

BIBLIOGRAPHY

Hartshorne, R., *The Nature of Geography* (1939).

Jong, G. de, *Chorological Differentiation as the Fundamental Principle of Geography* (1962).

Richthofen, F. von, *Aufgabe und Methoden der heutigen Geographie* (1883).

Christaller, Walter (1893-1969)

Ph.D., University of Erlangen, 1932. Intermittently employed as journalist, writer, and office worker. *Die zentralen Orte in Süddeutschland* (1933), *Die ländliche Siedlungsweise im Deutschen Reich* (1937), *Das Gesicht Unserer Erde* (1961). K. & R. Hottes and P. Schöller in *GBS*, 7 (1983), 11-16; K. Hottes and P. Schöller in *GZ*, 56 (1968), 81-84 (also 85-101); R. Hottes in *Geographisches Taschenbuch 1981/1982* (1981), 59-70 (abridged trans. by G. Weigend in *AAAG*, 73 [1983], 51-54); E. Wirth in *GZ*, 70 (1982), 293-297; obits. by B. Berry and C. Harris in *GR*, 60 (1970), 116-119; and P. Haggett in *GJ*, 136 (1970), 500.

Clark, Andrew Hill (1911-1975)

Ph.D., University of California, Berkeley, 1944. Rutgers University, 1946-51, University of Wisconsin, 1951-75. *The Invasion of New Zea-*

land by People, Plants and Animals: The South Island (1949); *Three Centuries and the Island* (1959); *Acadia: The Geography of Nova Scotia to 1760* (1968). *GOF* (1971); *60 Years of Berkeley Geography* (1983); obit. by D. Ward in *AAAG*, 67 (1977), 145–148.

Clark, William Arthur Valentine (b. 1938)

Ph.D., University of Illinois, 1964. University of Canterbury, 1964–66; University of Wisconsin, 1966–70; University of California, Los Angeles, 1970 to the present. *Residential Mobility and Public Policy* (1980); *Recent Research on Migration and Mobility* (1982); *Human Migration* (1986); "Residential Segregation in American Cities: A Review and Interpretation," *Population Research and Policy Review* (1988).

Claval, Paul Charles Christophe (b. 1932)

D.-ès-L., University of Paris, 1970. University of Besançon, 1960–72; University of Paris, 1972 to the present. *Régions, nations, grands espaces* (1968); *Les mythes fondateurs des sciences sociales* (1980); *La logique des villes* (1981); *La géographie humaine et économique contemporaine* (1984). *WWF.*

Climatology

Climatology fits mostly between meteorology and geography, but since weather and climate can be viewed as driving or at least affecting just about everything near or on the earth's surface, climate studies can be found within all of the physical, environmental, biological, and social sciences. Although there are no sharp boundaries among the disciplines, emphasis here is on studies by climatologists identified mostly with geography.

During the first half of the twentieth century the focus in climatology was largely on collection of standardized temperature and precipitation data, and their geographical interpretation as longterm "average" climate sets. The data were organized into regional climatologies with the assumption that natural vegetation indexed inputs of longterm weather (climate) within the biological realm. The classic examples are the climatic classification of Wladimir Köppen (with modifications by Glenn Trewartha) and Richard Russell's emphasis on variability in terms of climate years.

During the late 1940s Warren Thornthwaite continued the focus on the interactions of energy and water at the surface of the earth, introducing his concepts of potential evapotranspiration (PE) and the water budget, initially in terms of climatic classification. The basic water-budget model has been further developed and applied, especially in hydrology, as the basis for drainage-basin models. In agriculture and forestry it is used for crop models and irrigation and drainage scheduling. Geographers also continue to be active in the study of physical processes at and near the surface; physical climatology and boundary-layer climatology; urban climatology; local climatic differences from place to place; microclimatology; and statistical and probabilistic analyses of the occurrences of extreme events, first and last freeze dates, and frequencies of rainfall above critical thresholds.

Geographers also participate in the climatology of atmospheric circulations. In terms of geographical scales, studies include analyses of synoptic weather types and their properties and frequencies, especially as related to floods and droughts at a place or over a region; generalized descriptions of the dynamic circulation patterns that generate regional climates; and the dynamics of regional interactions between surface properties and circulations aloft.

Climatologists in geography departments have also contributed for many years to paleoclimatology (interpretation of past climates). With the very recent concerns about "greenhouse warming" and "earth systems change," climatologists are contributing to the modeling of global circulation patterns (GCMs). Much attention is focused on applications of GCM outputs to the environmental and economic impacts of regional patterns of warming, and increasing or decreasing precipitation.

BIBLIOGRAPHY

Carter, D. B., and J. R. Mather, "Climatic Classification for Environmental Biology," *Publications in Climatology*, 19 (1966), 305–395.

Mather, J. R., R. T. Field, L. S. Kalkstein, and C. J. Willmott, "Climatology: The Challenge for the Eighties," *PG*, 32 (1980), 285-292.

Oliver, J. E., and R. W. Fairbridge, eds., *Encyclopedia of Climatology* (1987).

Wilcock, A. A., "Köppen after Fifty Years," *AAAG*, 58 (1968), 12-28.

Yarnal, B., R. G. Crane, A. M. Carleton, and L. S. Kalkstein, "A New Challenge for Climate Studies in Geography," *PG*, 39 (1987), 465-473.

• *Robert A. Muller*

Cohen, Saul Bernard (b. 1925)

Ph.D., Harvard University, 1955. Boston University, 1952-65; Clark University, 1965-78; Queens College CUNY (president), 1978-85; Hunter College CUNY, 1986 to the present. *Geography and Politics in a World Divided* (1963); *Problems and Trends in American Geography* (1967); *Experiencing the Environment* (1976); *Resources and Human Networks* (1977). *GOF* (1977); *WWA*; *AAG Newsletter*, 24 (Jan. 1989), 4.

Colby, Charles Carlyle (1884-1965)

Ph.D., University of Chicago, 1917. University of Chicago, 1916-49. *Source Book for the Economic Geography of North America* (1921); *Land Classification in the United States* (1941); (ed.) *Geographic Aspects of International Relations* (1942). W. Calef in *GBS*, 6 (1982), 17-22; obits. by C. Harris in *AAAG*, 56 (1966), 378-382; and R. Harper in *GR*, 56 (1966), 296-297.

Colombia

Colombia ranks high in the recent trend of geographic development in the Latin American area, though well behind Brazil, a country in which the take-off for modernization in this context started three decades earlier, in the 1930s. If modernization in geography is associated with that process that includes professional organization, advanced training of geographers, and active participation in, and awareness of, current paradigms and trends of the discipline in the world, then Colombia shares with Venezuela, Mexico and Chile a healthy awakening that took place during the past twenty or thirty years. A remarkable parallelism occurs in the history of geography in these countries. That history can be typified by the early foundation of a geographic academy or society and a geographic or cartographic institute, both official. Then, college departments of social science with strong pedagogic commitments were established to prepare high school teachers in history and geography. During such an early stage—in which several Latin American countries still remain—every geography post is staffed with non-geographers, many of them retired military officers. Eventually, the take-off stage comes through university innovation prompted by enlightened, self-educated native geographers, or by a foreign scholar (such as Pierre Deffontaines in Brazil), or by native professionals who became geographers through graduate training overseas.

Starting geography en route to becoming a modern scientific discipline and a profession based on university-trained geographers was a rather tardy occurrence in Colombia. By and large, two governmental institutions, the Universidad Pedagógica y Tecnológica de Colombia (UPTC) at Tunja and the Instituto Geográfico "Agustín Codazzi" (IGAC), Bogotá, are credited as sponsors of decisive actions taken in this direction since the mid-1960s.

UPTC established geographic education at the college level in 1938, an academic innovation undertaken within the social science curriculum of its school of education, Colombia's first such school (founded at Tunja in 1928 as the Escuela Normal Superior by the German educator Julius Sieber). UPTC provides undergraduate geographic training with courses on physical, human and economic geography; cartography; theory and methodology; world regional geography; and research seminars. These are combined with history, economics, anthropology and sociology in a four-year licentiate (B.A.) degree program. Although a private university (Universidad Jorge Tadeo Lozano) in Bogotá established a school of geographic engineering in 1952, and although schools of education with social science departments have been established throughout Colombia, the leading role played by UPTC in the promotion of geographic teaching is widely acknowledged.

Such leadership was reaffirmed in 1984 when the Graduate Geography Program (EPG) was implemented in Bogotá as a joint project of UPTC and the IGAC. Angel Massiris, now a member of the IGAC's research staff, was the first Colombian fully trained within the country to get the Master's degree in geography, from the EPG in 1987.

The IGAC is undoubtedly the leading geographic institution in Colombia. Founded in 1935 by President Alfonso López, this institute has traditionally dealt with such functions as geodetic and topographic surveying, mapping, soil and resource inventory, and cadastral surveying. Its interest in geographic matters per se became clear in 1967 when the IGAC joined UPTC to host in Tunja the first open meeting of persons concerned with earth sciences, a meeting that concluded with the founding of the Association of Colombian Geographers (ACOGE). In 1972 the IGAC included geographic research as one of its major tasks by adding to its structure another specialized subsection. In 1988 the Inter-American Center of Photointerpretation (CIAF)—an independent agency primarily devoted to giving advanced technical instruction in photogrammetric engineering, remote sensing, geomorphology and regional surveying—was relocated within IGAC. This addition, plus the agreement with UPTC, makes it possible to foresee a strengthening of the IGAC's involvement in advanced teaching in the years ahead. The institute has a growing list of publications both in monograph and serial form. *Colombia Geográfica* (semiannual) and *Colombia: Sus Gentes y Regiones* (quarterly) are its principal periodicals. The IGAC's Director General, Alvaro González-Fletcher, is the president of the Colombian National Committees of the Pan-American Institute of Geography and History (PAIGH) and the International Geographical Union (IGU), of which Colombia is an active member.

The Association of Colombian Geographers (ACOGE), founded in 1967 at a meeting organized by Dieter Brunnschweiler and Héctor Rucinque in Tunja, has been the open forum for the profession and its liaison with the IGAC, UPTC and other institutions related to geography. Ten biannual conventions have been convened by the association, one of them international. A series of graduate seminars in the early 1970s sponsored by ACOGE with the aid of the U.S. Fulbright Program led to the establishment of the UPTC/IGAC Master's degree project. A group of American geographers, led by C. W. Minkel, contributed to the success of these seminars. ACOGE's quarterly journal, *Trimestre Geográfico*, is growing in importance as well as in quality.

Today, Colombian geographers are still limited in number. They usually cluster in small groups in university towns such as Tunja, Medellín, Armenia, Pasto, Manizales, Barranquilla, Montería and Chiquinquirá. The largest group is located in Bogotá, where the major sources of employment are found (government, the IGAC, the National University, the National Pedagogic University, Javeriana University, Jorge Tadeo Lozano University, and Distrital and Libre Universities). Because more individuals receive their Master's degrees at the EPG every year, those cells will increase in number and activities.

The new geography is being spread by Colombians who are getting graduate training at American and European universities; at the EPG; and at the University of Nariño, Pasto, which now has a program offering "specialist" degrees. These new geographers are likely to induce the founding of new undergraduate projects. Programs in "pure" geography are badly needed. The schools of the licentiate nature require academic upgrading and curriculum innovation.

To conclude, it is evident that geography is progressing at a good pace in Colombia. A new generation of geographers has entered the profession, well abreast of what is going on in their discipline and fully conscious of limitations and potentials as well as responsibilities.

Bibliography

Acevedo Latorre, E., "Las ciencias en Colombia; geografía, cartografía," Academia Colombiana de Historia, *Historia Extensa de Colombia*, 24 (1974), 1–284.

Delgado Mahecha, O., "Desarrollos recientes de la geografía colombiana," *Trimestre Geográfico*, no. 13 (1989), 3–19.

Mateus Cárdenas, C. R., *Evolución histórica de la geografía moderna en Colombia* (1981).

Rucinque, H. F., "Cincuenta años y siglos más de geografía en Colombia," *Colombia: Sus Gentes y Regiones* (August 1985), 4-15.

——— , "Formación de geógrafos en los niveles pre- y post-graduado en Colombia: Antecedentes y perspectivas," *X Congreso Colombiano de Geografía* (1988), 251-260.

——— , "Geography in Colombia," *PG*, 41 (1989), 218-220.

• *Héctor F. Rucinque*

Compage

An example of the many stillborn words or phrases that have been proposed for the geographer's lexicon, "compage" is an obsolete noun that Harvard geographer Derwent Whittlesey tried to resurrect in 1954 to apply to "the type of region differentiated as to human occupance." "The compage is," said Whittlesey, "something less than spatial totality; but it does include all of the features of the physical, biotic, and societal environments that are functionally associated with man's occupance of the earth." After the term was exhumed long enough to be used in Whittlesey's own study of Southern Rhodesia, it was quickly reinterred, and no one has yet been so disrespectful as to inquire about the circumstances of its passing or to suggest its reintroduction. *See also* Region and Regional Geography.

BIBLIOGRAPHY

Whittlesey, D. S., "The Regional Concept and the Regional Method," pp. 19-68 in *American Geography: Inventory and Prospect*, ed. by P. E. James and C. F. Jones (1954).

——— , "Southern Rhodesia—An African Compage," *AAAG*, 46 (1956), 1-97.

Coppock, John Terence (b. 1921)

Ph.D., University of London, 1960. University College London, 1950-65; University of Edinburgh, 1965-86. *The Changing Use of Land in Britain* (with R. Best) (1962); *An Agricultural Geography of Great Britain* (1971); *Recreation in the Countryside* (with B. Duffield) (1975); *Innovation in Water Management: The Scottish Experience* (with W. Sewell and A. Pitkethly) (1986). *WW.*

Cornish, Vaughan (1862-1948)

D.Sc., University of Manchester, 1901. No occupation (private means). *Waves of the Sea and Other Water Waves* (1910); *Geography of Imperial Defence* (1922); *The Great Capitals, An Historical Geography* (1923); *National Parks and the Heritage of Scenery* (1930). B. Waites in *GBS*, 9 (1985), 29-35; G. Crone in *DNB 1941-50* (1950), 179-180; E. Gilbert, *Vaughan Cornish (1862-1948) and the Advancement of Knowledge Relating to the Beauty of Scenery in Town and Country* (1965); A. Goudie, "Vaughan Cornish: Geographer," *TIBG*, 55 (1972), 1-16; anon. obit. in *GJ*, 111 (1948), 294.

Cotton, Charles Andrew (1885-1970)

D.Sc., University of New Zealand, 1915. Victoria University of Wellington, 1909-54. *Landscape as Developed by the Processes of Normal Erosion* (1941); *Geomorphology* (1942); *Climatic Accidents in Landscape Making* (1942); *Volcanoes as Landscape Forms* (1944). J. Soons and M. Gage in *GBS*, 2 (1978), 27-32; obit. by J. Marwick in *Proceedings of the Royal Society of New Zealand*, 99 (1971), 100-105.

Cressey, George Babcock (1896-1963)

Ph.D., University of Chicago, 1923; Clark University, 1931. Syracuse University, 1931-63. *China's Geographic Foundations* (1934); *Asia's Lands and Peoples* (1944); *How Strong Is Russia* (1954); *Crossroads—Land and Life in Southwest Asia* (1960). P. James and A. Perejda in *GBS*, 5 (1981), 21-25; obits. by T. Herman in *AAAG*, 55 (1965), 360-364; and P. James in *GR*, 54 (1964), 254-257.

Cuba

At the start of the nineteenth century, with Alexander von Humboldt, Felipe Poey, Esteban

Pichardo, and others, the first steps were taken towards the birth of modern geography in Cuba, although its active pursuit had decreased by the century's close as regards both practice and teaching. It was not until 1925 that university instruction in geography commenced. In 1926 more or less regular courses began to be given in secondary schools, and in 1927 an official chair was established in the University of Havana. In 1914 the Sociedad Geográfica, and in 1940 the Sociedad Espeleológica, were founded, both of which have been important centers for geographical knowledge of the country. In 1962 the Institute of Geography of the Cuban Academy of Sciences and the School of Geography of the University of Havana were created, the latter being raised to a faculty in 1978. A Faculty of Geography was also set up in 1977 in the Instituto Superior Pedagógico, which specializes in the training of teachers in the secondary schools.

In the first half of the twentieth century the following people stood out in the different fields of geography: C. M. Trelles, P. Cañas Abril, L. Marrero, E. Rodríguez Busto, and G. Barraqué. They have been followed by a new generation in which figure prominently J. L. Díaz, M. Acevedo, L. R. Díaz, J. Mateo, T. Ayón, L. Orbera, N. Montes, and E. Propín. Separate mention must also be made, for their important contributions, of Carlos de la Torre, Alfredo M. Aguayo, Salvador Massip, Sara Ysalgué, and Antonio Núñez Jiménez.

If the birth of Cuban geography dates from the beginning of the nineteenth century, it is only after 1962 that its consolidation as a science comes. Recent achievements have on the whole been associated with better knowledge and more rational exploitation of the natural environment, with overall assessment of natural and socioeconomic factors. The *Atlas Nacional de Cuba* (1970), the *Nuevo Atlas Nacional de Cuba* (1989), and comprehensive geographical studies of various provinces and municipalities are among the accomplishments. At the present time, geographic information systems (GIS) and automated cartography constitute the basic means for the assimilation and application of new knowledge, and reflect use of the most modern technologies on the part of Cuban geographers.

BIBLIOGRAPHY

Instituto de Geografía, Academia de Ciencias de Cuba, and Instituto Cubana de Geodesia y Cartografía, *Nuevo Atlas Nacional de Cuba* (1989).

Instituto de Geografía, Academia de Ciencias de Cuba, and Instituto de Geografía, Academia de Ciencias de la URSS (Institut Geografii Akademii Nauk SSSR), *Atlas Nacional de Cuba* (1970).

Massip, S., *Estudio Geográfico de la Isla de Cuba* (1925).

Núñez Jiménez, A., *Geografía de Cuba* (1954, latest ed. 1972).

——— , *Cuba, la Naturaleza y el Hombre* (1982).

Trelles, C. M., *Biblioteca Geográfica Cubana* (1920).

• *Gladstone Oliva Gutiérrez*
(translated by C. Julian Bishko)

Cultural Ecology

Cultural ecology is the study of the mutual relations of people and the environments (of places and resources) they occupy and use. This field of inquiry, which constitutes a major theme of geography, is shared between geographers and anthropologists. The term cultural ecology was coined by the anthropologist Julian Steward in 1937 to distinguish it from biological, human and social concepts of ecology. Steward defined it as "the study of the processes by which a society adapts to its environment. Its principal problem is to determine whether these adaptations initiate internal social transformations or evolutionary change. It analyzes these adaptations, however, in conjunction with other processes of change. Its method requires examination of the interaction of societies and social institutions with one another and with the natural environment" (Steward, 337).

Steward made a careful effort in his definition to avoid the environmental determinist trap. Indeed, many anthropologists, in their enterprise of explicating human responses to environmental conditions, formerly avoided the term geography itself, perhaps fearful of contamination by the taint of environmental determinism, a term they associated with the geographers they had read in the 1920s and 1930s. Geographers since the 1930s, guided greatly by the example of Carl Sauer, have also been careful about the way they have framed the question of "adaptation to environment."

Key concepts in cultural ecology are appraisal of environment (environmental perception); adaptation to the possibilities in the environment, given the repertoire of tools and techniques available to a given culture and the goals of its members; and functional coherence of a group's appraisal and management of its environment. In practice cultural ecologists have studied small populations at a local level. Their studies tend to be small-scale and concerned with non-western societies. Much of the concern for mutual relations centers on the intersection of two domains—the biophysical world and the lifeworld of human beings, individually and in groups. The main set of relations can be thought of as vertical ensembles of things occurring at a place (site characteristics), such as soils, vegetation, climate, water, crops, land use, human settlement; yet spatial or locational considerations (situational characteristics) can be important at any scale, from differences along the slope of a field to the location of a household in a global market economy.

Enduring subthemes in cultural ecology have been population pressure and human adaptation, the evolution of cultural systems, environmental perception and natural hazards, medical studies, and cultural ecology in larger contexts. Indeed, if anything can be said to represent a recent trend in cultural ecology, it has been to incorporate links between places, whether these be rural to urban, cash crop production to national circuits of exchange, or land ownership and human livelihood to multinational corporations and the international political economy. A new term has been proposed: political ecology.

Some of the leading practitioners of cultural ecology in geography are Nigel Allan, Piers Blaikie, James Blaut, Harold Brookfield, Karl Butzer, Claudia Carr, William Denevan, William Doolittle, Lawrence Grossman, John Hunter, Douglas Johnson, Robert Kates, Gregory Knapp, C. Gregory Knight, Kent Mathewson, Melinda Meade, James Newman, Bernard Nietschmann, Campbell Pennington, Philip Porter, Paul Richards, B. L. Turner II, Donald Vermeer, Michael Watts, Gene Wilken, and Ben Wisner.

Bibliography

Bennett, J. W., *The Ecological Transition: Cultural Anthropology and Human Adaptation* (1976).

Blaikie, P. M., and H. C. Brookfield, eds., *Land Degradation and Society* (1987).

Butzer, K. W., "Cultural Ecology," pp. 192-208 in *Geography in America*, ed. by G. L. Gaile and C. J. Willmott (1989).

Grossman, L. S., "Man-Environment Relationships in Anthropology and Geography," *AAAG*, 67 (1977), 126-144.

Porter, P. W., "Geography as Human Ecology: A Decade of Progress in a Quarter Century," *American Behavioral Scientist*, 22 (1978), 15-39.

Steward, J. H., "Cultural Ecology," pp. 337-344 in *IESS*, vol. 4 (1968).

• *Philip W. Porter*

Cultural Geography.

See Human Geography; Landscape; Settlement Geography.

Cumberland, Kenneth Brailey (b. 1913)

D.Sc., University of New Zealand, 1945. University of Canterbury, 1938-45; University of Auckland, 1946-78. *Soil Erosion in New Zealand* (1944); "Aotearoa Maori: New Zealand about 1780," *GR*, 39 (1949), 401-424; "A Land Despoiled: New Zealand about 1838," *New Zealand Geographer*, 6 (1950), 13-34; "'Jimmy Grants' and 'Mihaneres': New Zealand about 1853," *EG*, 30 (1954), 70-89. A. G. Anderson (ed.), *The Land Our Future* (Cumberland Festschrift) (1980).

Curry, Leslie (b. 1922)

Ph.D., University of New Zealand, 1959. University of Auckland, 1953-60; University of Maryland, 1960-63; University of Toronto, 1964 to the present. "Climatic Change as a Random Series," *AAAG*, 52 (1962), 21-31; "Central Places in the Random Spatial Economy," *Journal of Regional Science*, 7 (1967), 217-238; "A Spatial Analysis of Gravity Flows," *Regional Studies*, 6 (1972), 131-147; "Inefficiency of Spatial Prices Using the Thermodynamic Formalism," *Environment and Planning A*, 16 (1984), 5-16.

Cvijić, Jovan (1865–1927)

Doctorate, University of Vienna, 1893. University of Belgrade, 1893–1927. *Das Karstphänomen* (1893); *La péninsule balkanique* (1918); *Morphologie terrestre* (2 vols., 1924, 1926). M. Vasović in *GBS*, 4 (1980), 25–32; T. W. Freeman, *The Geographer's Craft* (1967), 72–100; obits. by L. Gallois in *AG*, 36 (1927), 181–183; and F. Machatschek in *PGM*, 73 (1927), 102–103.

Dalla Vedova, Giuseppe (1834-1919)

Laurea, University of Padua, 1864. University of Padua, 1872-75; University of Rome, 1875-1909. *Delle origini e dei progressi della geografia scientifica* (1868); "Il concetto popolare e il concetto scientifico della geografia," *BSGI*, 7 (1880), 5-27; *Scritti geografici (1863-1913)* (1914). F. Porena, "L'opera di Giuseppe Dalla Vedova," pp. ix-xxi in *Scritti . . . pubblicati in onore di Giuseppe Dalla Vedova* (1908); I. L. Caraci, "A sessant'anni dalla morte di Giuseppe Dalla Vedova," *Pubblicazioni dell'Istituto di scienze geografiche di Università di Genova*, 32 (1978); I. L. Caraci in *Dizionario Biografico degli Italiani*, 32 (1986), 53-54; obit. by R. Almagià in *BSGI*, 47 (1920), 31-50.

Danish Geographical Society. *See* Royal Danish Geographical Society.

Darby, Henry Clifford (b. 1909)

Ph.D., University of Cambridge, 1931. University of Liverpool, 1945-49; University College London, 1949-66; University of Cambridge, 1932-45, 1966-76. (ed.) *An Historical Geography of England before A.D. 1800* (1936); *The Domesday Geography of England* (7 vols., 1952-77); (ed.) *A New Historical Geography of England* (1973). *WW*; P. J. Perry, "H. C. Darby and Historical Geography," *GZ*, 57 (1969), 161-176; appreciations by R. Lawton, R. Butlin, and D. W. Meinig in *Journal of Historical Geography*, 15 (1989), 14-23 (H. C. D. bibliog. 5-13).

Davidson, George (1825-1911)

Graduated 1845 from Central High School, Philadelphia, an institution that offered university-level work. United States Coast Survey, 1845-95; University of California, Berkeley, 1898-1905. *Directory for the Pacific Coast of the United States* ("Coast Pilot") (1859, 4th ed. 1889); *Coast Pilot of Alaska* (1869); *The Alaska Boundary* (1902); plus monographs on discovery and exploration of the Pacific Coast. G. Dunbar in *GBS*, 2 (1978), 33-37; O. Lewis, *George Davidson: Pioneer West Coast Scientist* (1954); C. Davenport, "Biographical Memoir of George Davidson," National Academy of Sciences, *Biographical Memoirs*, 18 (1938), 189-217.

Davis, William Morris (1850-1934)

M.Eng., Harvard University, 1870. Harvard University, 1878-1912. *Elementary Physical Geography* (1902); *Geographical Essays* (1909); *Die Erklärende Beschreibung der Landformen* (1912); *The Coral Reef Problem* (1928). S. Judson in *DSB*, 3 (1971), 592-596; R. Beckinsale and R. Chorley in *GBS*, 5 (1981), 27-33; R. Daly, "Biographical Memoir of William Morris Davis," National Academy of Sciences, *Biographical Memoir*, 23 (1945), 263-303; R. Chorley, R. Beckinsale, and A. Dunn, *History of the Study of Landforms*, vol. 2, *The Life and Work of William Morris Davis* (1973); obit. by K. Bryan in *AAAG*, 25 (1935), 25-31.

Debenham, Frank (1883-1965)

B.A., University of Sydney, 1904. University of Cambridge, 1919-49. *Kalahari Sand* (1953); *The Way to Ilala* (1955); *Antarctica: The Story of a Continent* (1959); *Discovery and Exploration* (1960). Obits. by J. Steers in *GJ*, 132 (1966), 173-175; and anon. in *TIBG*, 40 (1966), 195-198.

Deffontaines, Pierre (1894-1978)

D.-ès-L., Facultés catholiques de Lille, 1932. Facultés catholiques de Lille, 1925-39; French Institute of Barcelona, 1939-64; University of Montpellier, 1964-67; plus concurrent professorial appointments in Canada, Latin America, and Spain. *Géographie humaine de la France*, vol. 2 (with J. Brunhes) (1926); *Les hommes et leurs travaux dans les pays de la Moyenne Garonne* (1932); *Géographie et religions* (1948), *Atlas aérien de France*, 5 vols. (with M.J.-B. Delamarre) (1955-64). R. Blais *et al.*, "Hommage à Pierre Deffontaines," *Acta Geographica*, 3rd series, no. 38 (1979), 4-24.

De Geer, Sten (1886-1933)

Doctorate, University of Uppsala, 1911. University of Stockholm, 1911-28; University of Göteborg, 1928-33. *Karta over Befolkningens Fördelning i Sverige* (1919); plus 3 papers in English in *Geografiska Annaler*: "On the Definition, Method and Classification of Geography" (1923), "The American Manufacturing Belt" (1927), and "The Subtropical Belt of Old Empires" (1928). T. W. Freeman in *The Geographer's Craft* (1967), 124-155; obits. by H. W:son Ahlmann in *Ymer*, 53 (1933), 441-445; and by H. Nelson in *Svensk Geografisk Årsbok* (1933), 185-192.

Demangeon, Albert (1872-1940)

D.-ès-L., University of Paris, 1905. University of Lille, 1904-11; University of Paris, 1911-40. *La plaine picarde* (1905); *Les Iles britanniques* (1927); *Problèmes de géographie humaine* (1942); *France économique et humaine* (2 vols., 1946-48). G. Parker in *GBS*, 11 (1987), 13-21; E. Franceschini in *DBF*, 10 (1965), col. 963; A. Perpillou in *LGF* (1975), 81-106; obits. by E. de Martonne in *AG*, 49 (1940), 161-169; and L. Febvre in *Annales d'histoire sociale*, 3 (1941), 81-89.

Denmark

The founding in 1876 of the Royal Danish Geographical Society under the leadership of Edvard Erslev (1824-1892) and the establishment of a chair in the natural sciences faculty at the University of Copenhagen in 1883 under Ernst Løffler (1835-1911) mark the institutionalization of modern geography in Denmark. The intellectual groundwork was laid by a physical geographer holding a chair in botany, Joachim Frederik Schouw (1789-1852). Schouw was an internationally respected physical scientist who was a pioneer in the field of plant geography. He was also, however, a politically engaged thinker with an overarching concern for the historically developing man/nature relationship. Schouw's social concerns led him to devote considerable energy to popular education and the development of progressive geography textbooks.

Erslev continued Schouw's broad historical interest and concern for popular education. His research on geographical perception foreshadowed modern humanistic approaches. Erslev's work was eclipsed by the natural science orientation to the subject promulgated by Løffler, who because of his concern to thus legitimize the "new" discipline academically, refused to be identified with Erslev. Danish geography has since been dominated by physical geography, at times in collaboration with the pioneering Danish biological ecologists, and supplemented by an anthropogeography emphasizing physical factors. Among the more prominent physical geographers are Martin Vahl (1869-1946), Niels Nielsen (1893-1981) and Axel Schou (1902-1971). On the human side one could name Hans Peter Steensby (1875-1920), Kaj Birket-Smith (1893-1977) (who founded the department of anthropology at the University of Copenhagen), and Gudmund Hatt (1884-1960). This tradition is continued today by the human ecologist Sofus Christiansen.

In the 1960s and 1970s a more social- and economics-oriented geography flourished in the wake of the Anglo-American quantitative revolution and the later development of Marxist political economy. At the University of Copenhagen these new ideas were pioneered by people such as the historical geographer Viggo Hansen (1913-1989) and the Marxist economic geographer Steen Folke. Social and economic geography, on the other hand, dominated the department at Aarhus University from its foundation in 1943 (under Prof. Johannes Humlum) to the point that physical geography eventually was moved to the geology department. The department established at the new Roskilde University Center in 1974, as an outgrowth of the University of Copenhagen, embodied a section continuing the traditional emphasis on the physical environment as well as a section oriented toward social and economic geography.

The 1980s have witnessed a suppression of these modern trends under a right-wing minister of education. The traditional physical orientation of the discipline has been strengthened at both the Roskilde and Copenhagen departments and the state's primary and secondary school curricula are also being revised to reflect this orientation. The phasing out of the Aarhus department began in 1985 and it will close in 1991.

There are geography departments at the University of Copenhagen, circa twenty-five faculty; Roskilde University Center, circa ten faculty; and the Royal School of Educational Studies (a postgraduate school), circa five faculty. The Copenhagen Business School employs three geographers in its Department for Traffic, Tourism and Regional Economy. *See also* Royal Danish Geographical Society.

BIBLIOGRAPHY

Asheim, B. T., "A Critical Evaluation of Postwar Developments in Human Geography in Scandinavia," *PHG*, 11 (1987), 333-353.

Christiansen, S., N. K. Jacobsen, and N. Nielsen, "Geografi," pp. 377-446 in *Københavns Universitet 1479-1979*, ed. by S. Ellehøj, vol. 13 (1979).

Hansen, V., "Reflections on Historical Geography," pp. 147-163 in *Geographers of Norden*, ed. by T. Hägerstrand and A. Buttimer (1988).

Matthiessen, C. W., "Videnskabelig geografi i Danmark," *Geografisk Notiser*, 45 (1987), 7-16.

Olwig, K. R., "Historical Geography and the Society/Nature 'Problematic': The Perspective of J. F. Schouw, G. P. Marsh and E. Reclus," *Journal of Historical Geography*, 6 (1980), 29-45.

——— , *Nature's Ideological Landscape* (1984).

• *Kenneth R. Olwig*

Derruau, Max Emile Léon (b. 1920)

D.-ès-L., University of Grenoble, 1949. University of Grenoble, 1948-52; University of Clermont-Ferrand, 1943-48, 1952-86. *La grande Limagne* (1949); *Précis de géomorphologie* (1956); *L'Europe* (1958); *Géographie humaine* (1976). *WWF.*

Dickinson, Robert Eric (1905-1981)

Ph.D., University College London, 1932. University College London, 1928-47; Syracuse University, 1947-58; University of Leeds, 1958-67; University of Arizona, 1967-75. *The Making of Geography* (with O. Howarth) (1933); *City, Region and Regionalism* (1947); *The Makers of Modern Geography* (1969); *Regional Concept: The Anglo-American Leaders* (1976). L. Pederson in *GBS*, 8 (1984), 17-25; obits. by T. W. Freeman in *GJ*, 148 (1982), 147-148; and W. G. East in *TIBG*, 8 (1983), 122-124.

Diet, Geography of. *See* Nutritional Geography.

Diffusion (Spatial Diffusion)

Spatial diffusion may be viewed as one of the most fundamental processes by which the evolution of landscape takes place. It concerns the spread, over both space and time, of phenomena outward from limited origins. If we view the most basic geographic process as the creation of the use and thus the identity of any piece of the earth's surface, and that the next or as basic a process is interaction or movement among these pieces of territory, then we can view spatial diffusion as the tendency or possibility for the "use" at any location to be extended to the surrounding area. In other words, if processes of using locations effectively lead to areal differentiation, then spatial diffusion tends to lead toward uniformity, as the more effective displaces the less effective. As an example, under conditions of very limited mobility and interaction extreme differentiation in language or in agricultural practices could occur, but as interaction increases, more effective practices will tend to diffuse and displace less effective practices, creating greater homogeneity.

The study of spatial diffusion is of long standing in biology/biogeography (*e.g.*, the spread of animal and plant species) and in anthropology/cultural geography (*e.g.*, the spread of languages, peoples, agriculture and other aspects of culture). Carl Sauer and his students were especially prominent in these kinds of studies. In these fields arguments about the relative role of diffusion versus independent invention are both lively and unresolved.

The formal modelling of spatial diffusion and its extension to other aspects of geography was introduced in the 1950s by the Swedish geographer Torsten Hägerstrand, who studied the spread of agricultural innovations. The Hägerstrand model of spatial diffusion was notable for its micro-scale behavioral approach, its focus on the importance of person-to-person contact in explaining the spread of innovations, and for its use of Monte Carlo simulation—that is, probabilistic approaches. But most important is the idea of diffusion as a basic geographic process by which spatial separation conditions interaction among people, and thus their ideas and practices. Subsequent work distinguished between (a) local or contagious diffusion, as of diseases, and spread down the urban hierarchy, as of television broadcasting, and (b) processes like the spread of rumors or disease that might not alter the landscape and processes like frontier settlement and the physical growth of the city. Also developed were more aggregate and deterministic models aimed at predicting the patterns of diffusion over space and time as a function of initial conditions. Lawrence Brown shifted the emphasis to the role of firms and agencies in propagating diffusion, and extended studies to other parts of the world. The deliberate use of techniques of diffusion for planned development has been widely studied, both in such positive aspects as the spread of acceptance of family planning, and in such negative effects as the subversion of local culture or an increase in dependency. In particular, Blaut has criticized "diffusionism" as ethnocentric, viewing the spread of "superior" Euro-American traits to a supposedly more primitive rest of the world.

Bibliography

Blaut, J. M., "Two Views of Diffusion," *AAAG*, 67 (1977), 343-349.

Brown, L., *Innovation Diffusion: A New Perspective* (1981).

Hägerstrand, T., *Innovation Diffusion as a Spatial Process*, trans. by A. Pred (1967, orig. pub. 1953).

Morrill, R., G. Gaile, and G. Thrall, *Spatial Diffusion* (1988).

Rogers, E., *Diffusion of Innovations* (1983).

• *Richard Morrill*

Dion, Roger (1896-1981)

D.-ès-L., University of Paris, 1934. Ecole Normale Supérieure, 1924-34; University of Paris, 1928-32; University of Lille, 1934-44; Paris and ENS, 1939-46; University of São Paulo, 1946-47; Collège de France, 1948-68. *Essai sur la formation du paysage rural français* (1934); *Le Val de Loire* (1934); *Histoire de la vigne et du vin en France* (1959); *Histoire des levées de la Loire* (1961). J. L. M. Gulley, "The Practice of Historical Geography: A Study of the Writings of Professor Roger Dion," *TESG*, 52 (1961), 169-183; obits. by N. Broc in *AG*, 91 (1982), 205-217; and by P. Gourou in *Journal of Historical Geography*, 8 (1982), 182-184.

Dresch, Jean Emmanuel (b. 1905)

D.-ès-L., University of Paris, 1941. University of Caen, 1943-45; University of Strasbourg, 1945-47; University of Paris, 1948-77; President of IGU, 1972-76. *Recherches sur l'évolution du relief dans le massif central du grand Atlas* (1941); *La Méditerranée et le Moyen-Orient* (with P. Birot) (2 vols., 1953-56); *Un géographe au déclin des empires* (1979); *Géographie des régions arides* (1982). *WWF*; A. Buttimer (ed.), *The Practice of Geography* (1983), 121-122, 130-132, 136, 138; biography and bibliography in "Désert et montagne au Maghreb," *Revue de l'Occident musulman et de la Méditerranée*, no. 41-42 (1987), 19-42.

Drygalski, Erich von (1865-1949)

Doctorate, University of Berlin, 1888. University of Berlin, 1899-1906; University of Munich, 1906-35. *Die Grönland-Expedition der Gesellschaft für Erdkunde zu Berlin 1891 bis 1893* (2 vols., 1897); *Zum Kontinent des eisigen Südes* (1904); *Der Einfluss der Landesnatur auf die Entwicklung der Völker* (1922); *Gletscherkunde* (with F. Machatschek) (1942). G. Tiggesbäumker in *GBS*, 7 (1983), 23-29; obits. by N. Creutzberg in *Erdkunde*, 3 (1949), 65-68; and W. Joerg in *GR*, 40 (1950), 489-491.

Dubois, Marcel (1856-1916)

D.-ès-L., University of Paris, 1884. University of Paris, 1885-1916. (with C. Guy) *Album géographique* (5 vols., 1896-1906); *Systèmes coloniaux et peuples colonisateurs* (1895); (with A. Terrier) *Un siècle d'expansion coloniale* (1902); *La crise maritime* (1910). Obit. by L. Gallois in *AG*, 25 (1916), 466; N. Broc, "Nationalisme, colonialisme et géographie: Marcel Dubois (1856-1916)," *AG*, 87 (1978), 326-336.

Dutch Geographical Society.

See Royal Dutch Geographical Society.

East Africa (Kenya, Tanzania, and Uganda)

Before the establishment of institutions of higher education in East Africa, geographical study and research was represented by official publications or by authors who were not professional geographers. The Survey Departments issued national atlases in 1942 (Tanganyika), 1959 (Kenya) and 1962 (Uganda, with assistance from Makerere College staff), and publications of the Geological Surveys included much of geographical interest. Influential concepts of regional geomorphology were advanced by E. J. Wayland during his government service as a geologist in Uganda, 1919–1939. A handbook of Tanganyika was published in 1930 (succeeded by two separate volumes in 1955 and 1958) and one for Uganda in 1935 (H. B. Thomas and R. Scott, replacing previous volumes of 1913 and 1920). Scholarly articles appeared in the *Tanzania* (formerly *Tanganyika) Notes and Records* from 1936 and in the *Uganda Journal* from 1934 (re-founded in 1946). Two articles in the *Geographical Review* (U.S.A.) by Clement Gillman, on a population map of Tanganyika (1936) and on a vegetation map (1949), are examples of remarkable single-handed studies by an expatriate official long resident in the country.

Departments of geography teaching at university level were founded in Kampala (Makerere College) in 1949, in Nairobi in 1961 and Dar es Salaam in 1965. At first staff were few and students many, but despite heavy teaching loads a corpus of geographical studies was built up, in part guided by the need for teaching material and field course data, increasingly supplemented by undergraduate dissertations. There was much scope for field recording, classification and interpretation, facilitated by the rapidly increasing supply of topographical maps based on air photography, which was also used directly. Photography was by the Royal Air Force, using post-war spare capacity, and mapping by the U.K. Directorate of Colonial (later Overseas) Surveys in support of the three survey departments. Similarly influential was the round of greatly improved and detailed censuses of 1957–1962. The *East African Geographical Review* was founded at Makerere in 1963 and the *Kenyan Geographer* in 1975. With the accelerating pace of development and change initiated by independence of the three countries in 1961–1963, research became more applied and geographers, increasingly of local origin, began to provide reports and advice or entered government service. This was institutionalized in Tanzania through the Bureau of Resource Assessment and Land Use Planning (BRALUP) founded in 1967 and renamed the Institute of Resource Assessment in 1983. In Nairobi, Professor S. H. Ominde initiated the Population Studies and Research Institute.

Although most of the newer universities are intended to promote practical studies in support of national development and despite the demonstrated relevance of the subject, not every university has a department of geography. The university departments of geography and their heads (in 1988) are given below, with selected previous and present staff members and their fields of interest: Makerere University: M. V. Mwaka, S. J. K. Baker (1947-67, regional), D. N. McMaster (land use), B. W. Langlands (historical, medical), A. O'Connor (economic); Nairobi: E. Mwagiru, W. T. W. Morgan (1956-67, regional), S. H. Ominde (1964-88, formerly Makerere, population, development), R. S. Odingo (land use), F. F. Ojany (geomorphology, regional), R. B. Ogendo (economic, regional), R. A. Obudho (urban); Kenyatta University: M. B. K. Dako, C. Nyamweru (geomorphology); Dar es Salaam University: W. F. L. Mlay, L. Berry (regional, development), B. Datoo (transport); I. R. A.: A. S. Kauzeni, A. Mascarenhas (1972-84).

Bibliography

Berry, L., ed., *Tanzania in Maps* (1971).

Langlands, B. W., *Inventory of Geographical Research at Makerere 1947-1972* (Makerere University, Department of Geography, Occasional Paper, no. 50, 1972).

McMaster, D. N., *A Subsistence Crop Geography of Uganda* (1962).

Morgan, W. T. W., *East Africa* (1973).

Ominde, S. H., *Land and Population Movements in Kenya* (1968).

Russell, E. W., ed., *The Natural Resources of East Africa* (1962, 2nd ed. W. T. W. Morgan 1969, rev. 1972).

• *W. T. W. Morgan*

Eckert, Max (1868-1938)

Dr.phil., University of Leipzig, 1895. Technical University of Aachen, 1907-37. *Die Kartenwissenschaft* (2 vols., 1921-25), *Kartenkunde* (1936); *Kartographie* (1939). R. Ogrissek, "Studium, Promotion und Lehrtätigkeit Max Eckerts an der Universität Leipzig im 19. Jahrhundert," *International Yearbook of Cartography*, 25 (1985), 139-158.

Economic Geography

Economic geography is concerned with areal variation in the livelihood activities of people; with regional patterns of production, distribution, and consumption of goods and services; with the spatial structure and development of economic systems; and with human use of earth resources. It differs from economics in being based not on economic theory but on the interpretation of spatial distributions of economic phenomena. The two fields are united by their common research interests in regional economics and in locational theory and analysis.

The resource and economic content of area has been an enduring geographic interest, expressed in the work of Herodotus and Strabo in antiquity and—with the revival of European study—in the writings of Sebastian Münster and others in the sixteenth century. The Dutch geographer Bernard Varen (Varenius) in the mid-seventeenth century wrote a regional guide to Japan and Siam for the information of Dutch merchants doing business in those distant lands. Alexander von Humboldt's work on New Spain (1811) is recognized as an early work in regional economic geography, while Carl Ritter is credited with initiating if not completing studies in its systematic aspects.

Practical concerns with opportunities for business and trade sparked popular interest in geography during the nineteenth century in Europe and the United States and formed one of the two lines of development leading to modern academic economic geography. Businessmen and merchants supported such geographical societies as that of Scotland and Manchester in Britain and the American Geographical Society in the United States, and "commercial geography" became a subject of instruction in business schools and high schools (and their equivalents) in the United States, Britain, and on the continent. Carl Zehden of the Handels-Akademie of Vienna published one of the first comprehensive commercial geographies in 1871; in Britain, Hugh Robert Mill's *Elementary Commercial Geography* appeared in 1888. John N. Tilden, writing that "commercial geography treats of the productions of the earth, and of their distribution; of the routes of commerce, and of markets and manufactures," wrote a textbook for American secondary and commercial schools that was published in 1890. The most famous contribution was that of the Scottish geographer George G. Chisholm, whose *Handbook of Commercial Geography*, first published in 1889, became a classic, eventually to appear in ten editions up to 1925 under his authorship; it appeared in still later editions rewritten and edited by L. Dudley Stamp.

A second and more direct line of development leading to modern academic economic geography emerged during the later nineteenth century from concerns of German and Austrian economists of the "new historicism" viewpoint. Maintaining that modern economies could be understood only in the context of geographically, historically, and politically unique national circumstances, they supported geographic study and teaching contributing to those economic understandings. The name "economic geography" appears to have been introduced in 1882 by W. Götz (a student of Friedrich Ratzel), who stressed the scientific nature of the study in contrast to the practical interests of commercial geography. But the new discipline as taught in German universities initially reflected the interests and purposes of economists.

It was through American economists trained in Germany that economic—as opposed to commercial—geography was introduced in the United States and became part of the economics curriculum in many major American universities beginning in the early 1890s. One of the first (1893) to offer it was the Wharton School of Finance and Commerce of the University of Pennsylvania, taught there by the economist Emory R. Johnson and geographers J. Paul Goode, J. Russell Smith, and Walter S. Tower, Johnson's students and colleagues. Erroneously, the Wharton group is frequently credited with developing academic economic geography in the United States and, through J. Russell Smith's *Industrial and Commercial Geography* (1913), producing the first textbook in the field. In reality, instruction was widely offered by German-trained American economists, many college texts preceded Smith's volume, and major philosophical and methodological statements defining economic geography had been authored by individuals with no connection with the University of Pennsylvania—notably Lindley Keasbey, then of Bryn Mawr College, by Lincoln Hutchinson of the University of California, and by Edward Van Dyke Robinson of the University of Minnesota. Economic geography was generally dropped from American economics curricula before World War I as no longer relevant to that changing discipline.

Absorbed into geography department programs, economic geography lost its direct connection to economics and focused on the concern then central to geographic inquiry: the theoretical relationship between human (in this case, economic) activity and the physical environment. Environmentalism remained a dominating orientation to the 1930s, when general reaction against its inconsistencies and illogic induced economic geographers to concentrate increasingly on regional variations in economic activity and on the delimitation of economic regions. An emphasis on specific commodities and their conditions of production and trade led, as well, to a fragmentation of economic geography and the rise of subdisciplines including agricultural, industrial, transportation, and resource geography, among others. A more theoretical systems-oriented economic geography based once again on the work of economists and on quantitative techniques emerged beginning in the 1950s. Most recently, domestic and international social consequences of economic systems and structures have fostered concern with socioeconomic spatial patterns and problems of development and underdevelopment. Marxist interpretations increased in number and kind during the 1980s, focusing particularly on the role of the market system in patterns of land use allocation and on problems of regional and international development and exchange.

Economic geography has become or remained an important research interest in all countries, though its development and present role is unique to each of them. In Germany, it has retained both the regional and systematic emphases early introduced by von Humboldt and Ritter, and has been marked by more extensive methodological discussions of aims and procedures than have other national schools. Contributions of German economists and geographers to theories of agricultural land use (J. H.

von Thünen), of plant location (Alfred Weber), urban size, spacing, and functional hierarchies (Walter Christaller), and spatial economic and urban organization (August Lösch) have influenced economic geographic thought in all countries. In France, the subject earlier expressed the strong regional emphasis of Vidalian (P. Vidal de la Blache) tradition and the definition given it by the work of Jean Brunhes (*La géographie humaine*, 1910). More recently, the systematic approach has received greater attention, and studies of factors of production and consumption, economic systems, transportation geography, and economic planning have been emphasized.

Building on a pre-revolutionary interest in economic geography (the term was first used by Lomonosov in the 1760s) and on work in economic regionalization attributed to Arsen'yev in the first half of the nineteenth century, Soviet geographers have made economic geography the leading branch of human geography and, following Leninist dictates, have concentrated on problems of regional distribution of economic activities, particularly on detailed studies of planned integrated economic regions. Nikolai N. Baranskiy, founder of the Department of Economic Geography at Moscow State University, was a leader in establishing the strong economic and planning orientation of Soviet geography. The same interests in geography as a tool in economic planning and development have motivated work in economic geography in many of the newer Third World universities and governmental agencies. *See also* Agricultural Geography; Industrial Geography.

BIBLIOGRAPHY

Berry, B. J. L., E. C. Conkling, and D. M. Ray, *Economic Geography* (1987), especially Chap. 2, pp. 17–35, "The Changing Nature of Economic Geography as a Field of Study."

Chisholm, M., *Geography and Economics* (1966), especially Chap. 2, pp. 4–28, "Relations between Geography and Economics."

Fellmann, J. D., "Myth and Reality in the Origin of American Economic Geography," *AAAG*, 76 (1986), 313–330.

Saushkin, Y. G., *Economic Geography: Theory and Methods* (1980).

Smith, R. H. T., E. J. Taaffe, and L. J. King, *Readings in Economic Geography: The Location of Economic Activity* (1968).

Wheeler, J. O., and P. O. Muller, *Economic Geography* (2nd ed., 1986), especially Chap. 1, pp. 2–16, "The Field of Economic Geography."

• *Jerome D. Fellmann*

Egypt

Egypt has a long tradition in geography. The marvellous astronomical orientation of the pyramids, the solar calendar discovered c.4540 B.C., and the achievements of such figures as Ptolemy and Eratosthenes in antiquity and Idrisi and Ibn Khaldun in the Middle Ages stand as witnesses of such a tradition. In modern Egypt, geography has always occupied a distinguished place in the curricula of general education. It is noteworthy that Rifaa el Tahtawy (1800–1873), one of the pioneers of modernization in Egypt, translated parts of Conrad Malte-Brun's *Géographie universelle* into Arabic around 1853. Egyptian students resumed their interest in the subject in the High Training College, which was founded in the 1870s. Geography was then combined with astronomy and geology on one hand and history and archaeology on the other. M. Ramzy (1871–1945) was a distinguished administrator and researcher in the H.T.C. who was interested in the origins and evolution of place names in Egypt. With the establishment of the first Egyptian University (later called Fuad I and then Cairo University) in 1925, Mustafa Amer (1896–1973) and M. Awad (1896–1972) initiated the foundation of the geography department. Amer and the Austrian scholar O. Menghin were interested in prehistoric archaeology. They excavated a Neolithic site at Maadi south of Cairo. Rizkana is now publishing a compendium on Neolithic cultures. S. Huzzayin, one of the first graduates of Cairo University, continued his postgraduate studies under Roxby and Fleure in Britain and became an authority on Pleistocene geography with his work, *The Place of Egypt in Pre-history* (1940). The Egyptian school of geography in the interwar period had historical and literary inclinations and was generally deterministic. After World War II, young graduates were sent to western universities, where they were quick to learn new specialties, and on their return they freed their subject from non-geographic irrelevancies and the obsolete deterministic views. They advanced new branches of geography, *e.g.*, geography of population and settlement (Ghallab and Abdel-Hakim), geography of land use (Nasr), and quantitative methods. This generation also conducted research in other Arab lands. Sayyad worked in the Sudan (1949), Ghallab published many works on the historical geography of the Levant (1951–1965), and Sherief submitted a thesis on Iraq. In 1947 a university was opened that specialized first on Sudanese and later on African studies. The newer generation was called upon to teach in newly opened schools and universities in the emerging Arab countries. The Egyptian school of geography is not confined to the twenty departments in the Egyptian universities but is extended to daughter departments widespread from Beirut to Khartoum and from the Atlantic to the Persian Gulf.

In 1875 Khedive Ismail founded the Société de géographie d'Egypte, one of the first societies of its kind in the world. The society played an important role in the exploration and mapping of equatorial Africa, in particular Ethiopia and the Sudan. Its *Bulletin*, issued annually in English and French, has recorded geographic knowledge and research for more than a century. Since 1965 it has been felt that an Arabic magazine should be added, as the number of geography graduates has increased in Egypt and other Arab countries. The society also published some selected geographical works, mostly in English and French. It houses the only ethnographic museum in Egypt. It organizes series of extramural geographical lectures, which are keenly attended by geographers and other social scientists. The tradition of exploration continued through the initiative of several individuals, the most notable of whom was Ahmed Hussanien (1889–1946), who with the support of King Fuad explored the western desert from the Mediterranean to Darfur in western Sudan in 1923.

Bibliography

Ezzat Abdel Karim, M., *Taarikh Al Taaleem fi Misr* (History of Education in Egypt) (1945).

Foucart, G., and A. Cattaui, *La Société de Géographie du Caire. Son Oeuvre (1875–1921)* (1921).

Hamdan, G., "Nahwa Madrassattin Arabiyattin fi Algographia" (Towards an Arabic School of Geography), *Miraatu Al Oloum Al Ijtimaiati* (Social Science Magazine), December 1964.

Toni, Y., "The 'Bulletin de la Société de Géographie d'Egypte': A Review of Its Volumes, 1875–1965," *Bulletin de la Société de géographie d'Egypte*, 39 (1966), 83–114.

• *M. S. Ghallab*

England. *See* United Kingdom.

Environment

In the broadest sense, the environment includes all external components that environ any object, but among geographers it most commonly refers to location and the aggregate of physical, climatic and biological elements that impinge on humanity. In view of the fact that few, if any, purely "natural" environments now exist in their pristine state, the term may be broadened to include features already modified by man, but a distinction between elements of human and non-human origin is usually implicit. At times, some geographers extend the concept to incorporate such components as social, cultural, and economic conditions, and the meaning of environment (and cognate terms such as milieu) naturally varies with context.

Further refinements of meaning may be indicated by a qualifying adjective. For instance, the objectively existing total environment may be distinguished from the operational or effective environment, and the perceptual and behavioral environments may be further differentiated. But, without such qualifications, the "environment" usually denotes the locational and biophysical setting of human life.

Bibliography

Bates, M., *The Human Environment* (1962).

Taylor, T. G., ed., *Geography in the Twentieth Century* (3rd ed., 1957).

Wagner, R. H., *Environment and Man* (2nd ed., 1974).

• *Gordon R. Lewthwaite*

Environmental (or Geographical) *Determinism*

Environmental determinism postulates that the "natural" environment, incorporating the locational, physical, climatic, and biological components of geography, governs the course of human history and the levels of cultural and economic achievement.

This "environmentalist" interpretation has deep roots in, at least, Western thought. It was adumbrated in the fifth century B.C. by Hippocrates, surfaced in the works of Bodin and Montesquieu, and came to extreme expression in the writings of social historians such as Cousin and Buckle. Furthered by the Lamarckian and Darwinian linkage of environment and evolution, it formed an influential strand of thought in the 1870–1920 period, when geography assumed its university role. Influential leaders, especially in America and Britain, sought to organize the discipline around a unifying search for man-environment relationships in an era when deterministic law seemed the touchstone of science. Thus the German geographer Friedrich Ratzel sought (though not without qualification) to elucidate laws of environmental relationship in the initial volume of *Anthropogeographie* (1882) and Ellen Churchill Semple advocated a related approach in America. There William Morris Davis defined geography as the study of relationships—usually of control and response—between physiography and ontography or organic life, and Ellsworth Huntington, while not ignoring Semple's stress on the influence of surface forms and routeways, focussed attention on the postulated impact of climate on civilization. In Australia, Griffith Taylor propounded "stop-and-go determinism": man could change the pace but not the direction of development.

This tendency to seek "geographic" determinants receded as the independent potency of human action was highlighted, as Vidal de la Blache developed the theme of "possibilism," and as other European geographers elucidated alternative concepts of the nature of geography—landscape study and areal differentiation in particular. As Carl Sauer and Richard Hartshorne elucidated these concepts in American geography, they called for explanatory analyses unbiased by any preferential assumption of environmental causation.

Yet the critical evaluation of environmental influence remains valid, in part transmuted and revived as the themes of ecosystem and environmental limitations surface in geography. Significantly, however, "environmentalism" today refers less to postulated physical controls than to the conservation movement, and "determinism" is commonly softened into "probabilism."

BIBLIOGRAPHY

Glacken, C. J., *Traces on the Rhodian Shore* (1967).

Lewthwaite, G. R., "Environmentalism and Determinism: A Search for Clarification," *AAAG*, 56 (1966), 1–23.

Taylor, T. G., ed., *Geography in the Twentieth Century* (3rd ed., 1957).

• *Gordon R. Lewthwaite*

Environmental Perception

"Environmental perception" constitutes a research tradition arising out of a concern for environmental hazards, and the way these are evaluated as people make decisions about locating their activities—for example, building houses on floodplains, or choosing to grow certain crops in drought-prone areas.

Although the term was coined in the 1960s, "environmental perception" has many precedents in geographic inquiry. Historical geographers, for example, have taken as a major theme the images of the new lands of North America and Australia in the nineteenth century, images generated by people forced to evaluate a new physical environment in varying degrees of ignorance, using former, and often inappropriate, categories. Some "perceptions" crossed the border from genuine enthusiasm to sheer propaganda to enhance land sales and financial gain.

Contemporary research has moved away from an earlier concern with perception *per se* to an evaluation of marginal, and increasingly fragile, environments under heavy pressure from human and animal populations. *See also* Behavioral Geography; Mental Maps.

BIBLIOGRAPHY

Downs, R., and D. Stea, eds., *Image and Environment* (1973).

Heathcote, R. L., *Back of Bourke* (1965).

Saarinen, T., *Perception of the Drought Hazard on the Great Plains* (1966).

• *Peter Gould*

Evans, Emyr Estyn (1905–1989)

D.Sc., University of Wales, 1939. Queen's University of Belfast, 1928–70. *Irish Heritage* (1942); *Mourne Country* (1951); *Irish Folk Ways* (1957); *The Personality of Ireland* (1973). *WW*; H. J. Fleure, "Emyr Estyn Evans: A Personal Note," pp. 1–7 in *Man and His Habitat: Essays Presented to Emyr Estyn Evans* (ed. by R. Buchanan, E. Jones, and D. McCourt) (1971); obit. by E. Jones in *GJ*, 156 (1990), 116–117.

Faucher, Daniel (1882–1970)

D.-ès-L., University of Grenoble, 1927. University of Toulouse, 1926–52. *Plaines et bassins du Rhône moyen* (1927); *Géographie agraire* (1949); *La vie rurale vue par un géographe* (1962); *L'homme et le Rhône* (1968). F. Taillefer in *LGF* (1975), 173–183; *France méridionale et pays ibériques* (Faucher festschrift) (2 vols., 1948–1949); obit. by L. Papy in *AG*, 80 (1971), 385–396.

Fawcett, Charles Bungay (1883–1952)

M.Sc., University of London, 1916. University of Leeds, 1919–28; University College London, 1928–49. *Frontiers* (1918); *Provinces of England* (1919); *Political Geography of the British Empire* (1933). T. W. Freeman in *GBS*, 6 (1982), 39–46; obits. by R. Buchanan in *GJ*, 118 (1952), 514–516; and H. Fleure in *Geography*, 37 (1952), 232–233.

Feminist Geography

Emerging in the early 1970s as an academic expression of the women's movement, feminist geography has been shaped by a fundamental concern with inequality between women and men and a recognition that research, teaching, and practice in geography need to pay attention to women and to gender relations. Early studies in North America described differences in male and female spatial behavior and environmental relations, establishing the significance of gender in geographic research. The first comprehensive review of this work was published by Zelinsky, Monk, and Hanson in 1982. Another early approach was to critique human geography for its failure to consider women and gender (Monk and Hanson, 1982).

By the mid-1980s, feminist geography had become international in scope and more complex in the questions it asked and the approaches it employed. British geographers, in particular, focused on developing theoretical frameworks for analyzing gender inequalities, emphasizing how women were disadvantaged by economic constraints imposed by capitalism and ideological constraints imposed by patriarchal social orders. They reviewed and applied this thinking in the book *Geography and Gender*, which was designed to introduce feminist geography into undergraduate education. In comparison with British geographers, Americans have paid more attention to addressing the implications of feminist research for revising existing geographic theories.

Going beyond approaches that emphasize structural constraints are studies that consider women as social agents whose actions reflect their own values and interests. Research on women's paid and unpaid work, their daily movements and longer term migrations, and their efforts to provide shelter and services for themselves and their families, has now been conducted in an array of world areas. A substantial body of work has appeared on Third World women and development issues, and, as the collection edited by Momsen and Townsend (1987) illustrates, this is being written by geographers from the areas discussed, as well as by scholars from developed countries. Increasing attention is also being paid to diversity among women on such bases as race and ethnicity, class, and life stage, as well as geographic context. For example, sessions on such themes were organized at British and U.S. meetings in 1989. A group of geographers in Barcelona is emphasizing the study of rural women; Amsterdam geographers are taking up the theme of diverse family forms and their implications for urban and regional planning. In these works attention to gender has required breaking down distinctions between social and economic geography and recognizing how ideologies about gender shape life.

A recurring feature in the publication of feminist geographic studies has been the preparation of special journal issues, including *Antipode* (1974, 1986), *Journal of Geography* (1978), *The Professional Geographer* (1982), *Cahiers de géographie de Québec* (1987), *Espace, Population, Société* (1989), *Geoforum* and *Documents d'Anàlisi Geogràfica* (forthcoming). Two atlases of American women have been published, as well as an international atlas.

Feminist geographers have also devoted attention to examining the status of women in the geographic profession and to forming organizations to support women geographers and feminist scholarship. By 1989, such groups operated in Australia, Britain, Canada, the Netherlands, Nordic countries, and the United States, with a German-speaking group incorporating geographers from Austria, Switzerland, and West Germany. A gender study group within the International Geographical Union, including African, Asian, and Latin American, as well as European, North American, and Australasian scholars, was established in 1988 and has initiated a program of international comparative research.

Bibliography

Momsen, J., and J. Townsend, eds., *Geography of Gender in the Third World* (1987).

Monk, J., and S. Hanson, "On Not Excluding Half of the Human in Human Geography," *PG*, 32 (1982), 11-23.

Seager, J., and A. Olson, *Women in the World: An International Atlas* (1986).

Women and Geography Study Group of the Institute of British Geographers, *Geography and Gender* (1984).

Zelinsky, W., J. Monk, and S. Hanson, "Women and Geography: Review and Prospectus," *PHG*, 6 (1982), 317-366.

• *Janice Monk*

Fenneman, Nevin Melancthon (1865-1945)

Ph.D., University of Chicago, 1901. University of Wisconsin, 1903-07; University of Cincinnati, 1907-37. *Physiography of Western United States* (1931); *Physiography of Eastern United States* (1938). B. Ryan in *GBS*, 10 (1986), 57-68; obits. by J. Rich in *AAAG*, 35 (1945), 180-189; and W. Bucher in *Proceedings of the Geological Society of America for 1945* (1946), 215-228.

Finland

Even before its institutionalization in the 1880s, geography in Finland had been a well-established tradition since the foundation of the university in 1640, where geography was taught together with mathematics and history. Two different conceptions of geography led to the founding of two societies in 1888. The first of these, the Geographical Society of Finland, was similar to a scientific academy, an elite organization of scholars from different disciplines, whose major aim was to emphasize Finland's national identity. Leading figures in the society were J. A. Palmen (1845-1919), professor of zoology, and K. E. F. Ignatius (1837-1909), historian, who published the first *Geography of Finland* (1881). The society's main achievements were the journal *Fennia* and the *Atlas of Finland*, the first national atlas in the world (1899; 5th ed. 1975–91). The second society was the Geographical Association; this was a collection of school geography teachers and amateur geographers under the leadership of ex-botanist Ragnar Hult (1857–1899), the first university geographer (docent 1890). The two societies were amalgamated in 1921.

Hult founded the department of geography at Helsinki University in 1893, firmly establishing geography as a university discipline, and teacher training got under way. There followed the appointment of J. E. Rosberg (1864–1932) as extraordinary (or associate) professor in 1901 (professor 1912–29), and J. G. Granö (1882–1956) as first assistant in 1902. The main subject was geomorphology. Another extraordinary professor, I. Leiviskä (1876–1953), appointed in 1921, also represented this field. Radical changes in Granö's thinking took place when he was professor at Tartu in Estonia (1919–23) and Helsinki (1923–26); he moved away from the geomorphology of Central Asia to the perceptual study of the landscape (see especially his *Reine Geographie*, 1929). In 1926 Granö moved to Turku as first professor there, but returned to Helsinki in 1945 (retiring in 1950). Granö was succeeded in Helsinki in 1927 by the palaeobotanist and later explorer of Patagonia, V. Auer (1895–1981). This strengthened the study of Quaternary geology, a trend also followed by L. Aario (b. 1906, professor 1951–69), who nonetheless later turned to human geography. Rosberg was followed by the geologist and geomorphologist V. Tanner (1881–1948, professor 1931–44), who made expeditions to Labrador.

Other professors before the expansion of the 1960s were Helmer Smeds (1908–1963) of the University of Helsinki, a human geographer and expert on Ethiopia; the geomorphologist Veikko Okko (b. 1912); the human geographer Stig Jaatinen (b. 1918); and Kaarlo Hildén (1893–1960), who was an anthropologist at the Helsinki School of Economics. Oiva Tuominen (b. 1909), an urban geographer, moved to Turku, as did the coastal geographer and historian of geography Olavi Granö (b. 1925). Uuno Varjo (1921–1986), an agricultural geographer, moved to Oulu, and Mauri Palomäki (b. 1931), whose interests are in urban and settlement geography, went to Vaasa. In Tampere the pioneer of quantitative geography Reino Ajo (1902–1974) worked outside the university. After 1960 the number of geographers increased so greatly that it is impossible to mention them in this brief essay.

In the 1960s there was a rapid expansion of graduate programs and growth of new departments (Oulu, Tampere, Vaasa, and Joensuu) outside Helsinki and Turku. Human and physical geography became separated, and regional geography was forgotten. There were numbers of new doctorates in both human and physical geography, while applied geography, with its own program of studies, attached increasing importance to planning. With the increasing number of university posts, geography as a university discipline predominated, and leadership of the Geographical Society passed almost entirely into the hands of university geographers. Departments started their own publications, and English superseded German as the language of publication. This expansion came to an end in the late 1970s. Then, in addition to the traditional areas, theoretical, epistemological, and humanist questions began to be discussed. It was a period of re-identification, and science policy spread to geography.

Today there are seven chairs of physical geography and thirteen chairs of human geography at Finnish universities, together with six chairs of economic geography at schools of economics. In the ten departments of geography the academic staff numbers approximately ninety.

Bibliography

Granö, O., "The Relationship between Intellectual Content and Institutionalisation in Finnish Geography," *Fennia*, 162 (1984), 9–20.

Hägerstrand, T., and A. Buttimer, eds., *Geographers of Norden* (1988).

Rikkinen, H., "Developments in the Status and Content of Geography in the Secondary Schools of Finland 1770–1977," *Fennia*, 160 (1982), 43–93, 313–383.

Rikkinen, K., "Ragnar Hult and the Emergence of Geography in Finland, 1880–1900," *Fennia*, 166 (1988), 3–192.

Yli-Jokipii, P., "Trends in Finnish Geography in 1920–1979 in Light of the Journals of the Period," *Fennia*, 160 (1982), 95–193.

——— , "Centenary Bibliography of Papers Published in the Journals of the Geographical Society of Finland, by Topic, 1888-1987," *Fennia*, 166 (1988), 193-285.

• *Olavi Granö*

Finland, Geographical Society of (Suomen Maantieteellinen Seura/Geografiska Sällskapet i Finland)

On April 6, 1888, the statutes of two separate geographical societies were formalized in Finland. The Finnish Geographical Association, founded primarily by Ragnar Hult, adopted the model of the Royal Danish Geographical Society and aimed at a broader membership. The Geographical Society of Finland, initiated by J. A. Palmen, recruited members who were directly concerned with research into the geography of Finland. The former established the journal *Geografiska Föreningens Tidskrift*, which subsequently became *Terra*; the latter initiated the research series *Fennia*. In 1921, the two organizations united as the Geographical Society of Finland. In addition to *Fennia* and *Terra*, they have also published *Acta Geographica* (1927-72). Members of the two societies collaborated in the production of the world's first national atlas, *The Atlas of Finland* (1899). The fifth national atlas is a joint venture with the National Survey Board. The society promotes research and teaching projects, intermittently awards *Fennia* medals, and actively encourages the Geographical Days, an annual gathering of research geographers.

BIBLIOGRAPHY

Rikkinen, K., "Ragnar Hult and the Emergence of Geography in Finland, 1880-1900," *Fennia*, 166 (1988), 3-192.

• *W. R. Mead*

Fischer, Theobald (1846-1910)

Dr.habil., University of Bonn, 1876. University of Kiel, 1879-83; University of Marburg, 1883-1910. *Studien über das Klima der Mittelmeerländer* (1879); *Länderkunde der südeuropäischen Halbinsel* (1893); *Mittelmeerbilder* (1906, 2nd ed. 1913). G. Glauert in *NDB*, 5 (1961), 205-206; obits. by H. Wagner in *PGM*, 56 (1910), 188-189; and K. Oestreich in *GZ*, 18 (1912), 241-254.

Fleure, Herbert John (1877-1969)

D.Sc., University of Wales, 1904. University of Wales, Aberystwyth, 1904-30; University of Manchester, 1930-44. *Human Geography in Western Europe* (1918); *The Corridors of Time* (with H. Peake) (10 vols., 1927-56); *A Natural History of Man in Britain* (1951). T. W. Freeman in *GBS*, 11 (1987), 35-51; E. Evans in *IESS*, 5 (1968), 494-495; J. Campbell, "Some Sources of the Humanism of H. J. Fleure," Oxford School of Geography, *Research Papers*, 2 (1972); obits. by E. Evans in *GJ*, 135 (1969), 484-485; and E. Bowen and A. Garnett in *Geography*, 52 (1969), 464-469.

France

I. An Old Geographic Tradition (sixteenth to nineteenth centuries)

Geography began to develop in France during the Renaissance. Information flowed toward Paris from all over the world. By the beginning of the eighteenth century, Paris was the main cartographic center in the world. French geographers of the mid-eighteenth century were famous. They were specialists in locations and did not try to explain patterns and distributions. Their best achievement was to produce wonderful maps out of imperfect records. With the advent of Harrison's chronometer, this kind of geography became obsolete.

Geography then experienced a crisis at a time when Paris was the best place to practice geography, and when new geographic curiosities were appearing: a taste for regional description; a will to produce better administrative divisions; a wish to build sounder cities; and, somewhat later, with the transportation revolution, the desire of fostering the division of labor. There were people with a capacity to mobilize these new orientations, like Volney, but the modernization of geography did not occur, since the prevailing philosophies did not raise major geographical issues. The main geographers, like Conrad Malte-Brun, adopted a conservative view of the discipline.

II. The French School of Geography (1870–1950)

Conditions changed by the last third of the nineteenth century. It was the time when the imperial expansion of Europe accelerated. Nationalisms were getting stronger. The democratic regimes were in search of new principles on which to build their national identities.

There were by then geographical issues of significance for philosophers. At the end of the eighteenth century, Herder had developed a conception of progress in which each people had its fate linked with the country where it settled. Darwinism recast the man/environment problem embodied in the Herderian paradigm in a more scientific setting. Hence the development, first in Germany, of a discipline focusing on the vertical relations between societies and milieus.

The French school of geography grew out of the teachings and works of three geographers. Elisée Reclus (1830–1905), a student of Carl Ritter, offered to the French public a modern description of the world's diversity. Emile Levasseur (1828–1911), a historian and demographer, introduced geography in the curricula of French secondary schools and focused attention on density maps. Paul Vidal de la Blache (1845–1918) provided French geography with its main themes and methods. The general architecture of his posthumous *Principes de géographie humaine* (1922) conformed with Ratzel's *Anthropogeographie*: the evolutionist point of view, the opposition between primitive and developed societies, and the role of circulation in the shaping of the modern world were similar. French geography was influenced by German geography, but it differed from it on major issues. For Vidal and for the French school, the aim of geography was more to explain the patterns of human densities throughout the world than to analyze their landscapes. Human densities reflected the diversity of environments; primitive people had to adapt themselves to natural conditions, but they did it actively, through the invention of tools and techniques. They were building *genres de vie*, ways of life, which varied according to the materials that nature offered, but became increasingly stable with progress. With civilization, *genres de vie* ceased to merely reflect the local milieus; they expressed the character of the societies that tried to impose them on new environments when they migrated. For Vidal, *circulation* gave an increasing autonomy to *genres de vie*.

With trade, the *genres de vie* could specialize in order to take advantage of local resources. Regions ceased to be isolated blocks; they became parts of complex spatial organizations. Their study was the main field of inquiry for the French School.

Through the development of social and economic complementaries, geographical entities developed; they often became political units and acquired a "personality." This kind of interpretation, first proposed by Michelet applying Herderian ideas to France, was the only way to explain the emergence of a unitary nation out of a bunch of different regions with different ethnic traditions. At this point, French geopolitics diverged from German geopolitics: race could not be central in a country built out of diversity! Vidalian geopolitics was well suited to preindustrial societies. The fact is best exemplified by Fernand Braudel relying on the characters of the Mediterranean lands to explain their history.

The *genre de vie* ceased to be a satisfactory tool when applied to complex urbanized and industrialized societies with a great variety of roles. Vidal was personally able to overcome this difficulty, but he developed his views at a late stage in his career. This part of his ideas was generally overlooked by his followers. In a way, the French School was poorer than the geography of Vidal de la Blache.

III. Crisis, Confusion and Recovery (1950 to the present)

The limitations of Vidalian geography began to appear disturbing during the 1940s. Some geographers tried to modernize the discipline—André Cholley, Maurice Le Lannou, and Maximilien Sorre, for instance. Eric Dardel's stimulating views on phenomenology and geography were overlooked at the time they were published (1952).

Marxism was then popular among the French intelligentsia. Some geographers, like

Jean Dresch and Pierre George, criticized French imperialism, fostered the study of city/countryside relationships, and stressed the role of industrialization in the transformation of Western societies. But they were unable to build a new scientific paradigm and to conceive geography as a social science.

From the early 1960s on, some people endeavored to provide new approaches and new structures for geography. Renewal came mainly after 1968. People began to explore economics, anthropology and history for new ideas, and to look for foreign experiences. German influences dwindled, while Anglo-American contacts grew.

The idiosyncracies of French geographers are much less than a generation ago, but some traditional attitudes have managed to survive, the emphasis on induction and description, for instance. The refusal to build causal theories was reformulated in terms of systems theory. But the renewal is real. French geography is now largely a social science. French geographers still start from density maps, but instead of describing *genres de vie*, they look for the complex role-setting of the urbanized and industrialized societies. The interest in regional organization is still alive, but more attention is devoted here than abroad to information flows as factors in shaping human patterns. Cities are conceived as machines devised in order to maximize social interactions; hence a new formulation of urban geography.

The study of perception is also central to the new orientations, but its psychological dimensions are thought less interesting than its cultural ones. It is impossible to understand the way people shape the earth without a view of their preferences, their dreams and their beliefs. Human space is differentiated between profane areas and sacred spaces. Hence the new interest in ethnogeographies.

Some of the original features of French geography have survived the process of modernization undergone during the past thirty years. There is no longer a French School of Geography, but French geography has retained some of its characteristics. *See also* Association of French Geographers; Paris Geographical Society.

Bibliography

Berdoulay, V., *La formation de l'école géographique française (1870–1914)* (1981).

Buttimer, A., *Society and Milieu in the French Geographic Tradition* (1971).

Claval, P., *Essai sur l'évolution de la géographie humaine* (Rev. ed., 1976).

Dunbar, G. S., *Elisée Reclus, Historian of Nature* (1978).

Meynier, A., *Histoire de la pensée géographique en France (1872–1969)* (1969).

Pinchemel, P., M.-C. Robic, and J.-L. Tissier, eds., *Deux siècles de géographie française: choix de textes* (1984).

• *Paul Claval*

Gallois, Lucien (1857-1941)

D.-ès-L., University of Paris, 1890. University of Lyon, 1889-93; University of Paris, 1893-1927. *Les géographes allemands de la Renaissance* (1890); *Régions naturelles et noms de pays* (1908); (ed.) *Géographie universelle* (planned with P. Vidal de la Blache). A. Meynier in *LGF* (1975), 25-33; obit. by E. de Martonne in *AG*, 50 (1941), 161-167.

Gambi, Lucio (b. 1920)

Laurea in Lettere, University of Rome, 1945. University of Messina, 1951-60; University of Milan, 1960-75; University of Bologna, 1975 to the present. *L'insediamento umano nella regione della bonifica romagnola* (1948), *Calabria* (1965), *Una geografia per la storia* (1973), *Milano* (1982).

Garrison, William Louis (b. 1924)

Ph.D., Northwestern University, 1950. University of Washington, 1950-60; Northwestern University, 1960-67; University of Illinois, Chicago, 1967-69; University of Pittsburgh, 1969-73; University of California, Berkeley, 1973 to the present. "The Analysis of Highway Networks," *Highway Research Board Proceedings* (1958), 1-14; "Spatial Structure of the Economy," *AAAG*, 49 (1959), 232-239, 471-482; "Values of Regional Science," Regional Science Association, *Proceedings*, 13 (1964), 7-10; "Transportation Technology," *Transportation Research*, 18A (1984), 267-276. *WWA*; *GOF* (1972).

Gautier, Emile-Félix (1864-1940)

D.-ès-L., University of Paris, 1902. University of Algiers from 1902. *Madagascar* (1902), *Le Sahara algérien* (1908), *L'Algérie et la Métropole* (1920), *Le Sahara* (1923). *Mélanges de géographie et d'orientalisme offerts à E.-F. Gautier* (1937); M. Larnaude in *LGF* (1975), 107-118.

Geddes, Patrick (1854-1932)

Student at Royal School of Mines, London, 1874-77. University College London, 1877-79; University of Edinburgh, 1880-89; University College Dundee, 1889-1919; University of Bombay, 1920-23; Scots College, Montpellier, 1924-32. *City Development* (1904); *Cities in Evolution* (1915); *Our Social Inheritance* (with V. Branford) (1919); *Life: Outlines of General Biology* (with J. Thomson) (2 vols., 1931). W. I. Stevenson in *GBS*, 2 (1978), 53-65; A. Geddes in *DNB 1931-1940* (1949), 311-313; H. Fleure, "Patrick

Geddes (1854–1932)," *Sociological Review*, 1 (1953), 5–13; obit. and appreciations in *Sociological Review*, supplement to vol. 24 (1932), 351–400.

Gellert, Johannes Fürchtegott (b. 1904)

Dr.phil., University of Leipzig, 1929; (habil.) University of Leipzig, 1937. Leipzig Institute for Colonial Geography, 1931–38; University of Leipzig, 1938–49; University of Halle, 1949–50; Pedagogical University of Potsdam, 1950–70. *Mittelbulgarien* (1937); *Physische Geographie von Deutschland* (1958); *Deutsche Demokratische Republik* (1977); *China* (1987).

Gender in Geography. *See* Feminist Geography.

Genre de Vie

French social scientists began to use the term *genre de vie* (way of life) as a descriptive device at the end of the eighteenth century; they contrasted hunters and gatherers or nomads with peasant farmers. Vidal de la Blache transformed the phrase into a fundamental category of geographic analysis.

The idea is to decipher the relations between man and nature through the listing of operations needed for exploiting the environment. The study of *genres de vie* gives geography an ecological dimension. *Genre de vie* also describes patterns of social relations and expresses social ideologies. *Genres de vie* differ from place to place, thus stimulating exchanges, fostering spatial organization, and explaining the emergence of political constructions. In this way, the study of *genres de vie* gives human geography a social dimension.

Genres de vie are standardized sets of roles observed in primitive or traditional societies; within them, nearly everybody has to participate in the same way in the exploitation of nature. In urbanized and industrialized societies, standardization disappears because of the increasing variety of productive occupations; geographers have to rely on the constitutive elements, the roles, to grasp the ecological bases and the social life of groups. Time geography, the modern form of *genre de vie* analysis, performs the same functions and offers similar advantages.

Bibliography

Claval, P., "Les trois niveaux d'analyse des genres de vie," pp. 73–87 in *Geographie des Menschen: Dietrich Bartels zum Gedenken*, ed. by G. Bahrenberg *et al.* (1987).

Sorre, M., "La notion de genre de vie et sa valeur actuelle," *AG*, 57 (1948), 97–108, 193–204 (English translation, "The Concept of Genre de Vie," pp. 399–415 in *Readings in Cultural Geography*, ed. by P. L. Wagner and M. W. Mikesell, 1962).

Vidal de la Blache, P., "Les genres de vie dans la géographie humaine," *AG*, 20 (1911), 193–213, 289–304.

• *Paul Claval*

Gentilcore, R. Louis (b. 1924)

Ph.D., University of Maryland, 1950. Indiana University, 1950–53, 1954–56; University College London, 1953–54; California State University, Los Angeles, 1956–58; McMaster University, 1958–89. *Canada's Changing Geography* (1967); *Geographic Approaches to Canadian Problems* (1971); *Studies in Canadian Geography: Ontario* (1971); *Ontario's History in Maps* (with G. Head) (1984). *CWW.*

Geoarchaeology

Physical geographers have long contributed to identification of events in Quaternary history and the resolution of related environmental changes, with geographers A. Penck and E. Brückner establishing the model of fourfold Pleistocene glaciation in 1909. Culture-historical geographers have also shown interest in "reconstructing" natural environments before human modification, such as R. Gradmann in Germany and C. O. Sauer and his students in North America. Geographers such as H. J. Fleure and P. Deffontaines collaborated with archaeologists or historians to understand changing human settlement or subsistence patterns. With the development of more systematic interdisciplinary excavation projects since the 1950s, geographers and other earth scientists have sharpened their analytical methodologies to participate more fully in the search for archaeological sites, in regard to specific landform units and sediments; in the interpretation of past human activities within specific sites; in recognizing abandoned agricultural landforms, such as raised fields and field terraces, in regard to potential productivity; and in assessing the mosaic of regional resources and potential environmental impact of a particular prehistoric group or historical community. This is a well-established international branch of applied geography that has significantly influenced the ways in which archaeologists work and interpret their data. Both methods and ideas from spatial theory and systematic human geography have also been transferred to archaeology with considerable success, and some archaeologists now work within a geographical paradigm.

BIBLIOGRAPHY

Butzer, K. W., *Environment and Archeology* (1964, 2nd ed. 1971).

——— , *Archaeology as Human Ecology* (1982).

Denevan, W. N., K. Mathewson, and G. Knapp, eds., *Pre-Hispanic Agricultural Fields in the Andean Region* (1987).

Turner, B. L., and P. Harrison, eds., *Pulltrouser Swamp* (1983).

Vita-Finzi, C., *The Mediterranean Valleys* (1969).

Wagstaff, J. M., ed., *Landscape and Culture* (1987).

• *Karl W. Butzer*

Geographic Information Systems

A Geographic Information System (GIS) may be defined as a method, normally digitally based, for the encoding, storage/management, retrieval, manipulation/analysis, and display of spatially addressable data. All operational GISs incorporate specific hardware and software for data input, storage, processing, and display of computerized maps. The incipient field of GIS may be traced, to a large extent, to activities in computer cartography and the initial developments at the Laboratory for Computer Graphics and Spatial Analysis at Harvard University. Here, it was the software package SYMAP, developed by Howard T. Fisher and his colleagues in the 1960s, that first enabled geographers, planners, and landscape architects, among others, to easily encode spatial data and produce thematic computer maps, including choropleth, isarithmic, and proximal types. Other systems developed during the 1960s were the Canadian Geographic Information System (1964), the New York Landuse and Natural Resources Information System (1967), and the Minnesota Land Management Information System (1969). DIDS (The Domestic Information and Display System), a state-of-the-art statistical mapping system designed for decisionmaking in the federal government during the late 1970s, produced both choropleth and bivariate maps, but for a variety of reasons was abandoned in 1983.

Two types of geographic information systems have emerged during the past fifteen years: vector-based and raster-based. Vector-based systems represent spatial data as strings of x-y coordinate pairs in a Cartesian coordinate system. Thus a railway line, which on a traditional manuscript map is a continuous feature produced with pen and ink, is represented in digital format with a set of x-y coordinate pairs, called an ARC. Alternatively, the raster-based systems represent the surface of the earth as a matrix of grid cells, or pixels (picture elements). Each of the individual grid cells depicts a constant area, such as 10m × 10m (SPOT imagery), 30m × 30m (Landsat thematic mapper imagery), or 79m × 79m (Landsat multispectral scanner imagery), and is referenced with both a row-column number and a value or category of the attribute (*e.g.*, land-cover type).

The significant feature that distinguishes GISs from basic computer mapping systems—such as SYMAP—or computer-aided design systems (CAD) is the capability for spatial analysis and modelling. Both C. Dana Tomlin and Joseph K. Berry have formalized the concept of a distinct map algebra and fundamental map analysis techniques. Additionally, P. A. Burrough's *Principles of Geographical Information Systems for Land Resources Assessment* (1986), which has become the standard text in the field of GIS, details the relationship between map algebra and spatial modelling. Three classes of transformation functions have been identified for analysis and modelling: point, region, and neighborhood operators. For instance, an elementary point function might simply *add* together, cell by cell in raster-mode, two of the overlays in the database, while a complex neighborhood function, SCAN, uses one layer (*e.g.*, major street intersections) and searches another layer (*e.g.*, housing density) in order to create yet a third layer (*e.g.*, average housing density within 0.5 km of major street intersections). Dana Tomlin's grid-based Map Analysis Package (MAP), created from the principles of fundamental operators, remains one of the major GISs used in academic environments. Another GIS package designed principally for pedagogic applications is IDRISI, written by Ron Eastman of Clark University.

Commercial activity in the field of GIS has, for the most part, remained ahead of developments in academia and has attracted some of the brightest minds away from academic careers. The most visible of the private sector systems is the vector-based software ARC/INFO, designed by Environmental Systems Research Institute (ESRI) of Redlands, California. ARC/INFO has now been installed in hundreds of university, government, and private-sector sites throughout the world. ESRI, with its rapid expansion over the past five years, now sponsors a yearly conference on GIS, a widely circulated publication (*ARC News*), and runs a series of national and international workshops on ARC/INFO. Other significant private sector efforts include Intergraph, designer of IMAP; Tydac Technologies, which innovated the SPANS (Spatial Analysis System), a quadtree-based GIS; and ERDAS, a combined GIS/image processing system.

The dissemination of knowledge in the field of GIS has been primarily through conferences and conference proceedings. Prominent among these are the International Symposia on Automated Cartography (AUTO-CARTO) series, which were initiated during the early 1970s. By 1983, with AUTO-CARTO 6 held in Ottawa, Canada, many of the papers had focused specifically on GIS or LIS (Land Information Systems). Additionally, two organizations, URISA (Urban and Regional Information Systems Association) and SORSA (Spatially-Oriented Referencing Systems Association), were organized in order to emphasize various aspects of GIS. Recently, conferences on GIS have proliferated, with the GIS/LIS'88 & 89 series in the United States, the National Conference on GIS in Canada, and the International Symposia on Spatial Data Handling (1984 in Zurich, 1986 in Seattle, and 1988 in Sydney). There have also been AUTO-CARTO conferences in Japan (1986) and an International Workshop on GIS in China. Although McHarg's 1969 *Design with Nature* was a seminal work in early GIS development (suitability analysis), it was not until 1986 that the first textbook (by Burrough) appeared. Finally, the first scholarly journal on this topic, the *International Journal of Geographic Information Systems*, edited by J. T. Coppock and E. K. Anderson, was inaugurated in 1987. Two other related journals, *The American Cartographer* and *Surveying and Mapping*, are now in the process of title changes to *Cartography and GIS* and *Surveying and GIS*. Because of the need to gather and process increasingly larger volumes of spatial information in many disciplines, it is a certainty that the field of GIS will continue its rapid growth through the end of this century. Recently, the establishment of the National Center for Geographic Information and Analysis (NCGIA) by the National Science Foundation at a consortium of universities (State University of New York at Buffalo, University of California, Santa Barbara, and the University of Maine) has assured that research and development in the field of GIS will be adequately addressed by the academic community.

Bibliography

Berry, J. K., "GIS: Learning Computer-Assisted Map Analysis," *Journal of Forestry*, 84, no. 10 (1986), 39–43.

Burrough, P. A., *Principles of Geographical Information Systems for Land Resources Assessment* (1986).

Dangermond, J., and L. K. Smith, "Geographic Information Systems and the Revolution in Cartography: The Nature of the Role Played by a Commercial Organization," *American Cartographer*, 15 (1988), 301–310.

McHarg, I., *Design with Nature* (1969).

Smith, T. R., S. Menon, J. Star, and J. E. Estes, "Requirements and Principles for the Implementation and Construction of Large Scale Geographic Information Systems," *International Journal of Geographic Information Systems*, 1 (1987), 13–31.

Tomlin, C. D., "Digital Cartographic Modelling Techniques in Environmental Planning," Unpublished Ph.D. dissertation, Yale University, 1983.

• *Robert B. McMaster*

Geographical Association

The British association of geography teachers, the Geographical Association, was founded 20 May 1893 at Christ Church, Oxford. The prime mover was B. B. Dickinson of the Rugby School, who had circulated the following message in January 1893: "I am trying to start an Association of Public School Masters and others interested in the subject, to promote the Study of Geography by means of lectures given as part of the school work and illustrated by lantern-slides." At the May meeting Halford Mackinder proposed a more formal organization than the one Dickinson had in mind. Dickinson served as secretary for the first seven years, but even more important in shaping the character of the association were his successors, who were university men—A. J. Herbertson, secretary from 1900 until his death in 1915, and H. J. Fleure, secretary from 1917 to 1947. The name of the association's journal, *The Geographical Teacher*, which commenced publication in 1901, was changed to *Geography* in 1927, because of "the involvement in the Association's affairs of a growing body of university geographers who, prior to the formation of the Institute of British Geographers, had no journal or platform of their own from which either to present the results of research or express opinions" (Lewis). University geographers have always been a distinct minority in the association but have provided leadership. The 1970s saw a swing toward greater concern with teaching, especially at the secondary level.

Bibliography

Fleure, H. J., "Sixty Years of Geography and Education: A Retrospect of the Geographical Association," *Geography*, 38 (1953), 231–266.

Lewis, G. M., "Association with a Mission," *Geographical Magazine*, 52 (1980), 455, 457.

Warrington, T. C., "The Beginnings of the Geographical Association," *Geography*, 38 (1953), 221–230.

Geographical Cycle

The Harvard geomorphologist William Morris Davis developed a theoretical cycle to explain the evolution of landforms from "youth" through "maturity" to "old age." This cycle, which he called the "geographical cycle" or "cycle of erosion," is also referred to as the "Davisian cycle." It is the very backbone of Davis's descriptive geomorphology, which dominated landform study in the first half of the twentieth century. Davis and his disciples constantly tinkered with this "normal" cycle in order to explain landforms that developed in areas of the world where certain erosional agents were dominant, *e.g.*, arid, glacial, and karst landscapes. Although such cyclical or evolutionary notions were also introduced into plant ecology and even into human geography, they have largely been discarded by geographers of the present day, or perhaps we should say that geographers now feel that they should hide their natural interest in such simplistic schemes by using greater subtlety or more elaborate circumlocutions.

Bibliography

Chorley, R. J., R. P. Beckinsale, and A. J. Dunn, *The History of the Study of Landforms*, vol. 2 (1973).

Davis, W. M., "The Rivers and Valleys of Pennsylvania," *National Geographic Magazine*, 1 (1889), 183–253.

——— , "The Geographical Cycle," *GJ*, 14 (1899), 481–504.

Geographical Society of New South Wales

The Geographical Society of New South Wales was founded in Sydney in 1927 under the leadership of Griffith Taylor, who became its first president. Membership has continued to include a broad cross-section of the community and is currently over 400. Until the founding of the Institute of Australian Geographers in 1959, this was the strongest Australian geographical society, and the society still has as its patron by tradition the Governor of New South Wales. Current publications include *Australian Geographer* (two issues per year since 1928) and the *Newsletter of the Geographical Society of New South Wales*. Apart from regular lecture meetings and its publications, the society organizes geographical study tours throughout the world.

Bibliography

Biddle, D., "The Australian Geographer, 1928–1988," *Australian Geographer*, 19 (1988), 3–6.

Powell, J. M., "Geographical education and its Australian heritage," *Australian Geographical Studies*, 26 (1988), 214-230.

• *R. L. Heathcote*

Geomorphology

Geomorphology is the science concerned with the form of the earth's surface, the forces shaping it, and their dynamic evolution through time. The spherical form of the earth is the province of geophysics. The word *geomorphology* was first proposed in 1888 by W. J. McGee, who attributed it to J. W. Powell but later reclaimed it for himself. Other similar words were proposed—*geomorphography* and *geomorphogeny*—but these see only occasional use today.

Despite the late appearance of a specific name for surface studies, significant writings on the topic had been in existence since ancient times, for man lives on the surface. The subject was an important component of what was to become geology before that science had a name, and much eighteenth-century writing about the earth was most accurate and perceptive about surface form and process.

The science is a difficult one, for when it is considered within conventional geological timescales—from millions to hundreds of millions of years—erosion dominates, and the very evidence for what has happened is missing. The present evidence of weathered rocks and sediments in transit to the ocean is a miniscule skin compared to the thousands of meters of rock that usually has been removed.

The earlier emphasis on description of form was replaced after World War II with intensive work on surface processes of weathering, erosion, and transportation—often with respect to very short time scales. More recently still, research has been conducted on Quaternary changes of climate and the new paradigm of plate tectonics. The twentieth century has also seen field work conducted in all the earth's major climatic regions, and this had led to a much greater variety of explanations being put forth for landforms—many of them based on regional variations in climate and the mix of surface processes it controls. Remote sensing of this and other planets, the digital representation of topography, real time monitoring of surface processes, and vastly improved computational devices—all these, together with existing methodology, seem likely to revolutionize the operation of the subject within the next century.

A number of journals are devoted entirely to the field: *Zeitschrift für Geomorphologie* (1925-1943, new series begun 1957), *Revue de géomorphologie dynamique* (1950), *Earth Surface Processes and Landforms* (1976; *Landforms* was added in 1981), and *Geomorphology* (1987). A *Journal of Geomorphology* began in the United States in 1938 and issued five volumes before becoming a casualty of wartime pressures in 1942. In addition, geomorphologists publish in a wide array of journals in geography, geology, and related sciences. The international abstracting service now known as *GeoAbstracts* was begun by the English geomorphologist Keith Clayton in 1960 as *Geomorphological Abstracts*.

The formal institutional basis of the subject is limited: As a discipline it is taught within both geography and geology programs, and practitioners usually owe their professional allegiance to one of those fields. In Great Britain the British Geomorphological Research Group, founded in 1960 by D. L. Linton, although affiliated with the Institute of British Geographers, is responsible for the journal *Earth Surface Processes and Landforms*, runs conferences, and has issued a series of Technical Bulletins (published by GeoBooks) on various aspects of geomorphological practice. It makes the D. L. Linton Award. In the United States there are geomorphological groups within both the Association of American Geographers and the Geological Society of America. Annual honor awards are made by each society for work in geomorphology: the G. K. Gilbert and the Kirk Bryan awards, respectively. An annual conference on a selected topic in geomorphology is run by an informal steering committee in the eastern United States and is known as the Binghamton Symposium in Geomorphology. It was named for the city (and university) in New York State where the first meetings were held. In October 1989 it met for the twentieth time, in Carlisle, Pennsylvania, on the topic of Appalachian geomorphology. In 1985 the first International Conference on Geomorphology was held in Manchester, England. The conference met again in Frankfurt, Germany, in 1989 and appears likely to continue regular meetings.

Two comprehensive modern textbooks are *Geomorphology* by R. J. Chorley, S. A. Schumm, and D. E. Sugden (1984) and *Earth's Changing Surface* by M. J. Selby (1985). The *Encyclopedia of Geomorphology* (1968) is still an excellent reference source, and *GeoAbstracts* provides a useful bibliographic resource for current journal articles in all branches of the subject. *See also* Physical Geography.

BIBLIOGRAPHY

Chorley, R. J., R. P. Beckinsale, and A. J. Dunn, *The Study of Landforms, or the History of the Development of Geomorphology* (2 vols., 1964-73; a third volume is in press).

Fairbridge, R. W., ed., *Encyclopedia of Geomorphology* (1968).

Gregory, K. J., *The Nature of Physical Geography* (1985).

Higgins, C. G., "Theories of Landscape Development: A Perspective," pp. 1-28 in *Theories of Landform Development*, ed. by W. N. Melhorn and R. C. Flemal (1976).

King, C. A. M., ed., *Landforms and Geomorphology: Concepts and History* (1976).

Tinkler, K. J., *A Short History of Geomorphology* (1985).

• *Keith Tinkler*

Geopolitics. *See* Political Geography.

George, Pierre Oscar Léon (b. 1909)

D.-ès-L., University of Paris, 1936. University of Lille, 1946-48; University of Paris, 1948-77. *Fin de siècle en Occident, déclin ou métamorphose* (1982); *Géopolitique des minorités* (1984); *L'immigration en France* (1986); *Les hommes sur la terre, la géographie en mouvement* (1989). *WWF*; A Buttimer (ed.), *The Practice of Geography* (1983), 123-125, 128-129, 131-134, 136-138.

Geosophy

Although the term "geosophy" or its German equivalent was employed in the nineteenth century by Friedrich Marthe and Patrick Geddes, modern usage usually derives from John K. Wright's presidential address to the Association of American Geographers in 1946, in which he defined geosophy as "the study of geographical knowledge from any or all points of view." "Sophogeography" would be the geography of knowledge, and "historical geosophy" would be what is customarily called "the history of geography." In later years Wright declared that he was not fond of the term geosophy. "What I meant by it," he said in 1965, "was study of the study of geography, i.e. *geography-sophy*." Although the term has been used sparingly, the basic idea is deeply imbedded in the works of many of Wright's spiritual descendants, including David Lowenthal, Leslie Heathcote, and Martyn Bowden. The Wright denkschrift, *Geographies of the Mind*, edited by Lowenthal and Bowden, has the subtitle, *Essays in Historical Geosophy in Honor of John Kirtland Wright*.

BIBLIOGRAPHY

Dunbar, G. S., "Geosophy, Geohistory, and Historical Geography: A Study in Terminology," *Historical Geography*, 10 (Fall 1980), 1-8.

Lowenthal, D., and M. Bowden, eds., *Geographies of the Mind* (1976).

Wright, J. K., "Terrae Incognitae: The Place of the Imagination in Geography," *AAAG*, 37 (1947), 1-15.

Gerasimov, Innokentii Petrovich (1905-1985)

Doctorate, Leningrad State University, 1936. USSR Academy of Sciences, 1936-85; Moscow State University, 1936-56. *Pochvy Tsentral'noy Evropy* (1960); *Osnovy Pochvovedeniya i geografii pochv* (with M. Glazovskoy) (1960) (published in English as *Fundamentals of Soil Science and Soil Geography*, 1965); *Geomorfologicheskaya Karta SSSR* (with A. Aseev) (1987). R. P. Zimina and Ya. G. Mashbits in *GBS*, 12 (1988), 83-93; B. Lewytzkyj, ed., *Who's Who in the Soviet Union* (1984), 105.

Germany

I. On the Development of Geographical Thought before 1890

The decades from 1890 to the present can, of course, never be understood without understanding the development that preceded them. Thus the most significant achievements in the history of geographical thought from the time of antiquity to 1890 will be briefly cited here:

a. Since antiquity geography has been a science of the present, with an increasingly important "historical element" (Carl Ritter), beginning in the seventeenth century, which makes the present more comprehensible.

b. From ancient times until the deaths of Humboldt and Ritter in 1859 geography was the science of the spatial circumstances of Mankind.

c. Since antiquity geography dealt with the science of nature only to a limited degree. It featured only general description: blue sky, aridity, rainy season, desert, and so on.

d. Since the eighteenth century a meaningful consideration of nature presented the greatest problem, which was solved by Alexander von Humboldt by means of perfected instrumentation. This enabled him to include nature to a much greater degree than was done in all the previous periods.

e. Consequently, Humboldt's solution of the problem of physical geography, which included helping and supporting Man with the application of physical geography in the spirit of the Enlightenment, could have unleashed the greatest revolution in geographical thought and forced a change of paradigm, if only his geography could have been understood.

f. After the deaths of Humboldt and Ritter, geographers in Germany gave up the instrumentation of physical geography (and also Humboldt's quantification in the field of human geography by means of scientific statistics), and developed geomorphology, to join the ranks of the then-triumphant natural sciences. German geographers were no longer fully aware of the importance of Humboldt's geography. It was in no way connected with the development that was to follow, even with respect to the important results in human geography. It was a sacrifice of creative geography that is without equal in the entire history of geographical thought.

g. After 1859 geography in Germany underwent its greatest revolution, due to the unilateral development and promotion of geomorphology, whose debatable position in the system of geography was recognized at first only by Friedrich Marthe. His criticism, however, was fruitless.

II. 1890-1905

From this time on geography was represented at all German universities. Because chairs of geography were available but there weren't enough geographers to fill them, and because historians, mathematicians, and high school teachers, among others, were appointed, a frequently humorous chaos in methodology ensued. The organization of the infrastructure—periodicals, geographical societies, publishers, professional meetings, and publications—was surprisingly good; everything was excellent and internationally exemplary. Younger geographers, such as Albrecht Penck, developed important excursions for their students, besides giving lectures and exercises. Friedrich Marthe, Hermann Wagner, Ferdinand von Richthofen, and Alfred Hettner attempted to straighten out the methodological muddle. Richthofen's formula—geography as the "science of the surface of the earth and the objects and phenomena causally connected with it"—finally shaped a first consensus that was profoundly far-reaching into the twentieth century. He ascribed to physical geography (different from Humboldt's approach to pure natural geography, which always included Man) the surface of the earth as an object, while he saw Man solely in an environmental deterministic relationship to and with a dependence upon it. After 1899 this was criticized by Richthofen's student, Otto Schlüter, but at first his criticism went unnoticed. The ruling geomorphology, which was always pursued more by trained geologists as a geographical subdiscipline, found its proponent in young Albrecht Penck, whose textbook was epoch-making (*Morphologie der Erdoberfläche*, 2 vols., 1894). Hettner, who emphasized the unity of physical and human geography, considered as "the vital point" the "reciprocity" between Man and Nature; thus Schlüter's approach, which was attacked by Hettner, did not gain ground. Geography of this era was on its way to becoming a general geography accentuated by natural science, with a tendency towards a general earth science with an exaggerated emphasis on geomorphology without measuring instruments, and with a great broadening of the field of observation through colonial geography (after 1884/1885) and scientific travels supported by the prevailing imperialism. The environmental determinism in Richthofen's and Hettner's system and in Ratzel's *Anthropogeographie* (2 vols., 1882-1891), although to all suspect and by all rejected, yet silently tolerated, is noticeable. Richthofen was the most renowned geographer of this era because of his travels and his work *China* (5 vols., 1877-1912; with atlas). Ratzel's manifold activity achieved greater success with the general public than at the university. While Hettner had expanded his position of power since 1895 through his *Geographische Zeitschrift*, young Albrecht Penck had already won remarkable international recognition on the basis of his *Morphologie* and his works on regional geography.

III. 1905-1918

The first generation of modern geographers entered the universities during this period. Penck became the leading geographer of the period, next to whom Hettner built up his position. The greatest achievements were hardly understood: the formulation for the first time of an antideterministic, workable geography of Man by Otto Schlüter; Alfred Rühl's very modern approaches with respect to the geography of Man, which were far ahead of their time; and Ewald Banse's concept of cultural continents. Schlüter was unappreciated by Hettner; Rühl was stymied by Penck, who had been responsible for Rühl's appointment with respect to the development of economic geography; and the pioneering approach of Siegfried Passarge in his *Physiologische Morphologie* (1912) went unnoticed. It was Passarge especially who picked up on Humboldt's instrumentation and who advocated the examination of small sections of terrain, which he evaluated in series of maps. By his determination of different forces, Passarge contradicted William Morris Davis, who, as an exchange professor in Germany, had suddenly arrived at an unexpected breakthrough in morphological thought in geography. German translations by the geographers Gustav Braun and Alfred Rühl facilitated Davis's breakthrough, but after World War I powerful criticisms arose that were tinged with anti-American tones.

As a result, the truly creative program of the younger geographers did not prevail, while the meanwhile-perfected structure of intrigues of geography was playing its fateful role.

IV. 1919-1932: Era of the Weimar Republic

At the beginning of this period there existed criticism, frequently justified, of the Treaty of Versailles because of the loss of the right of self-determination of nations, and for other reasons. This criticism came from antidemocratic forces as well as from true democrats. At the end of this period there was the crisis in the world economy, which created an army of unemployed; both phenomena paved the way for Adolf Hitler. Already at this time inhumanity, racism, and terrorism became apparent and could not be controlled by Germany's constitution, which was too idealistic. On the other hand, an amazingly high cultural level was reached in this republic.

The geomorphologists continued to attain results as before, which were very significant at that time. There was, for example, the discovery of the "trumpet valley" by Carl Troll in 1924, and the first systematic approaches to a climatic morphology in 1926 (at the meeting in Düsseldorf). Hettner's methodology, emphasizing the unity of geography, reached a very shaky consensus on account of the closed nature of his school, which said that geography by its very character was regional geography. It dealt with the influences of Nature on Man (a quite empty formula) and the influence of Man on Nature. Thus a determinism began to prevail that would continue to assure the unity of Hettner's geography and leave it unchallenged. Hermann Lautensach, who was denied an academic chair, was creating the most exemplary regional geography.

V. 1933-1945: Geography under the National Socialist Regime

Many geographers—for example, Ewald Banse, Walter Behrmann, Albrecht Burchard, Erwin Scheu, and Hans Schrepfer—became laden with guilt in the Third Reich. The majority went along with the regime, but a few—for example, Leo Waibel and Alfred Rühl—offered resistance in their own way. Helmut Anger's courageous attempt to warn Hitler before the campaign in Russia stands out.

After official pressure that met with no resistance, geomorphology was criticized for the first time, and the concept of geography shifted towards the geography of Man. Environmental determinism was ideologically not accepted. The pushing back of geomorphology was hardly effective, because the institutional structure of university geography had remained intact. It was even undercut by pointing to the "importance to the war" of this segment of the discipline in a publication by Hans Mortensen. The most significant results were attained by Hermann Lautensach and Albert Kolb in regional geography, by Alfred Rühl in economic geography, by Leo Waibel, and—not least—by Walter Christaller. The most important approaches, again, did not come to fruition.

VI. 1945-1967

After catastrophic losses in libraries and personnel, including the majority of the talented young geographers, Germany had to begin anew under the most difficult circumstances. There were only a few geographers who could assume any responsibility after 1945. One of the most eminent among them, Carl Troll, had even become known abroad. Hettner was now no longer being mentioned. He moved into the shadows, but his thinking continued to be carried on by his numerous and influential school. Social geography moved to center stage, especially through the work of Hans Bobek. Works on human geography had begun to become important during the Weimar Republic; even Christaller, finally, was now being looked upon with great respect. The significance of geomorphology, however, remained uncontested, and climatic geomorphology was successful under the leadership of Julius Büdel and Herbert Louis. Troll's chiefly natural-science orientation attained international recognition through his significant output (*Die tropischen Gebirge*, 1959, and *Karte der Jahreszeitenklimate*, 1963). The concept of ecology, at first limited by Troll to natural geography, began to become important to Man. Beginning in 1952, Lautensach discussed his "Geographische Formenwandel" and pointed out the fruitfulness of this concept for regional geography by means of excellent works (*e.g.*, *Die Iberische Halbinsel*). Also, in 1963 three outstanding geographers demonstrated the effectiveness of their conceptions of regional geography (Herbert Wilhelmy, *Die La Plata Länder*; Albert Kolb, *Ostasien*: and Oskar Schmieder, *Die Neue Welt*). The increasing unrest among the young geographers was consistent with a growing criticism where the concept of geography was involved. It would have seemed appropriate to have continued now with Schlüter or, finally, with Rühl and Christaller, and with the many problems that had been left over. And this certainly was done, to some extent. But while fewer internal problems were involved this time, it was the stimulus from American geographers that increased the pressure on the status quo.

VII. 1968-1969 to the Present

Unnoticed and without being overheard, a new scientific revolution was simmering. It broke out at the annual meeting (Geographentag) in Kiel in 1969 and almost ran through open doors. It was able to succeed only in part, because the young revolutionaries were unable to take over the institutes, the academic chairs, and the established organization. Thus it had to stop before the gates of the institutes.

When the revolution was proclaimed, an earlier basic problem (involving the works of Johan Georg Kohl, Otto Schlüter, Alfred Rühl, and others) became clear, of which the younger geographers had been unaware. In 1967 Dietrich Bartels presented his habilitation thesis (*Zur wissenschaftstheoretischen Grundlegung einer Geographie des Menschen*, 1968). The bibliography of this essay, so fundamental for the revolution that was soon to break out, gave proof of the previously unknown influence of English, Scandinavian, and, above all, American geographical thought. Bartels's views were accepted at a meeting of young geographers (October 1967) and systematically disseminated and elaborated in a catalog containing twelve points. At the Kieler Geographentag the professional associations of geography students forced a general meeting for the first time. It was a great surprise that it was not Bartels who succeeded in the following months but the Munich school of social geography. Later Bartels accused the Munich geographers of lacking theory. In Munich the way had been prepared since 1966 by Wolfgang Hartke and Karl Ganser. However, Franz Schaffer and Karl Ruppert, social geographers from Munich, presented their ideas at a decisive moment and, after a meeting in November 1968, began their victory march in the school as well.

Since 1945 traditional geography has proved itself in field research and has also given proof of its ability to alter the setting of goals. This geography, like Humboldt's geography, wants to assist Man. There is now a pluralism in the universities that tolerates traditional geography alongside the most modern approaches. Obviously this pluralism is influenced by the revolution in quantification and terminology. The split between natural and cultural geography is becoming increasingly apparent, and could lead to new concepts and thereby promote further development of Germany's geography. *See also* Berlin Geographical Society; Germany, East.

Bibliography

Beck, H., *Geographie: Europäische Entwicklung in Texten und Erläuterungen* (1973).

———, "Die Geschichte der Reisen: Grundzüge und Perspektiven eines Teilgebietes der Historie der Geographie," *Praxis Geographie*, 18 (1989), 6–10.

Hard, G., *Die Geographie: Eine wissenschaftstheoretische Einführung* (1973).

Wirth, E., "Dietrich Bartels 1931–1983," *GZ*, 72 (1984), 1–22.

———, "Geographie als moderne theorieorientierte Sozialwissenschaft," *Erdkunde*, 38 (1984), 73–80.

• *Hanno Beck*
(translated by Ursula Willenbrock Martin)

Germany, East (German Democratic Republic)

After World War II departments of geography existed in the universities of Berlin, Leipzig (closed 1969), Halle (merged with Leipzig in 1969), Greifswald, Jena (closed 1969), Rostock (closed 1969), and at the Technical University of Dresden. The famous old Justus Perthes Institute in Gotha, founded in 1785, still flourishes as the VEB Hermann Haack Institute. At Leipzig the Deutsches Institut für Länderkunde has become an institute for geography and geoecology. In the 1950s geography departments were established in the teacher training colleges of Potsdam and Dresden. Economic geography was introduced in the Berlin School of Economics and also at the Leipzig School of Commerce, where there was a chair in geography before 1945. In 1953 the Geographical Society of the German Democratic Republic was founded, and its journal, *Geographische Berichte*, has been published since 1956. But the most important periodical is still *Petermanns Geographische Mitteilungen*, which has been published in Gotha since 1855. After World War II, and under the conditions of the new anti-fascistic democratic order, new departures in academic teaching and research became necessary. This meant an abrupt break with the mistaken traditions of the past and an attempt to create a new geography in terms of a materialistic world view.

In 1946 the study of geography was allowed as a subsidiary subject. Since 1948 geography has been a principal subject for students, and in the same year a special program was introduced that leads to the academic degree of Diplom-Geograph, which qualifies graduates for jobs in applied geography. Since 1952 there has been a unified plan for the study of geography, compulsory for all academic institutions, but unfortunately this has led to the disappearance of some of the subfields of geography. Anthropogeography and the old economic geography (Wirtschaftsgeographie) were replaced by a new Marxist economic and political geography. The old regional geography, or Länderkunde, has been replaced by a new regional geography consisting of regional economic and regional physical geography. After 1969, when a comprehensive university reform was undertaken, traditional fields of research like geomorphology and historical geography lost their former significance. One aim of this reform was to improve academic education and research to serve better the demands of life outside the universities by forming larger departments and concentrations of scholars. In the field of physical geography, landscape ecology, introduced by Ernst Neef, has become the major research topic. In economic geography one of the most important research tasks is to find an efficient territorial structure in order to improve the production process of the society. Within the past ten years remote sensing has become another important topic of geographical research. GIS is going to be developed in some departments. Only the Halle

University geography department has a program in the history of geographical thought. *See also* Gotha Geographical-Cartographic Institute.

BIBLIOGRAPHY

Haase, G., "Ziele und Aufgaben der geographischen Landschaftsforschung in der DDR," *Geographische Berichte*, 22 (1977), 1-19.

Lichtenberger, E., "The German-Speaking Countries," pp. 156-184 in *Geography since the Second World War*, ed. by R. J. Johnston and P. Claval (1984).

Lüdemann, H., "Geographische Forschung in der DDR: Entwicklung und Perspektiven," *PGM*, 124 (1980), 97-103.

Mohs, G., "Globale Probleme und regionale Entwicklung im Blickfeld der Geographie," *Geographische Berichte*, 31 (1986), 1-12.

Schmidt, G., "Geoökologisches Systemverständnis in der Geographie," *Ibid.*, 33 (1988), 261-271.

• *Max Linke*

Ghana

Geography has a long history as a subject taught in the curriculum of Ghanaian schools. During the colonial era the geography of the Gold Coast (renamed Ghana in 1957), West Africa, and the world in general (with an emphasis on the British Isles) was studied. The syllabus during this early period concentrated on uncoordinated descriptions of the features of particular places. Thus the names and lengths of major rivers, heights of major mountains, location of specific towns and countries, products of various countries, and the capitals of the more well-known countries were memorized by pupils. Ingenious teachers composed songs to help their pupils remember the names and locations of the major towns and physical features of the Gold Coast.

Geography assumed a new status with the establishment of the country's universities—the University of Ghana in 1948, the University of Science and Technology in 1961, and the University of Cape Coast in 1971. The orientation of geography in the University of Cape Coast has been toward the training of teachers. All three universities offer a wide variety of courses: physical geography, human geography, regional geography, and practical geography, including mapwork, surveying, and statistics.

To further interest in geography, the Gold Coast Geographical Association (later the Ghana Geographical Association) was formed on April 19, 1955, under the direction of the Geography Department of the University College of Gold Coast (now the University of Ghana). Membership includes university lecturers, teachers in the schools and colleges, and students. The association encourages the formation of geographical societies in the secondary schools.

Geography is now taught all through the six years of primary education. Emphasis is on the geography of Ghana, but knowledge of phenomena of special geographical interest is not neglected. Specific topics on the geography of other countries are also studied. At the secondary level there is a classification of the geographical knowledge imparted—physical and human geography, regional geography of Ghana and West Africa, and practical geography (mainly mapwork), with surveying and statistics at the more advanced secondary level. In the first three years of secondary education geography is compulsory, and after the fourth year it is optional. In 1987 almost 33 percent of the secondary school students who were preparing for their Ordinary Level certificates were taking geography, and the corresponding figure at Advanced Level was 22 percent.

The present state of teaching of geography, especially at the university level, has a practical orientation. Teaching and research emphasize the applied aspects of geography, aiming at the solution of specific problems. Geography is thus seen as not merely an academic discipline but as an applied discipline that can aid in solving the practical problems of life. This practical orientation of geography in Ghana influenced the redesignation of the Department of Geography of the University of Ghana as the Department of Geography and Resource Development in 1989.

With the changing perception of the nature of the field, professional geographers are increasingly finding employment in non-teaching positions, where they can directly apply the knowledge they acquired in their school and university education. The Environmental Protection Council (EPC)—whose basic objective is formulating plans, policies, and strategies for the conservation of resources and of ensuring the sustainability of development—is one institution that employs geographical expertise. The Department of Geography and Resource Development in the University of Ghana has been collaborating with the EPC and other national agencies in the drafting of an Environmental Action Plan for Ghana. Geographers in Ghana have also been playing an increasingly important role in census taking.

The Danish Development Aid Organization is now assisting the University of Ghana in establishing a Remote Sensing Unit in the Department of Geography and Resource Development. This unit, to be set up in 1990, will analyze satellite photographs for the purpose of making an inventory of natural resources and monitoring resource utilization and environmental degradation. The department is also assisting the Accra Metropolitan Authority in drawing up a master plan. A major population research unit, the Population Impact Project, has been established in the department. Its objectives include dissemination of research findings through publications, seminars, and briefings. It also makes relevant population information available to policy makers and other leaders.

• *George Benneh*

Gilbert, Edmund William (1900-1973)

B.Litt., University of Oxford, 1928. University of Reading, 1926-36; University of Oxford, 1936-67. *The Exploration of Western America* (1933); *Brighton* (1954); *The University Town in England and West Germany* (1961); *British Pioneers in Geography* (1972). T. W. Freeman in *GBS*, 3 (1979), 63-71; G. Robinson and J. Patten, "Edmund W. Gilbert and the Development of Historical Geography," *Journal of Historical Geography*, 6 (1980), 409-419; obits. by anon. in *GJ*, 140 (1974), 176-177; and T. W. Freeman in *Geography*, 59 (1974), 68.

Gilbert, Grove Karl (1843-1918)

B.A., University of Rochester, 1862. Cosmos Hall, 1863-69; Ohio State Geological Survey, 1869-71; Wheeler Survey, 1871-74; Powell Survey, 1874-79; U.S. Geological Survey, 1879-1918. *Report on the Geology of the Henry Mountains* (1877); *Lake Bonneville* (1890); *Glaciers and Glaciation, Alaska* (1904); *The Transportation of Debris by Running Water* (1914). P. James in *GBS*, 1 (1977), 25-33; W. Davis, "Biographical Memoir, Grove Karl Gilbert, 1843-1918," National Academy of Sciences, *Memoirs*, vol. 21, no. 5 (1927); S. Pyne, *Grove Karl Gilbert: A Great Engine of Research* (1980); E. Yochelson (ed.), "The Scientific Ideas of G. K. Gilbert," Geological Society of America, *Special Paper*, 183 (1980).

Gillman, Clement (1882-1946)

Graduated from Technical University of Zurich (ETH), 1905. Tanganyika railway service, 1905-37; water consultant to Tanganyika government, 1938-40. "A Population Map of Tanganyika Territory," *GR*, 26 (1936), 353-375; "White Colonization in East Africa," *GR*, 32 (1942), 585-597; "A Short History of the Tanganyika Railways," *Tanganyika Notes and Records* (1942), 14-56; "A Vegetation-Types Map of Tanganyika Territory," *GR*, 39 (1949), 7-37. B. Hoyle in *GBS*, 1 (1977), 35-41; B. Hoyle, *Gillman of Tanganyika 1882-1946: The Life and Work of a Pioneer Geographer* (1987); E. Weigt, "Clemens Gillman und die neuere geographische Erforschung Ostafrikas," *Erdkunde*, 3 (1949), 193-199.

Glacken, Clarence James (1909-1989)

Ph.D., Johns Hopkins University, 1951. University of California, Berkeley, 1952-76. *The Great Loochoo: A Study of Okinawan Village Life* (1956); "Changing Ideas of the Habitable World," pp. 70-92 in *Man's Role in Changing the Face of the Earth* (ed. by W. Thomas) (1956); "Count Buffon on Cultural Changes of the Physical Environment," *AAAG*, 50 (1960), 1-21; *Traces on the Rhodian Shore* (1967). *GOF* (1980); C. Glacken, "A Late Arrival in Academia," pp. 20-34 in *The Practice of Geography* (ed. by A. Buttimer) (1983).

Golledge, Reginald George (b. 1937)

Ph.D., University of Iowa, 1966. University of Canterbury, 1961–63; University of British Columbia, 1965–66; Ohio State University, 1966–77; University of California, Santa Barbara, 1977 to the present. (ed.) *Environmental Knowing* (with G. Moore) (1976); (ed.) *Behavioral Problems in Geography Revisited* (with K. Cox) (1982); (ed.) *Proximity and Preference* (with J. Rayner) (1982); *Analytical Behavioural Geography* (with R. Stimson) (1987). *GOF* (1988); M. Parfit, "Mapmaker Who Charts Our Hidden Mental Demons," *Smithsonian*, 15 (May 1984), 122–131.

Gotha Geographical-Cartographic Institute

Founded in Gotha in 1785, this famous German geographical publishing house was long known as the Justus Perthes Anstalt after its founder, but it is now called the VEB Hermann Haack Geographisch-Kartographische Anstalt Gotha. Perthes began to publish geographical literature in 1793, and the first atlas appeared in 1809. Between 1817 and 1823 the first edition of the famous *Stieler Atlas* was published, named after the cartographer Adolf Stieler (1775–1836). The first school atlas appeared in 1821, and through the initiative of Emil von Sydow (1812–1873) the first wall map was produced in 1838. Heinrich Berghaus (1797–1884) planned a monumental *Grosser Atlas der aussereuropäischen Erdteile* but only produced a partial *Atlas von Asia* (1832–1837). Of paramount importance was Berghaus's *Physikalischer Atlas*, a thematic atlas to accompany Humboldt's *Kosmos*, which was issued in separate plates between 1838 and 1848 (2nd ed. 1849–52, 3rd ed. 1886–92). The entry of August Petermann (1822–1878) in 1854 was a milestone in the history of the firm. In 1855 he started the journal *Mitteilungen aus Justus Perthes' geographischer Anstalt*, renamed for him after his death, and now, as *Petermanns Geographische Mitteilungen*, it is still one of the most important geographical periodicals in the world. Another ongoing publication is the *Geographisches Jahrbuch*, initiated by Ernst Behm (1830–1884) in 1866. In 1897 Hermann Haack (1872–1966) joined the firm. Two years later he founded the *Geographischer Anzeiger*, which soon became the leading periodical for teachers of geography and lasted until 1944. After 1945 the firm continued its work, but became a state-owned enterprise in 1952. Its publications include journals, atlases, wall maps, and, since 1964, scientific geographical textbooks. Of particular importance are the *Haack Grosser Weltatlas* (1966), a two-volume historical atlas, and also the national atlas of the German Democratic Republic (1981).

BIBLIOGRAPHY

Köhler, F., *Gothaer Wege in Geographie und Kartographie* (1987).

Linke, M., M. Hoffmann, and J. A. Hellen, "Two Hundred Years of the Geographical-Cartographical Institute in Gotha," *GJ*, 152 (1986), 75–80.

Suchy, G., ed., *Gothaer Geographen und Kartographen* (1985).

• *Max Linke*

Gottmann, Jean Iona (b. 1915)

D.-ès-L., University of Paris, 1970. Institute for Advanced Study, Princeton, New Jersey, 1942–61; University of Oxford, 1968–83. *A Geography of Europe* (1950, 4th ed. 1968); *La politique des états et leur géographie* (1952); *Megalopolis* (1961); *La città invincibile* (1984). *WW*; *GOF* (1982); (anon.) "The Talk of the Town—French Geographer," *The New Yorker*, 37 (2 December 1961), 52–54; J. Patten, "Jean Gottmann: An Appreciation," pp. xi–xv in *The Expanding City: Essays in Honour of Professor Jean Gottmann* (ed. by J. Patten) (1983).

Goudie, Andrew Shaw (b. 1945)

Ph.D., University of Cambridge, 1972. University of Oxford, 1976 to the present. *Duricrusts of Tropical and Subtropical Landscapes* (1973); *Environmental Change* (1977, 1983); *The Human Impact* (1981, 1986); *Geomorphological Techniques* (1981). *WW*.

Gould, Peter Robin (b. 1932)

Ph.D., Northwestern University, 1960. Syracuse University, 1960–63; Pennsylvania State University, 1963 to the present. *Spatial Organization* (with R. Abler and J. Adams) (1971); *Mental*

Maps (with R. White) (1974); *The Geographer at Work* (1985); *Fire in the Rain: The Democratic Consequences of Chernobyl* (1989). *GOF* (1971); C. Browning (ed.), *Conversations with Geographers* (1982), 26-39.

Gourou, Pierre (b. 1900)

D.-ès-L., University of Paris, 1936. University of Montpellier, 1940-41; University of Bordeaux, 1942-45; Free University of Brussels, 1936-40, 1945-70; Collège de France, 1947-70. *Les paysans du delta tonkinois* (1936); *Les pays tropicaux* (1947); *Pour une géographie humaine* (1973); *Riz et civilisation* (1984). J. Gallais, "L'évolution de la pensée géographique de Pierre Gourou sur les pays tropicaux (1935-1970)," *AG*, 90 (1981), 129-150.

Gradmann, Robert (1865-1950)

Doctorate, University of Tübingen, 1898. University of Tübingen, 1909-18; University of Erlangen, 1919-34. *Das ländliche Siedlungswesen des Königsreichs Württemberg* (1913); *Die städtischen Siedlungen des Königsreichs Württemberg* (1914); *Süddeutschland* (2 vols., 1931). K. Schröder in *GBS*, 6 (1982), 47-54; R. Gradmann, *Lebenserinnerungen* (ed. by K. Schröder) (1965); obits. by O. Berninger in *PGM*, 95 (1951), 187-190; F. Metz in *Die Erde*, 2 (1950-1951), 333-338; and F. Huttenlocher in *Erdkunde*, 5 (1951), 1-6.

Granö, Johannes Gabriel (1882-1956)

Doctorate, University of Helsinki, 1911. University of Tartu, 1919-23; University of Turku, 1926-45; University of Helsinki, 1908-19, 1923-26, 1945-50. *Reine Geographie* (1929); *Die geographische Gebiete Finnlands* (1931); *Mongolische Landschaften und Ortlichkeiten* (1941). O. Granö in *GBS*, 3 (1979), 73-84; A. Paasi, "Connections between J. G. Granö's Geographical Thinking and Behavioural and Humanistic Geography," *Fennia*, 162 (1984), 21-31; obit. by K. Hildén in *Terra*, 68 (1956), 3-13.

Granö, Olavi Johannes (b. 1925)

Ph.D., University of Helsinki, 1955. University of Helsinki, 1948-52; Helsinki School of Economics, 1954-58; University of Turku, 1958 to the present (Chancellor since 1984). "Maantiede ja tieteen kehityksen ongelma" (Geography and the Problem of the Development of Science), *Terra*, 89 (1977), 1-9; "External Influence and Internal Change in the Development of Geography," pp. 17-36 in *Geography, Ideology and Social Concern* (ed. by D. Stoddart) (1981); "Maisematutkimus maantieteen traditiona" (Landscape Studies as a Geographical Tradition), *Terra*, 94 (1982), 7-12; "The Spread of the Peneplain Concept to Finland: An Example of a Paradigm's Relation to an Institutional Structure," *Striae*, 31 (1987). *International Who's Who*.

Haase, Günter (b. 1932)

Ph.D., University of Leipzig, 1962; (habil.) Technical University of Dresden, 1969. University of Leipzig, 1956–59; Technical University of Dresden, 1959–62; Geographical Society of the DDR, 1962–69; Academy of Sciences of the DDR (Leipzig), 1969 to the present. "Landschaftsökologische Detailuntersuchung und naturräumliche Gliederung," *PGM*, 108 (1964), 8–30; "Sedimente und Paläoböden im Lössgebiet," *PGM Ergänzungsheft* 274 (1970); "Struktur und Gliederung der Pedosphäre in der regionischen Dimension," *Beiträge zur Geographie*, 29 (1978); "Zur Ableitung und Kennzeichnung von Naturpotentialen," *PGM*, 122 (1978), 113–129.

Hägerstrand, Stig Torsten Erik (b. 1916)

Fil.Dr., University of Lund, 1953. University of Lund, 1953–82. *The Propagation of Innovation Waves* (1952); *Innovationsförloppet ur korologisk synpunkt* (1953) (trans. by A. Pred as *Innovation Diffusion as a Spatial Process* [1967]); *Migration and Area* (1957). *Who's Who in Scandinavia*; *Vem är Det*; T. Hägerstrand, "In Search for the Sources of Concepts," pp. 238–256 in *The Practice of Geography*, ed. by A. Buttimer (1983).

Haggett, Peter (b. 1933)

Ph.D., University of Cambridge, 1969. University College London, 1955–57; University of Cambridge, 1957–66; University of Bristol, 1966 to the present. *Locational Analysis in Human Geography* (1965); *Geography: A Modern Synthesis* (1972); *Spatial Diffusion* (1981); *Spatial Aspects of Epidemics* (1986). *WW*; *GOF* (1982); C. Browning (ed.), *Conversations with Geographers* (1982), 40–56.

Hamelin, Louis-Edmond (b. 1923)

Doctorat d'état, University of Paris, 1975. Laval University, 1951–78; University of Three Rivers (Trois-Rivières, Québec) (Rector of the University), 1978–83. (co-editor) *Atlas du monde* (1967); *Canada: A Geographical Perspective* (1973); *The Canadian North/Le Nord canadien* (1988). *WWA*; *CWW*; *Dictionary of International Biography*; *Contemporary Authors*.

Hard, Gerhard (b. 1934)

Dr.phil., University of Saarbrücken, 1962; (hab.) University of Bonn, 1969. University of Bonn, 1963–70; Free University of Berlin, 1970–71; Teachers College of the Rhineland (Bonn),

1971-77; University of Osnabrück, 1977 to the present. *Die "Landschaft" der Sprache und die "Landschaft" der Geographen* (1970); *Die Geographie: Eine wissenschaftstheoretische Einführung* (1973); *Prozesswahrnehmung in der Stadt* (1981); *Selbstmord und Wetter, Selbstmord und Gesellschaft* (1988).

Hare, Frederick Kenneth (b. 1919)

Ph.D., University of Montreal, 1950. McGill University, 1945-64; Kings College London, 1964-66; Birkbeck College London, 1966-68; University of British Columbia (President), 1968-69; University of Toronto, 1969-84. *The Restless Atmosphere* (1953); *On University Freedom* (1967); *Climate Canada* (with M. Thomas) (1974, 1979). *WW*; *CWW*; *GOF* (1985).

Harris, Chauncy Dennison (b. 1914)

Ph.D., University of Chicago, 1940. Indiana University, 1939-41; University of Nebraska, 1941-43; University of Chicago, 1943-84. (ed.) *International List of Geographical Serials* (with J. Fellmann) (1960, 1971, 1980); *Cities of the Soviet Union* (1970); *Guide to Geographical Bibliographies and Reference Works in Russian or on the Soviet Union* (1975); *Bibliography of Geography* (2 vols., 1976, 1984). *WWA*; *GOF* (1971, 1986); A. Rodgers, "The Contributions of Chauncy Harris to Geographical Studies of the Soviet Union: An Appreciation," pp. 1-9 in *Geographical Studies on the Soviet Union: Essays in Honor of Chauncy D. Harris* (ed. by G. Demko and R. Fuchs) (1984).

Hart, John Fraser (b. 1924)

Ph.D., Northwestern University, 1950. University of Georgia, 1949-55; Indiana University, 1955-67; University of Minnesota, 1967 to the present. *The Southeastern United States* (1967); (ed.) *Regions of the United States* (1972); *The Look of the Land* (1974); *The Land That Feeds Us* (1991). *WWA*; *GOF* (1972, 1987).

Hartke, Wolfgang (b. 1908)

Dr. phil., University of Berlin, 1932; (habil.) University of Frankfurt, 1938. University of Frankfurt, 1946-52; Technical University of Munich, 1952-73. *Kulturgeographische Wandlungen in Nordostfrankreich* (1932); *Das Land Frankreich als sozialgeographische Einheit* (1963). *Wer Ist Wer?*; *Zum Standort der Sozialgeographie: Wolfgang Hartke zum 60. Geburtstag* (1968); A. Buttimer, "An Interview with Wolfgang Hartke," *The Practice of Geography* (1983), 225-237; C. Borcherdt, "Wolfgang Hartke zum 80. Geburtstag," *Erdkunde*, 42 (1988), 1-6.

Hartshorne, Richard (b. 1899)

Ph.D., University of Chicago, 1924. University of Minnesota, 1924-40; University of Wisconsin, 1940-70. "Geographic and Political Boundaries in Upper Silesia," *AAAG*, 23 (1933), 132-153; *The Nature of Geography* (1939); "The Franco-German Boundary of 1871," *World Politics*, 2 (1950), 209-250; *Perspective on the Nature of Geography* (1959). *WWA*; *GOF* (1972, 1978, 1979, 1986).

Harvey, David (b. 1935)

Ph.D., University of Cambridge, 1961. University of Bristol, 1961-69; Johns Hopkins University, 1969-86; University of Oxford, 1987 to the present. *Explanation in Geography* (1969, 4th ed. 1978); *Social Justice and the City* (1973, 1978); *The Urbanisation of Capital* (1985), *Consciousness and the Urban Experience* (1985). *WW*; *GOF* (1972, 1983); J. Paterson, *David Harvey's Geography* (1984).

Haushofer, Karl (1869-1945)

Doctorate, University of Munich, 1913 (habil. 1919). University of Munich, 1921-39. *Das japanische Reich in seiner geographischen Entwicklung* (1921); *Geopolitik des Pazifischen Ozeans* (1924); *Weltpolitik von heute* (1934); *Weltmeere und Weltmächte* (1937). H. Heske and R. Wesche in *GBS*, 12 (1988), 95-106; H.-A. Jacobsen, *Karl Haushofer, Leben und Werk*

(2 vols., 1979); P. Schöller, "Die Rolle Karl Haushofers für Entwicklung und Ideologie nationalsozialistischer Geopolitik," *Erdkunde*, 36 (1982), 160–167; H. Heske, "Karl Haushofer: His Role in German Geopolitics and Nazi Politics," *Political Geography Quarterly*, 6 (1987), 135–144.

Hazards Geography

Hazards geography is concerned with the investigation of human responses to events, systems, and people that threaten human safety, emotional security, and material well-being, and is part of a broader multidisciplinary field of hazards research that brings together investigators and managers from a wide variety of fields and professions. The range of hazardous phenomena that have received the most detailed analysis by geographers includes environmental extremes such as earthquakes, tropical storms, floods, and drought, as well as catastrophic failures of technology such as nuclear power station accidents and toxic chemical spills. Responses to socioeconomic, political, or cultural tensions and other social hazards have received less scrutiny.

Environmental hazards have long been incidental objects of study in many subfields of geography, but the emergence of a systematic focus on hazards dates from the classic investigations of floods that were carried out by Gilbert White and his students at the University of Chicago during the 1950s and 1960s. White's work revealed that changes in the human use of floodplains are the primary factors involved in burgeoning flood losses. His work also uncovered the interactive nature of physical and human contributions to hazard and the role of risk perception in hazard management decisions. From this perspective, hazards are continuously variable phenomena—joint products of changeable societies and environmental fluxes that call forth changing streams of adjustments from affected populations. Subsequent studies of many types of hazard have confirmed the interpretive value of White's "human ecological" approach to the analysis of natural and technological hazards, particularly in developed societies. Human ecological models have also proven valuable to hazard managers because they often reveal previously unperceived points of intervention and suggest new modes of societal response.

In recent years a second "political economy" approach to hazards geography has emerged to complement the human ecological one. The political economy approach emphasizes the importance of social structures and historical factors as conditioners of individual responses to hazard. Its proponents draw mainly on evidence from the Third World to demonstrate that large numbers of people have little choice among the range of responses to hazard that is theoretically available. Such societies are increasingly vulnerable to loss because growing poverty and powerlessness undermine existing coping strategies and prevent the adoption of better alternatives. Analysts are now beginning to explore the many different contexts that affect the interpretation of specific hazard events and particular hazards systems.

BIBLIOGRAPHY

Bogard, W., *The Bhopal Tragedy: Language, Logic, and Politics in the Production of a Hazard* (1989).

Emel, J. and J. R. Peet, "Resource Management and Natural Hazards," pp. 49–76 in *New Models in Geography: The Political-Economy Perspective*, ed. by J. R. Peet and N. Thrift (1989).

Mitchell, J. K., "Hazards Research," pp. 410–424 in *Geography in America*, ed. by G. L. Gaile and C. J. Willmott (1989).

Palm, R., *Natural Hazards: An Integrative Framework for Research and Planning* (1990).

Platt, R. H., "Floods and Man: A Geographer's Agenda," pp. 28–68 in *Geography, Resources, and Environment*, vol. 2, *Themes from the Work of Gilbert F. White*, ed. by R. W. Kates and I. Burton (1986).

• *James K. Mitchell*

Herbertson, Andrew John (1865–1915)

Ph.D., University of Freiburg, 1898. University of Oxford, 1899–1915. (trans.) *American Life* by Paul de Rousiers (1892); *Man and His Work* (with F. D. Herbertson) (1899); "The Major Natural Regions," *GJ*, 25 (1905), 300–312. L. Jay in *GBS*, 3 (1979), 85–92; E. Gilbert *et al.*, "A. J. Herbertson Centenary Issue," *Geography*, 50 (1965), 313–372; obits. by H. Mackinder *et al.*, *The Geographical Teacher* 8 (1915–16), 143–146; and H. Beckit in *GJ*, 46 (1915), 319–320.

Hettner, Alfred (1859-1941)

Doctorate, University of Strassburg, 1881. University of Leipzig, 1891-97; University of Tübingen, 1897-99; University of Heidelberg, 1899-1928. *Die Oberflächenformen des Festlandes* (1921), *Der Gang der Kultur über die Erde* (1923), *Die Geographie, ihre Geschichte, ihr Wesen und ihre Methoden* (1927), *Vergleichende Länderkunde* (4 vols., 1933-34). E. Plewe in *NDB*, 9 (1971), 31-32; E. Plewe in *GBS*, 6 (1982), 55-63; G. Pfeifer (ed.), *Alfred Hettner . . . Gedenkschrift zum 100. Geburtstag* (1960); E. Plewe and U. Wardenga, *Der junge Alfred Hettner* (1985).

Hewes, Leslie (b. 1906)

Ph.D., University of California, Berkeley, 1940. University of Oklahoma, 1932-45; University of Nebraska, 1945-74. "The Northern Wet Prairie of the United States," *AAAG*, 42 (1952), 24-50; *The Suitcase Farming Frontier* (1973); "The Great Plains One Hundred Years after Major John Wesley Powell," pp. 203-214 in *Images of the Plains* (ed. by B. Blouet and M. Lawson) (1975); *Occupying the Cherokee Country of Oklahoma* (1978). *WWA*; *GOF* (1973); *60 Years of Berkeley Geography.*

Historical Geography

Historical geography is that branch of geography that is concerned with reconstructing the geographical aspects of the human past. Historical approaches in geography have traditionally been associated with regional geography in France, Latin America, and throughout the English-speaking world, and with settlement geography in Scandinavia, the Mediterranean countries, and throughout the German and Slavic-language worlds. In most of Africa and Asia, historical perspectives have been associated with colonialism and have remained of secondary importance to current development needs. Historical geography also has retained close association with the history of geography, with geographical exploration and discovery, and with historical cartography. The internationalization of historical interests has resulted in the publication since 1975 of the periodical, the *Journal of Historical Geography.*

Some geographers have claimed that all geography ultimately is historical geography and simply involves the retrospective application of current concepts and techniques to the study of past areas and conditions. Indeed, many important contributions to historical geography continue to be made by geographers who combine interests in contemporary regions and settlements with a concern for how these places evolved.

Yet the geographical study of past human occupance also involves sensitivity to the historical context of the people, place, and topic under consideration. Historical geographers generally subscribe to the view that the history of a place and its people is embedded in its geography and that geographical processes and patterns need to be understood within their environmental and social contexts. Some historical geographers have placed particular emphasis on reconstructing geographies of past periods from written or printed documents in the form of synchronic cross-sections that convey the characteristics of places at particular times. Others have been concerned with exploring particular themes or topics diachronically through time, from even prehistory to the present, by combining techniques from archeology, documentary interpretation, and landscape observation. Research has been enhanced most recently by advances in conceptual modelling, in quantitative and computer techniques, and in textual analysis.

Historical geography in North America has traditionally focussed on three principal topics: rural settlement and agriculture, regional distinctiveness and identity, and cultural origins and diffusion, studied within the contexts of European colonization and subsequent frontier expansion. During the past fifteen years, however, scholars have paid increasing attention to the major processes of change during the nineteenth and twentieth centuries, most notably large-scale immigration, urbanization, and industrialization. These recent trends in topic and technique suggest an increasing concern with linking processes and patterns discerned in the past with the unfolding geography of the present, thus contributing to the creation of a more historically informed human geography.

BIBLIOGRAPHY

Baker, A. R. H., ed., *Progress in Historical Geography* (1972).

Baker, A. R. H., and M. Billinge, eds., *Period and Place: Research Methods in Historical Geography* (1982).

Darby, H. C., ed., *A New Historical Geography of England* (1973).

Gibson, J. R., ed., *European Settlement and Development in North America* (1978).

Mitchell, R. D., and P. A. Groves, eds., *North America: The Historical Geography of a Changing Continent* (1987).

Sauer, C. O., *Land and Life: A Selection from the Writings of Carl O. Sauer* (1963).

• *Robert D. Mitchell*

History of Geography

During the past century the writing of the history of geography has substantially changed in content and method. These changes are primarily associated with the professionalization of geography as a university discipline.

Before 1890, except in Germany (where university chairs in geography were established earlier than elsewhere), the history of geography was written largely by librarians, learned amateurs, classicists, Biblical scholars, and retired military officers. As the professionalization of geography spread from country to country some of the older forms, such as the history of cartography, the history of exploration and discovery, and the history of classical geographic thought, tended to shear off from the disciplinary agendas of professional geographers.

With a few honorable exceptions (among whom John K. Wright in the U.S. and J. N. L. Baker and Eva G. R. Taylor in Great Britain had pride of place in the English-speaking world), few professional geographers of the first half of this century were competent research specialists in the history of geography. As ordinarily practiced by professional geographers, the history of the discipline tended to be cast in a Whiggish mode, a "rise and progress" ambience suited to legitimating the discipline, providing fodder for celebratory occasions such as presidential addresses, or supporting through selective historical reference one's own position concerning a current professional "ism." Such habits have not entirely disappeared, but in a number of countries since the 1950s more systematic and sophisticated forms of teaching and research have emerged, especially in Canada, the Federal Republic of Germany, France, Great Britain, Italy, Scandinavia, Spain, the U.S.S.R., and the United States.

Of all the possible genres, studies of individual scholar-practitioners have proven the most popular, and geographical biographers have attained increased skill in using archival and manuscript collections to augment and correct the received tradition and the printed word. A large-scale international effort to reappraise leading contributors to geographic thought (whether or not by persons identifiable as geographers) was begun in 1977 under the auspices of the International Geographical Union's Commission on the History of Geographical Thought (established 1968) through its "biobibliography" series. In the past two decades there has also been a rising interest in studies of the institutionalization and professionalization of geography in various countries and of the spread of ideas among the various national traditions. By contrast, the history of geographical instruction has been little cultivated, though here again there are distinguished exceptions.

More recently, there have been sophisticated attempts to enlarge the history of geographic thought to uncover linkages to the more general climate of thought in which it is embedded. Studies of particular ideas about places and specific environments were in part a by-product of the environmental cognition and phenomenological currents within geography in the late 1960s and the 1970s. Similarly, a few studies have systematically examined more general notions of space and location among groups of people who were not self-identified geographers. Beyond the use of new types of source materials, some scholars have employed metahistorical constructs and theories (Marxist or other) in the history of geographic thought, and others have explored the uses and limits of oral history in text and filmed interview. Since 1981 the *History of Geography Newsletter* (now the *History of Geography Journal*) has served as a barometric register of these kinds of changes.

At present, only a handful of scholars in each country have chosen the history of geography as a primary research task. The published results of their investigations have provided raw materials for a far richer quality of courses in the history of geography than those earlier taught in geography departments. No useful English-language general text has emerged to incorporate these newer research approaches and replace the "textbook chronicles" of the immediate post-World War II generation. The scholarly journals are, however, increasingly open to articles in the field and a growing number of useful books are available for collateral reading in such courses.

The appearance in 1986 of the first volume of a large-scale international series on the history of cartography promises a reintegration in the 1990s of that area of inquiry with the mainstream of the history of geography. We may also expect further intensification of the discovery and use of manuscript and archival materials for both biographical and institutional studies, as well as a broader examination of printed sources outside those produced by self-identified geographers. The isolation of historians of geography from scholars making methodological advances in other disciplinary areas is beginning to be breached, though cautionary voices have been raised against the uncritical adoption of extra-disciplinary models, such as the paradigm concept in the history of science. On the other hand, at least one major work by a geographer, Clarence Glacken's *Traces on the Rhodian Shore*, has been acclaimed by historians of science internationally.

The future will undoubtedly realize the promise of a richer history of geography as geographers become better acquainted with developments in social and cultural as well as intellectual history and the history of science. Until the major graduate departments develop the capacity to train successor scholars in the field, however, the history of geography is likely to remain for some time to come a primary interest of a few devoted individuals rather than a core professional field for the discipline as a whole.

BIBLIOGRAPHY

Dunbar, G. S., ed., *The History of Modern Geography: An Annotated Bibliography of Selected Works* (1985).

Glacken, C. J., *Traces on the Rhodian Shore: Nature and Culture in Western Thought from Ancient Times to the End of the Eighteenth Century* (1967).

Harley, J. B., and D. Woodward, eds., *The History of Cartography* (1986—).

International Geographical Union, Commission on the History of Geographical Thought, *Geographers: Biobibliographical Studies* (1977—).

Johnston, R. J., and P. Claval, eds., *Geography since the Second World War: An International Survey* (1984).

Martin, G. J., ed., *History of Geography Newsletter* (now *History of Geography Journal*) (1981—).

• *William A. Koelsch*

Ho, Robert (1921-1972)

M.A., King's College London, 1950. University of Malaya, Singapore, 1948-59, Kuala Lumpur, 1959-65; Australian National University, 1965-72. *Farmers of Central Malaya* (1967); (ed.) *Studies of Contemporary Thailand* (with E. Chapman) (1973). G. Cho in *GBS*, 1 (1977), 49-54; obit. by T. Bahrin in *TIBG*, 59 (1973), 155-157.

Holmes, James Macdonald (1896-1966)

Ph.D., University of Glasgow, 1934. University of Sydney, 1929-61. *Atlas of Population and Production for New South Wales* (1931); *The Geographical Basis of Government, Specially Applied to New South Wales* (1944); *Soil Erosion in Australia and New Zealand* (1946). J. Powell in *GBS*, 7 (1983), 51-55; obit. by W. Maze in *Australian Geographer*, 10 (1966-68), 220-221.

Hooson, David John Mahler (b. 1926)

Ph.D., University of London, 1955. University of Glasgow, 1954-56; University of Maryland, 1956-60; University of British Columbia, 1960-66; University of California, Berkeley, 1966 to the present. "The Distribution of Population as the Essential Geographical Expression," *CG*, 17 (1960), 10-20; *The Soviet Union—People and Regions* (1966); "National Cultures and Academic Geography in an Urbanizing Age," pp. 157-178 in *The Expanding City: Essays in Honour of Jean Gottmann* (ed. by J. Patten) (1983), with autobiographical sketch on pp. 160–163; "The Soviet Union," pp. 79–106 in *Geography since the Second World War* (ed. by R. Johnston and P. Claval) (1984). *GOF* (1980).

Human Geography

Viewed from an orbiting satellite, the most conspicuous features of the earth are the division between land and water and the colors and configurations of the continents. At ground level, an observer may see roads, cultivated fields, factories, and other features of human activity. The study of the location, distribution, and environmental context of such features is the task of human geography. This effort is essentially a search for order or explanatory principles. Human geographers seldom entertain thought that what they study is random, accidental, or a product of forces beyond human control. For example, during a trip from a suburban airport to the center of a large American city a traveler encounters residential, commercial, and industrial districts plus land devoted to recreational and governmental activities. This distribution is a reflection of zoning regulations, property values and taxes, commuting patterns, and a host of other influences that are products of human choice, regulation, and tradition. In this case, human geography fosters understanding and prediction that is not available to physical geographers devoted to laws of nature or to scholars trained in disciplines that do not focus on the locational or distributional aspects of human accomplishment and behavior.

Human geography is well established in many colleges and most universities. It is usually cultivated by scholars committed to particular rather than general approaches to this vast field. Although textbooks often cover a wide array of human-geographical topics, research is usually more sharply focused on specific problems or areas. For this reason, human geography tends to be supplanted by more explicit labels, such as urban, rural, and transportation geography. The division of effort among human geographers also reflects the departments or disciplines that have evolved in response to the particular interests of scholars devoted to aspects of the human condition. In the United States especially the broad realm of human geography includes several well-recognized subfields, such as economic, historical, social, and political geography. Scholars working in these subfields are usually well aware of the methods and findings of the neighboring enterprises of economics, history, sociology, and political science. Cultural geographers enjoy a similar complementary relationship with anthropology, although the concept of culture is so broad and all-encompassing that cultural geography may be regarded as a synonym for human geography. The same can be said of anthropogeography, a term employed frequently to describe the pioneering work on the connections between natural environment and human activity that was undertaken from the closing decades of the nineteenth century until about the time of World War I.

As is true of most scholarly disciplines, human geography has experienced shifts of emphasis and debate about its objectives and accomplishments. The first generation of human geographers (c. 1900–30) sought to discover a causal relationship between human activities and elements of physical geography, such as climate, landforms, vegetation, and soil. Later scholars abandoned this inquiry in favor of distributional studies that led eventually to the formulation and testing of appropriate theories, models, and hypotheses. In time, geographers also tended to shift from an initial conception of nature and culture as separate entities or opposing forces to a view that most of the environment of mankind has been modified to some degree and hence is a product of both natural and human-induced forces. In studies of drainage, deforestation, erosion, and climatic modification the old distinction between human and physical geography could not be maintained. Investigation of such altered environments, neither purely natural nor artificial, has long been a major focus of concern. Most human geographers are in fact students of built environments and cultural as opposed to natural landscapes.

Much of the effort of human geographers, especially since the 1960s, has been devoted to the refinement of theories designed to explain the size and spacing of settlements, the zonation of land use, resource management strategies, and the diffusion of innovations. Other large research themes include the discordant relationship between the world political map and more complicated patterns of linguistic and religious distributions, and the global contrast between relatively rich and poor countries. Researchers have also been inclined in recent years to devote themselves to problems of environmental perception, and the distinction between appearance and reality has become a major preoccupation. Studies of landscape taste, residential preferences, and human awareness of and response to droughts, floods, and other hazards appear frequently in the geographical literature.

The last comprehensive work on human geography by a single author was accomplished by the French scholar Maximilien Sorre (3 vols., 1947-52). More recently, attention has been directed to aspects of or problems inherent in particular human activities or habitats. The isolation experienced by individual scholars with specific areal or topical interests is overcome to some extent by professional organizations such as the Association of American Geographers and the International Geographical Union. The cause of human geography is also served by periodical publications of both eclectic and specialized character. The persisting mission of human geography is to discover the logic or pathology that is revealed in the record of human use and misuse of the earth. *See also* Historical Geography; Population Geography; Settlement Geography; Social Geography.

Bibliography

Goudie, A. S., *The Human Impact on Natural Environment* (1986).

Jordan, T. G., and L. Rowntree, *The Human Mosaic: A Thematic Introduction to Cultural Geography* (4th ed., 1986).

Mikesell, M. W., "Tradition and Innovation in Cultural Geography," *AAAG*, 68 (1978), 1-16.

Sorre, M., *Les fondements de la géographie humaine* (3 vols., 1947-52).

Stoddard, R. H., D. J. Wishart, and B. W. Blouet, *Human Geography: Peoples, Places, and Cultures* (2nd ed., 1986).

Zelinsky, W., ed., "Human Geography: Coming of Age," *American Behavioral Scientist*, 22 (1978), 3-167.

• *Marvin W. Mikesell*

Humanistic Geography

Coined by geographer Yi-Fu Tuan in 1976, the term humanistic geography describes a diverse body of research emphasizing the role of human experience and meaning in understanding peoples' relationship with geographical environments. Recognizing that human involvement with the geographical world is complex and multilayered, humanistic geographers seek to describe and interpret human awareness and action as they both create and are created by such geographical qualities as place, space, nature, landscape, home, journey, region, dwelling, and built environment.

The reasons for the development of humanistic geography are complex and arise from both inside and outside the discipline. In the early 1970s, some geographers, following a trend in other social sciences, reacted against the dominant positivist approach, which emphasized *a priori* theory, empirically measurable evidence, mathematical modelling, and statistical analysis. A first criticism was that positivist geography ignored or denied qualities of human experience and meaning that could not be measured—for example, a sense of place, a feeling of at-homeness, or heightened experiences of nature. A second criticism was that positivist research devalued the role of individuals in affecting the course of events; people were assumed to be passive, interchangeable units that could be counted and aggregated and whose actions could be predicted and explained. A third criticism was that positivist geography, emphasizing control and pragmatic utility, could readily be used to manipulate human behavior and thereby undermine human freedom. At the same time, a growing interest in "environmental perception" within the positivist tradition itself led to research on themes dealing with the geographical behaviors of individuals and groups—for example, studies of cognitive mapping; territoriality; and environmental images, attitudes and preferences.

By the late 1970s, geographers began to complement the dominant positivist research on environmental perception and behavior with qualitative, descriptive approaches that were labelled "humanistic geography." Some of these researchers sought to resurrect and adapt the approaches of earlier geographers like Paul Vidal de la Blache, Carl Sauer, and J. K. Wright, while others turned to new philosophical traditions like phenomenology, hermeneutics, existentialism, idealism and pragmatism. Because these ways of working are very much different from each other, the term humanistic geography fell out of favor in the late 1980s, replaced by more exact descriptions like phenomenological geography, hermeneutical geography, and post-structural geography. Like the earlier humanistic geography, however, these approaches continue to hold in common an emphasis on human experience, meaning, interpretation and values as the essential foundation for understanding human behavior and events in the geographical world.

BIBLIOGRAPHY

Jackson, P., and S. J. Smith, *Exploring Social Geography* (1984).

Ley, D., and M. Samuels, eds., *Humanistic Geography: Prospects and Problems* (1978).

Relph, E., *Place and Placelessness* (1976).

Rowntree, L. B., "Orthodoxy and New Directions: Cultural/Humanistic Geography," *PHG*, 12 (1988), 575-586.

Seamon, D., and R. Mugerauer, *Dwelling, Place and Environment: Towards a Phenomenology of Person and World* (1985).

Tuan, Yi-Fu, "Humanistic Geography," *AAAG*, 66 (1976), 266-276.

• *David Seamon*

Humlum, Johannes Peter Christian Nicolai (b. 1911)

Ph.D., University of Copenhagen, 1942. University of Copenhagen, 1936-43; University of Aarhus, 1943-81. *Landsplanlægning og Egnsplanlægning i Danmark* (1962); *Landsplanlægningsproblemer* (1966); *Kulturgeografisk Atlas, Danmark* (1975-76); *Kulturgeografi l og 2/Atlas of Cultural and Economic Geography* (with H. Thomsen) (8th ed., 1977-78). J. Humlum, "Peregrination and Irrigation," pp. 164-189 in *Geographers of Norden* (ed. by T. Hägerstrand and A. Buttimer) (1988).

Hungary

Geography has a long history in Hungary, as a subject in schools and as a genre of writing and publishing. Maps have been designed, printed and published by Hungarian mapmakers ever since the sixteenth century. The rise of academic geography, on the other hand, was a long and slow process.

By the early twentieth century Budapest's Peter Pazmany University had a chair/institute of geography and one of geology; the primary emphasis in the institute of geography was on physical aspects of the discipline. It was not until after World War I that human geography occupied a place in university teaching. A chair of economic geography was established in the early 1920s in the semi-independent College of Economic Sciences of Budapest University. Although geography played a part in the life of Pécs and Szeged universities, the leadership was clearly vested in the two principal personalities in Budapest, Eugene Cholnoky in the physical and Paul Teleki in the economic-political aspects of geography. The Hungarian Geographical Society, founded in 1872, published its own journal—and continues to do so still—and encouraged popularization of geography.

Brief mention should be made also of the role played by Hungarians in geographical exploration in the nineteenth century and the first half of the twentieth century: Armin Vámbéry and Aurel Stein in Inner Asia, Samuel Teleki in East Africa, and others.

Geography in Hungary underwent profound changes in the years following World War II. The domination of the country by a one-party system modelled on that of the Soviet Union resulted in a "remodelling" of the entire system of higher education and advanced research. In higher education this meant an impressive increase in the number of university chairs and departments of geography and a corresponding increase both in the numbers of faculty teaching geography and students at the undergraduate and graduate levels specializing in the various aspects of the field.

The change has been even more impressive in research. Following the Soviet approach, as was the case in other countries that came under the dominant Soviet influence after 1945-48, it was the Hungarian Academy of Sciences that took the lead. Long regarded as the guardian of traditional values in letters and the arts, the reorganized, post-1948 academy established, following the Soviet model, a number of semi-autonomous institutes, including the Institute for Geographical Research, which now consists of two centers, one in Budapest and one in Pécs. These institutes, endowed with substantial funds and employing an impressive number of active researchers, many of whom also hold appointments in higher education, have put Hungarian geographical research "on the map." Their publications, all including foreign language summaries and quite a few also published in languages other than Hungarian, enjoy wide circulation. A measure of the current standing of the discipline may be illustrated by the fact that two geographers are members of the Hungarian Academy of Sciences and one is a Vice-President of the International Geographical Union.

The two principal geographical periodicals are *Földrajzi Közlemények* (Geographical Journal), published by the Hungarian Geographical Society since 1872, and *Földrajzi Ertesítö* (Geographical Review), published by the Institute of Geography, Hungarian Academy of Sciences, since 1952.

BIBLIOGRAPHY

Bernát, T., and G. Enyedi, *A Magyar Mezögazdaság Termelési Körzetei* (1961).

Kish, G., "Geography in Hungary," *PG*, 37 (1985), 89-90.

Markos, G., *Magyarország Gazdasági Földrajza* (1962).

Pécsi, M., and B. Sárfalvi, *The Geography of Hungary* (1964).

Sárfalvi, B., ed., *The Changing Face of the Great Hungarian Plain* (1971).

Teleki, P., F. Koch, and L. Kádár, *A Gazdasági Elet Földrajzi Alapjai* (2 vols., 1936).

• *George Kish*

Huntington, Ellsworth (1876-1947)

Ph.D., Yale University, 1909. Yale University, 1907-15, 1919-47. *The Pulse of Asia* (1907), *Civilization and Climate* (1915), *The Character of Races* (1924), *Mainsprings of Civilization* (1945). *WWA*; G. Martin, *Ellsworth Huntington, His Life and Work* (1973); G. Martin in *DAB*, supp. 4 (1974), 412-414; O. Spate in *IESS*, 7 (1968), 26-27; obits. by S. Visher in *AAAG*, 38 (1948), 38-50; and S. Van Valkenburg in *GR*, 38 (1948), 153-155.

Imhof, Eduard (1895–1987)

Diploma, Swiss Federal Institute of Technology (ETH), 1919. Swiss Federal Institute of Technology (ETH), 1920–65. *Kartographische Geländedarstellung* (1965, pub. in English as *Terrain Representation*, 1973); (ed.) *Atlas der Schweiz* (1965–78); *Thematische Kartographie* (1972); *Die grossen kalten Berge von Szetschuan* (1974). Obit. by E. Meynen in *Geographisches Taschenbuch 1987/1988* (1988), 158–164.

India. *See* South Asia.

Indonesia

The development of geography as a discipline is a rather recent phenomenon in Indonesia. It is true that during the past 100 years a wide variety of amateurs and professionals, even scientists, contributed heavily to the geographical knowledge of the East Indies. The Royal Dutch Geographical Society and the Treub Society for Scientific Research of the (Netherlands) East and West Indies, for instance, organized several large-scale expeditions to the interior and distant parts of the archipelago, contributing vastly to the knowledge thereof. And, certainly, over the years, the activities of various governmental institutions had important geographical components. But, as there was no academic teaching in geography, all these contributions did not lead to a flourishing discipline of geography.

Two early governmental initiatives were of great importance for the future development of geography in Indonesia. In 1910 the Encyclopaedic Bureau was established by the Department of Internal Affairs. In the eleven years of its existence the bureau published a series of thirty-six thorough regional studies under the innovative leadership of L. van Vuuren, who later held the chair of human geography at Utrecht University from 1927 to 1945. In 1920, a year before the Encyclopaedic Bureau was closed for budgetary reasons, the Cartographic Division in the Topographical Service welcomed its first geographer. As a consequence, this division became the successor to the Encyclopaedic Bureau as the main center of geographical information on Indonesia. Its first geographer was Dr. Samuel van Valkenburg, who later became a professor of geography at Clark University in the United States. He was succeeded in the Cartographic Division by P. B. Kessel, and after Kessel's untimely death by A. J. Pannekoek. The latter was responsible for the joint publication in 1938 by the Topographical Service and the Royal Dutch Geographical Society of the *Atlas van Tropisch Nederland*.

In 1947 the first geographical institute in Indonesia was founded by the head of the Topographical Service, A. Kint. This institute was to function within the Cartographic Division. The human geographer F. J. Ormeling, a future president of the International Cartographic Association, was appointed in 1948 to be its first head. The institute spearheaded the development of geography in Indonesian universities, in particular in Jakarta, Yogyakarta and Bandung. Institute staff members such as H.Th. Verstappen, the present first vice-president of the IGU, delivered lectures in the universities on geography, cartography and the interpretation of aerial photographs, but it was only in 1959 that through a direct link with the Pajajaran University this first geographical institute could reach academic status.

Since independence (1949) two newly established governmental institutions have had a decisive influence on the further development of geography in Indonesia—the Bakosurtanal and the Directorat Land Use. The Bakosurtanal is responsible for the coordination and planning of all mapping activities in Indonesia. The beginnings of this organization date back to 1951, when it was decided to bring together in a so-called "mapping-council" the secretary-generals of all government departments that were involved in mapping activities, but the present Bakosurtanal was not formed until the late 1960s. It now consists of two main divisions: one for topographic mapping and one for thematic mapping (of natural resources). In order to train the personnel required, the Bakosurtanal established in cooperation with the Universitas Gadjah Mada (UGM, Yogyakarta) a specialized training center for remote sensing applications and integrated mapping surveys (PUSPICS, Yogyakarta).

Land use mapping has been since 1960 the responsibility of the Directorat Land Use, headed by Dr. I Made Sandy. The legend used links up easily with the World Land Use Survey Classification developed by the IGU. To limit costs it was decided not to use aerial photographs or remote sensing but to rely on experienced manpower. At present 2,000 surveyors are employed, half of them academics, of whom 20 hold M.Sc. and Ph.D. degrees. By 1983, 12,000 sheets, most of them black-and-white, had been prepared.

As in so many countries, the teaching of geography in Indonesian universities started with geography as a minor subject to a major in another discipline, *e.g.*, agriculture. As a major study it entered the Indonesian universities in the mid-1950s.

The only full-fledged geographical faculty in Indonesia at present is the Fakultas Geografi of the Universitas Gadjah Mada (UGM) at Yogyakarta. It started in 1950 as an institute for the training of geography schoolteachers but soon expanded in scope. Under the leadership of Kardono Darmoyuwono, who later became the head of the thematic division of Bakosurtanal, the institute acquired the status of an independent faculty by 1963. From that year to the mid-1980s the number of full-time staff members increased from sixteen to sixty-five and the number of students to 550. In 1980 the faculty was granted PEMBINA (*i.e.*, center of excellence) status, which means that the faculty has the right to offer specialized postgraduate (S-2) courses to staff members of other universities and teacher training colleges. New academic teaching programs in rural and regional planning have been organized in cooperation with the regional planning authorities, Utrecht University and ITC Enschede (International Institute for Aerospace Survey and Earth Sciences in the Netherlands).

The second academic geographical institute in Indonesia, which was established by the Topographical Service in 1947 and which reached academic status with Pajajaran University in 1959, developed more slowly. The situation changed in 1967 when the institute became a part of the Universitas Indonesia at Jakarta and more decisively after 1978 when the institute moved to the grounds of that university. In the mid-1980s this geographical institute had forty staff members, of whom thirteen were full time, and close to 200 students.

In addition to the two universities of Yogyakarta and Jakarta a number of teacher-training colleges offer instruction in geography. Since 1964, these IKIP colleges have been independent institutions at university level. Geography is taught, to an average of 300 students each, in eleven IKIPs: Bandung, Jakarta, Jayapura, Malang, Medan, Padang, Semarang, Solo, Surabaya, Ujung Padang and Yogyakarta.

The current status of geography in Indonesia is promising. Twice a year the Fakultas

Geografi at Yogyakarta publishes the *Indonesian Journal of Geography*, with the financial support of the Ministry of Research. Indonesian geographers have a professional association, the Ikatan Geograf Indonesia, founded in 1959. Indonesia has been a member of the IGU since 1952. Indonesian geographers have growing contacts with various countries. Within their own country they contribute to many important fields, such as integrated mapping, natural resources, population studies, coastal research and management, environmental research and regional planning.

BIBLIOGRAPHY

Ormeling, F. J., and D. G. Montagne, "Het Geografisch Instituut van de Topografische Dienst te Batavia," *Tijdschrift van Koninklijk Nederlands Aardrijkskundig Genootschap*, 68 (1950), 1-20.

Verstappen, H.Th., and S. Hadisoemarno, "De Ontwikkeling van de Geografie in Indonesia," *Geografisch Tijdschrift*, 20 (1986), 446-454.

• *J. A. van Ginkel*

Industrial Geography

This major part of economic geography seeks to explain (a) the locational decisions and mobility of the production activities of firms and industries, and (b) urban and regional differences in the growth, decline and evolution of industrial specializations.

The primary elements of the "locational problem" in which a firm seeks to minimize its costs of access to markets, materials, and labor while maximizing the advantages from agglomeration were first defined by Alfred Weber (1929). This theoretical tradition applied to manufacturing firms is still a major influence on neoclassical location theory. Nevertheless, much empirical inquiry in industrial geography has reflected the convergence of regional development and industrial location theory: the geographic extent of production and information bonds between enterprises, the origin of industrial clustering, and the impact of agglomeration economies continue to be important fields of research, even though the particular research questions and the methods of data generation and analysis continue to evolve.

In response to a number of international economic trends, industrial geography has changed dramatically over the past twenty-five years:

1. The economic significance and complexity of multinational corporations have been addressed in a variety of ways—organizational and decision models have been sought to explain the operations of these multilocational and multifunctional firms, political-economic assessments have been made of the impacts of foreign direct investment on "host economies," and research has explored the allocation of labor-intensive phases of production to low-wage countries.
2. The world industrial map has been redrawn by virtue of the emergence of newly industrialized countries and the rise of Japan as a major industrial power. Intensified competition has thrust western economies into a period of restructuring involving the out-migration and closure of uncompetitive plants, and negative regional changes have been investigated, often through the use of Marxian analysis. Other enquiries have analyzed the location of firms in previously non-industrial regions.
3. The deindustrialization of advanced economies features large employment shifts to "white collar" jobs in the services, a process that has expanded the field of industrial geography. New intellectual products such as computer software have replaced an exclusive concern with tangible manufactured products, and the impact of computer systems on production, management, and communications systems promises to be an area of research growth. The location of "office industries" (such as financial services and the head offices of corporations, and activities linked to them) in the central cities of metropolitan regions has stimulated the analysis of contact systems as a means of understanding these choices.
4. Currently, industrial geography is strongly concerned with the locational impacts of technological change, and with the locational conditions that favor industrial innovation. This has led to locational analysis of the growing managerial, scientific, and technical producer services. The product life-cycle

model has been used to explain the location of R&D labs and the locational diffusion of production within large firms. Recently, however, this work has been balanced by attention to small firms as important sources of product innovation.

BIBLIOGRAPHY

Dicken, P., *Global Shift: Industrial Change in a Turbulent World* (1986).

Massey, D., and R. Meegan, *The Anatomy of Job Loss* (1982).

Smith, D. M., *Industrial Location: An Economic Geographical Analysis* (2nd ed., 1981).

Thomas, M. D., "Growth Pole Theory, Technological Change and Regional Economic Growth," *Regional Science Association Paper*, 34 (1975), 3-25.

Tornqvist, G., *Contact Systems and Regional Development* (Lund Studies in Geography, Series B, 35, 1970).

Weber, A., *The Theory of the Location of Industries*, trans. by C. J. Friedrich (1929).

• *John N.H. Britton*

Innis, Harold Adams (1894-1952)

Ph.D., University of Chicago, 1920. University of Toronto, 1920-52. *The Fur Trade in Canada* (1930); *The Cod Fisheries* (1940); *Empire and Communications* (1950); *The Bias of Communication* (1951). D. Creighton, *Harold Adams Innis: Portrait of a Scholar* (1957); G. Dunbar, "Harold Innis and Canadian Geography," *CG*, 29 (1985), 159-164; V. Berdoulay and R. Chapman, "Le possibilisme de Harold Innis," *CG*, 31 (1987), 2-11; I. Parker, "Harold Innis as a Canadian Geographer," *CG*, 32 (1988), 63-69.

Institute of Australian Geographers

The Institute of Australian Geographers Inc. is the national professional organization for geography in Australia. The governing council held its first meeting in February 1959, representing a membership drawn exclusively from university teaching staff. Subsequently membership has been broadened to include geographers from other educational sectors and from business, and is currently over 350. Griffith Taylor was the first president (1959-60). Current publications are *Australian Geographical Studies* (two issues per year, since 1963), which replaced the earlier *Australian Geographical Record* (five numbers, 1959-64), and the *IAG Newsletter* (two issues per year, since 1978).

BIBLIOGRAPHY

Powell, J. M. "Geographical education and its Australian heritage," *Australian Geographical Studies*, 26 (1988), 231-236.

Walmsley, D. J., and J. E. Hobbs, "AGS—25 years on," *Ibid.*, 26 (1988), 3-4.

• *R. L. Heathcote*

Institute of British Geographers

The Institute of British Geographers was founded in 1933 by academic geographers (73 founding members) who felt that more publication outlets were needed for academic papers. The institute holds annual conferences in early January and publishes two important journals, *Transactions* (established 1935) and *Area* (1969), analogous to the *Annals* and *Professional Geographer* of the Association of American Geographers, as well as occasional monographs. Although initiated partly as a reaction to the publication policies of the Royal Geographical Society at the time, the IBG has enjoyed good relations with the RGS, which provides the academic group with office space in London.

BIBLIOGRAPHY

Steel, R. W., *The Institute of British Geographers: The First Fifty Years* (1983).

Stoddart, D. R., "Progress in Geography: The Record of the I.B.G.," *TIBG*, 8 (1983), 1-13, and other articles in this "special issue of *Transactions* to mark the fiftieth anniversary of the Institute."

International Geographical Congress

The first international geographical congress was held in Antwerp, Belgium, in 1871. It was called the Congrès des Sciences Géographiques, Cosmographiques et Commerciales and came about largely through the initiative of Charles Ruelens of the Royal Library in

Brussels. The existence since 1922 of an international union of geographical societies and national committees has provided greater continuity and more consistent planning in the inter-congress years. Before World War II only two congresses were held outside Europe (United States 1904 and Cairo 1925), but about half of the postwar congresses have been convened on other continents. The two world wars created long hiatuses between congresses (1913–25 and 1938–49), but since 1952 the International Geographical Congress has met regularly every four years, typically in a one-week meeting in late August, with pre- and post-congress symposia and field trips in the countryside and smaller cities. Publication of congress proceedings has been the responsibility of the convenors and often occurs belatedly. Other publications spawned by the congresses include excursion guides and other works on the geography of the host country. Congresses also provide the occasion for summaries of the geographical activities of the inter-congress years.

BIBLIOGRAPHY

Leconte, P., "Histoire de l'Union Géographique Internationale et des Congrès Internationaux de Géographie," *IGU Newsletter*, 10 (1959), 3–20, 43–69.

Pinchemel, P., ed., *La géographie à travers un siècle de congrès internationaux* (1972).

International Geographical Union

The International Geographical Union is composed of the national committees from the member countries. It was founded in Brussels in 1922 at a meeting of the International Research Council (Conseil International de Recherches). One of the chief functions of the IGU is to provide leadership and planning for periodic congresses. In fact, such congresses had been held since 1871, but the union insured greater continuity and stability through a permanent executive committee. The IGU has spawned numerous commissions and working groups (since 1924), pre- and post-congress symposia (since 1952), and regional meetings in the inter-congress years (since 1955), and has published a *Newsletter* since 1950. Commission reports have been published sporadically since 1928. The Commission on the History of Geographical Thought has sponsored an annual volume, *Geographers: Biobibliographical Studies*, since 1977.

BIBLIOGRAPHY

Leconte, P., "Histoire de l'Union Géographique Internationale et des Congrès Internationaux de Géographie," *IGU Newsletter*, 10 (1959), 3–20, 43–69.

Martonne, E. de, "Brief History of the International Geographical Union," *IGU Newsletter*, 1 (June 1950), 3–5.

Pinchemel, P., ed., *La géographie à travers un siècle de congrès internationaux* (1972).

Ireland

Irishmen played a significant part in the exploration of our world during the second half of the nineteenth century, and early in the twentieth century the most famed Irish explorer was Ernest Henry Shackleton (1874–1922). Around 1900 geography was nevertheless poorly represented at home within the Irish educational curriculum, although the Dublin firm of Sullivan Brothers was publishing a successful series of internationally used geographical texts. Their *Geography Generalised* reached its seventy-first edition in 1887 and it was still in print in 1914. The first Irish institution devoted to the furtherance of geography was the Irish Geographical Association, founded in 1918 by Grenville Arthur James Cole (1859–1924). The association was chiefly concerned with the improvement of geography within Irish schools, but it did also press for geography to receive university recognition. In 1923 it was claimed that the newly independent Irish Free State shared with Albania and Latvia the doubtful distinction of being one of only three European nations where geography could not be studied as a university discipline.

The Irish Geographical Association died in 1928, but three years later a Lectureship in Geography was founded in Trinity College (the University of Dublin). That post became a Readership in 1944 and between 1936 and 1949 these positions were held by Thomas Walter Freeman (1908–1988), a geographer widely known for his research in Irish geography and the history of geographical thought. Between 1959 and 1978 chairs of geography and full departments in the subject were founded in Trinity College and in the colleges of the National University of Ireland in Cork, Dublin, Galway, and Maynooth. By 1981 there were 37 geographers at work in the Republic's universities (the total has since declined slightly) and there are about 1,600 students studying the subject at either undergraduate or graduate level. In Irish schools geography is compulsory for most students, up to the age of 15.

The Royal Irish Academy, Ireland's premier learned society, has seven geographers among its members, and the academy inaugurated the Irish National Committee for Geography in 1963. The academy published the *Atlas of Ireland* in 1979 and is currently publishing the *Irish Historic Towns Atlas*. The two principal specialist Irish geographical societies are the Geographical Society of Ireland (founded 1934, membership 220) and the Association of Geography Teachers of Ireland (founded 1961, membership 300). The Society has published *Irish Geography* annually since 1944. Noteworthy features of Irish geography are the unusually strong emphases upon physical geography and historical geography.

BIBLIOGRAPHY

Alexander, R. W., and D. A. Gillmor, eds., *Geography in Education in the Republic of Ireland* (1989).

Fahy, G., "Geography and Geographic Education in Ireland from Early Christian Times to 1960," *Geographical Viewpoint*, 10 (1981), 5-30.

Gillmor, D. A., "Geography in the Republic of Ireland," *PG*, 40 (1988), 103-106.

Glasscock, R. E., "Geography in the Irish Universities, 1967," *Irish Geography*, 5 (1968), 459-468.

Herries Davies, G. L., "The Making of Irish Geography, II: Grenville Arthur James Cole (1859-1924)," *Irish Geography*, 10 (1977), 90-94.

———, *This Protean Subject: The Geography Department in Trinity College Dublin 1936-1986* (1986).

• *Gordon L. Herries Davies*

Ireland, Geographical Society of

The Geographical Society of Ireland was founded in Dublin in 1934 for "the furtherance of geography in all its branches, particularly the geography of Ireland." The society has no premises of its own and its lectures and symposia are held in Trinity College Dublin, in the house of the Royal Irish Academy, or in University College Dublin. In recent years occasional meetings have been held in such other cities of the Republic as Cork, Galway, and Limerick. The society no longer holds the field-weeks that featured prominently in its program in earlier days, but field excursions continue to be organized during the summer. The society's journal *Irish Geography* (founded 1944) now appears in two issues each year and is devoted to the publication of geographical research relating to Ireland. The society's library is housed within the Freeman Library of the Department of Geography in Trinity College Dublin. The membership stands at 220 and consists largely of university geographers and schoolteachers.

BIBLIOGRAPHY

Herries Davies, G. L., ed., *Irish Geography: The Geographical Society of Ireland Golden Jubilee 1934-1984* (1984).

• *G. L. Herries Davies*

Isachsen, Fridtjov Eide (1906-1979)

M.A., University of Oslo, 1929. University of Oslo, 1931-76. *Norge Vart Land* (1937); *Verdens geografi* (2 vols., 1939); *Bosteder og arbeidssteder i Oslo* (with T. Sund) (1942). H. Myklebost in *GBS*, 10 (1986), 85-92; obits. by L. Hertsberg in *Svensk Geografisk Årsbok*, 55 (1979), 100-102; and H. Myklebost in *Norsk Geografisk Tidsskrift*, 34 (1980), 1-8.

Isard, Walter (b. 1919)

Ph.D., Harvard University, 1943. Harvard University, 1949-53; Massachusetts Institute of Technology, 1953-56; University of Pennsylvania, 1956-79; Cornell University, 1979-89. *Location and Space-Economy* (1956); *Methods of Regional Analysis* (1960); *Spatial Dynamics and Optimal Space-Time Development* (1979). *WWA*; *GOF* (1985).

Israel

The geography of Palestine, or Eretz-Israel, a land held sacred by the three major monotheistic religions, has been studied since ancient times. But geography as an academic discipline is comparatively young in this country. It started in 1948, after the establishment of the State of Israel, with the opening of the first Department of Geography, headed by Professor David H.K. Amiran, at the Hebrew University of Jerusalem.

There is an abundant literature by nineteenth-century western explorers and travellers on the geography of the Holy Land. In the late nineteenth century and early twentieth century, the new Jewish population becoming established in Eretz-Israel produced its own scholars, who published the results of their many researches in the various branches of knowledge concerning this land. Such studies became more numerous during the British Mandate (1918–48), with the great development of historical and archaeological study of Eretz-Israel on one hand and that of the earth sciences on the other.

At first, the Department of Geography at the Hebrew University emphasized regional geographical research, the impact of physical conditions on developments in settlement, and the practical applicability of geographical research. This kind of research was based on studies in physical geography, geomorphology and climatology, together with a solid cartographic foundation, including field work. The department was part of the Faculty of Natural Sciences. In time, the major emphasis, both in research and curriculum, shifted to human and social geography, and the department became part of the Faculty of Social Sciences, maintaining a branch in the Natural Sciences Faculty within the Institute of Earth Sciences, while some of its academic staff became active in the Faculty of the Humanities as teachers and researchers.

The Department of Geography at the Hebrew University of Jerusalem has laid the foundations for the development of geography in Israel. A large proportion of the academic staff of the existing departments of geography at the five universities of Israel, as well as teachers in various colleges and high schools in Israel, are graduates of the department in Jerusalem. Some of the academic staff have graduated from universities abroad, particularly American and English universities.

During the sixties and seventies geography in Israel was greatly influenced by developments originating in the West. Systematic geography became the main subject of study. Quantitative methods came into use, while planning and urban problems became major issues. In the eighties, the curriculum was extended to include social issues, such as urban and economic geography, humanistic issues and ecological problems. The study of Eretz-Israel has always occupied a central place, whether as subject matter by its own merit or as a laboratory for studying systematic geographic subjects in general.

There are now in Israel departments of geography in five universities: the Hebrew University of Jerusalem, Tel-Aviv University, the Haifa University, Ben-Gurion University of the Negev at Beer-Sheva, and Bar-Ilan University in Ramat-Gan near Tel-Aviv (a religious university). The academic staff in these five departments numbers around seventy; there are about 1,000 B.A. students, 100 M.A. students, and 25 Ph.D. students.

The majority of M.A. students and most Ph.D. students study in Jerusalem. The greatest part of the B.A. graduates are employed in teaching or in the service of the government and other public institutions. A considerable number of M.A. graduates are part of the planning staffs of governmental departments and municipal and regional councils. In 1960 the Israel Geographical Association was founded. By 1989 it counted some 350 members. The association holds an annual meeting, usually in December.

Bibliography

Atlas of Israel (3rd ed., 1985).

Atlas of Jerusalem (1973).

Geography Research Forum (journal published by the Department of Geography, Ben-Gurion University of the Negev, Beer-Sheva).

Horizons, Studies in Geography (*Okafim*) (journal in Hebrew issued irregularly by the Department of Geography, University of Haifa, since 1975).

Studies in the Geography of Israel (*Mehkarim*) (serial publication in Hebrew issued irregularly since 1959 by the Department of Geography, Hebrew

University of Jerusalem, and the Israel Exploration Society).

Waterman, S., "Not Just Milk and Honey—Now a Way of Life: Israeli Human Geography since Six Day War," *PHG*, 9 (1985), 194–234.

• *Yehoshua Ben-Arieh*

Italian Geographical Society (Società Geografica Italiana)

Founded in Florence in 1867 with 163 original members, the Italian Geographical Society is one of the oldest active geographical societies in the world. Begun in a period of great interest in exploration and colonization, the society placed particular emphasis on studies of the regions of Africa that the Italians were colonizing, in addition to studies of the home territory.

The society's scientific and cultural activities are manifest in its publications: the *Bollettino della Società Geografica Italiana*, published without interruption since 1868 and containing articles by the most important Italian geographers; the *Memorie della Società Geografica Italiana*, a series of 42 monographs; and the *Bibliografia geografica della regione italiana*, an annual survey of geographical publications relating to Italy.

Apart from its publishing ventures, the society also concerns itself with the organization of national and international congresses, conferences, excursions, and other scientific initiatives for spreading geographical knowledge and methods.

In its seat, a Renaissance villa in the center of Rome, the society boasts a library of more than 250,000 volumes and a substantial collection of maps, periodicals, and manuscripts. Of great importance for the history of Italian exploration are the archives of the society, the photo archives, and the museum.

The society maintains close relations with the National Research Council and other Italian governmental and public institutions and is now preparing for the celebration of the fifth centenary of the discovery of America, to be held in 1992. For the coordination of its scientific initiatives, the society is now being equipped with a modern computerized geographical and cartographic information center.

Bibliography

Taberini, A., and C. Cerreti, *Società Geografica Italiana* (1988).

• *Graziella Galliano*

Italy

University teaching of geography in Italy started in 1859, and by the end of the century schools of geography had been formed in all major universities. The most authoritative geographer of that period was Giuseppe Dalla Vedova (1834–1919), generally considered as the father of Italian geography, since his influence was similar to that of Ritter in Germany, to whom he could be considered ideologically related.

A deep influence was also exerted by Giovanni Marinelli (1846–1900), who introduced the positivistic perspective in geographical research. His son Olinto (1872–1926) was undoubtedly the most outstanding of the numerous young scholars at the beginning of this century. He was very young when he got a university chair, and although he did not live very long, he made a lasting impression on geographical research in Italy. Unlike the scholars who preceded him, Olinto Marinelli was not a theorist but developed new practical models of research. He conceived geography as a whole, as a synthetic science, whose object was the study of the distribution of physical phenomena on the earth's surface and the formulation of general laws.

In the 1920s and '30s, Italian geography felt the effects of fascist ideology, to which the studies on colonial geography and geopolitics of the time are to be related. In order to maintain some independence from the directives of the regime, some Italian geographers, such as Roberto Almagià, Giuseppe Caraci, and Alberto Magnaghi, among others, preferred to concentrate on subjects that did not so easily lend themselves to political exploitation, such as historical geography and the history of exploration. Human geography, on the other hand, unless it confined itself to works of merely descriptive or popular character, often had to tailor its research to the requirements of fascist propaganda—*e.g.*, studies of the depopulation of mountain areas, rural settlements, and, above all, land reclamation.

From 1950 onward, Italian geographers resumed lively contacts with geographers in other European countries. The teachings of Paul Vidal de la Blache had a large audience in Italy. Therefore, in spite of the efforts of the older geographers such as Roberto Almagià, Renato Biasutti, and Aldo Sestini to conserve the character of geography as a unitary science, the prevailing trends favored human and regional geography.

In the 1960s, as a consequence of the deep renewal in all cultural fields and the swift transformations of social and economic structures, Italian geographers directed their interests more towards economic studies. Some of the outstanding economic geographers of that period were Dino Gribaudi, Ferdinando Milone, and Umberto Toschi. By the beginning of the 1970s, the number of chairs of geography increased, and a greater degree of differentiation and specialization was introduced into Italian geography. After Lucio Gambi's criticism of the traditional attitude, the uncertainty of the conception of geography as either a natural or a social science was resolved by the variety of new methodologies and research fields. The new ideas that are being introduced and widely accepted no longer imply a hierarchy of research subjects but their coexistence on the same level of importance as well as their interrelation. Besides the traditional human geography, which has been affected by the influence of Benedetto Croce's historicism, quantitative geography and functionalism have become prominent. Lively theoretical debate alternates with practical and applied research. Important work is being done in the new branches of geographical research dealing with ecology and perception.

The development of Italian geography has been aided in substantial ways by the activities of the geographical societies, among which special mention should be given to the Italian Geographical Society (Società Geografica Italiana), founded in 1867 on the initiative of a group of politicians led by Carlo Cattaneo, and the Society for Geographical Studies (Società di Studi Geografici), as well as the publication of periodicals, such as the *Bollettino della Società Geografica Italiana* (since 1868), *Rivista Geografica Italiana* (since 1893), and *Universo* (since 1920). The Italian Geographical Congresses (Congressi Geografici Italiani) should also be mentioned; twenty-five conventions have taken place since 1892. Since 1977 Italian geographers have had a professional organization, the Association of Italian Geographers (Associazione dei Geografi Italiani [AGEI]).

Geography is quite well represented in Italian universities, although it is dispersed in various departments or faculties. The faculties with the largest numbers of chairs of geography are the faculties of arts, philosophy, economics, political science, and, in physical geography, the faculties of mathematical, natural, and physical sciences. The Italian university geographers now number about three hundred.

Bibliography

Baldacci, O., *Correnti del pensiero geografico contemporaneo* (1972).

Caldo, C., *Il territorio come dominio: la geografia italiana durante il fascismo* (1982).

Caraci, I. L., *La geografia italiana tra '800 e '900 (dall'unità a Olinto Marinelli)* (1982).

Celant, A., and A. Vallega, eds., *Il pensiero geografico in Italia* (1984).

Corna Pellegrini, G., and C. Brusa, eds., *La ricerca geografica in Italia, 1960–1980* (1980) (also published in a condensed, English version).

Società Geografica Italiana, *Un sessantennio di ricerca geografica italiana* (1964).

• *Ilaria Luzzana Caraci*

Jaatinen, Stig (b. 1918)

Ph.D., University of Helsinki, 1950. Helsinki School of Economics, 1959–68; University of Helsinki, 1951–57, 1968–82. *The Human Geography of the Outer Hebrides* (1957); (ed.) *Atlas över Skärgårds-Finland* (1969); "Geographical Research in the Archipelagoes of Finland," *Fennia*, 162 (1984), 81–101; *Ålands Kulturlandskap—1700-talet* (1989). *Kuka kukin on* (Who's Who in Finland); S. Jaatinen, "The Marchland Theme: A Landmark in a Geographer's Life," pp. 112–129 in *Geographers of Norden* (ed. by T. Hägerstrand and A. Buttimer) (1988).

Jackson, John Brinckerhoff (b. 1909)

B.A., Harvard University, 1932. Owner and editor of *Landscape* magazine, 1951–68; University of California, Berkeley, 1967–77; Harvard University, 1969–77. *Landscapes* (ed. by E. Zube) (1970); *American Space, the Centennial Years, 1865–1876* (1972); *The Necessity for Ruins* (1980); *Discovering the Vernacular Landscape* (1984). D. Meinig (ed.), *The Interpretation of Ordinary Landscapes* (1979), 210–244.

James, Preston Everett (1899–1986)

Ph.D., Clark University, 1923. University of Michigan, 1923–45; Syracuse University, 1945–70. *An Outline of Geography* (1935); *Latin America* (1942, 5th ed. 1985); (ed.) *American Geography: Inventory and Prospect* (with C. Jones) (1954); *All Possible Worlds: A History of Geographical Ideas* (1972, 2nd ed. with G. Martin, 1981). *GOF* (1970, 1979, 1984); G. Martin in *GBS*, 11 (1987), 63–70; D. Robinson, "On Preston E. James and Latin America: A Biographical Sketch," pp. 1–101 in *Studying Latin America: Essays in Honor of Preston E. James* (ed. by D. Robinson) (1980); obits. by G. Martin in *AAAG*, 78 (1988), 164–175; and R. Jensen in *JG*, 85 (1986), 273–274.

Japan

As in other old civilizations, geography has existed in Japan from early times, whether in the form of knowledge of the earth's surface or in the sense of consideration of spatial configurations. Since the beginning of the nineteenth century, after the introduction of Western geography, either through material obtained directly at the source or through Chinese writings, the term *chiri* (literally, the "logic" or "pattern" of the earth's surface) had begun to be used and its usage became common after the Meiji Restoration, especially after the establishment of the compulsory educational system in 1872. Geography came to be made much of as workable material for the encouragement of nationalism and, especially, as an effective means of driving the Japanese people to further modernization efforts by contrasting the situation in prosperous Western countries with the abject conditions in Asian countries that had fallen under Western domination. The necessity for the formation of geography teachers resulted in the establishment of geography institutions of higher education and the publication of geographical textbooks of advanced level.

The authors of important geographical writings appearing between the 1880s and the 1920s, such as Kanzo Uchimura (1861–1930), Inazo Nitobe (1862–1933), Shigetaka Shiga (1863–1927), Tsunesaburo Makiguchi (1871–1944), and Michitoshi Odauchi (1875–1954), were broadly trained in the agricultural sciences but also dealt with the political and social themes. While the practitioners themselves were Meiji-period nationalists, they nonetheless came to adopt a critical stance with regard to the official policies of imperialist Japan after World War I, since the brand of nationalism that they professed differed from the dominant ideological trends of that time. While the geographical works they published exercised considerable influence, they were not academic geographers in the strict sense of the term.

The establishment of chairs of geography at higher normal schools took place in 1895, and the chairs of geography at imperial universities were created in 1907 in Kyoto with the chairmanship of Takuji Ogawa (1870–1941) and in 1911 in Tokyo with the chairmanship of Naomasa Yamasaki (1870–1929). These two figures, both with backgrounds in geology, exercised strong influence in academic circles, together with Goro Ishibashi (1877–1946) and Taro Tsujimura (1890–1983), their respective successors at Kyoto and Tokyo. In contrast to the school of geography of the Imperial University of Tokyo, which was oriented toward physical geography and quantitative studies, the geographical school of the Imperial University of Kyoto was human-geography oriented. Besides these, at the Tokyo Bunrika University (the present University of Tsukuba) a newly formed department of geography became very active in studies in regional geography under the leadership of Keiji Tanaka (1885–1975) and Kanzo Uchida (1888–1969).

During the ultranationalist and militarist period in Japan, most of the academic geographers tried to retreat to the ivory tower and to maintain a politically neutral or value-free position. But when the geopolitical movement gained currency during the second half of the 1930s and the first half of the 1940s, a certain number of geographers were actively involved, especially Saneshige Komaki (1898–), who was well known also for his work in prehistoric geography and for his geographical school of Kyoto.

The situation of Japanese geography after World War II was characterized by (a) an increase in the number of geographers working in higher education and research establishments, consequent upon the institutional reforms, and (b) increased facilities for communication and scientific intercourse on both the domestic and international levels, which have made for fewer differences among the various geographical schools in Japan than is the case with geographical studies in other countries. For instance, the "quantitative revolution" that broke out in the second half of the 1950s and the neopositivist paradigm are accepted by the majority of Japanese geographers. "Radical-humanist" perspectives have come into vogue, and a certain number of geographers continue their tentative researches of these orientations. In contrast to the firm establishment of geography in schools and academies, which constitutes an institutional safeguard of Japanese geography, geography has not yet obtained a strong position in the fields of application such as planning and marketing.

Bibliography

Isida, R., *Nihon ni okeru kindaichirigaku no seiritu* (1984).

Komaki, S., *Senshichirigaku kenkyu* (1937).

Makiguchi, T., *Jinsei chirigaku* (1903).

Takeuchi, K., "Japan," pp. 235–263 in *Geography*

since the Second World War, ed. by R. J. Johnston and P. Claval (1984).

Takeuchi, K., and H. Nozawa, "Recent Trends in Studies on the History of Geographical Thought in Japan—Mainly on the History of Japanese Geographical Thought," *Geographical Review of Japan*, 61 (1988), 59-73.

Tsujimura, T., *Chikeigaku* (1923).

• *Keiichi Takeuchi*

Jefferson, Mark Sylvester William (1863-1949)

M.A., Harvard University, 1898. Eastern Michigan University, 1901-39. *Recent Colonization of Chile* (1921); *The Rainfall of Chile* (1921); *Peopling the Argentine Pampa* (1926); "The Law of the Primate City," *GR*, 29 (1939), 226-232 (reprinted in *GR*, 79 [1989], 226-232). G. Martin, *Mark Jefferson: Geographer* (1968); obits. by I. Bowman in *GR*, 40 (1950), 134-137; and S. Visher in *AAAG*, 39 (1949), 307-312.

Jobberns, George (1895-1974)

D.Sc., University of New Zealand, 1936. University of Canterbury, 1937-60. "The Raised Beaches of the North East Coast of the South Island of New Zealand," *Transactions of the New Zealand Institute*, 59 (1928), 508-570; "Geography and National Development," *New Zealand Geographer*, 1 (1945), 5-18; "Geography," pp. 215-220 in *Science in New Zealand* (ed. by F. Callaghan) (1957); "Of Many Things," *New Zealand Geographer*, 15 (1959), 1-17. W. Johnston in *GBS*, 5 (1981), 73-76; M. McCaskill (ed.), *Land and Livelihood: Geographical Essays in Honour of George Jobberns* (1962); obit. by K. Cumberland in *New Zealand Geographer*, 31 (1975), 1-5.

Johnson, Douglas Wilson (1878-1944)

Ph.D., Columbia University, 1903. Massachusetts Institute of Technology, 1903-07; Harvard University, 1906-12; Columbia University, 1912-44. (ed.) *Geographical Essays* by W. M. Davis (1909); *Topography and Strategy in the War* (1917); *Shoreline Processes and Shoreline Development* (1919); *Battlefields of the World War* (1921). R. Chorley in *DSB*, 7 (1973), 143-145; W. Bucher, "Biographical Memoir of Douglas Wilson Johnson," National Academy of Sciences, *Biographical Memoirs*, 23 (1947); obits. by J. Wright in *GR*, 34 (1944), 317-318; and A. Lobeck in *AAAG*, 34 (1944), 216-222.

Johnston, Ronald John (b. 1941)

Ph.D., Monash University, 1967. Monash University, 1964-66; University of Canterbury, 1967-74; University of Sheffield, 1974 to the present. *Geography and Geographers: Anglo-American Human Geography since 1945* (1979, 3rd ed. 1987); (ed.) *Dictionary of Human Geography* (with others) (1981, 2nd ed. 1986); *On Human Geography* (1986); (ed.) *A World in Crisis* (with P. Taylor) (1985, 1989). *GOF* (1982).

Jones, Emrys (b. 1920)

Ph.D., University College Wales, Aberystwyth, 1947. University College London, 1947-50; Queen's University of Belfast, 1950-59; London School of Economics, 1959-84. *Social Geography of Belfast* (1960); *Towns and Cities* (1965); *An Introduction to Social Geography* (with J. Eyles) (1977); *Metropolis* (1990). *WW*; *Who's Who in Western Europe*.

Jones, Llewellyn Rodwell (1881-1947)

Ph.D., University of London, 1925. London School of Economics, 1919-45. *North England: An Economic Geography* (1921); *North America* (with P. Bryan) (1924); *The Geography of London River* (1931). M. Wise in *GBS*, 4 (1980), 49-53; obits. by S. Wooldridge in *GJ*, 110 (1947), 258; and *Geography*, 32 (1947), 138.

Jordan, Terry Gilbert (b. 1938)

Ph.D., University of Wisconsin, 1965. Arizona State University, 1965-69; University of North Texas, 1969-82; University of Texas, 1982 to the present. *Texas Graveyards* (1982); *American Log Buildings* (1985); *The European Culture Area* (1973, 2nd ed. 1988); *The American Backwoods Frontier* (with M. Kaups) (1989). *WWA*; *GOF* (1984).

Juan Sebastián Elcano Institute of Geography (Instituto de Geografía Juan Sebastián Elcano, Madrid, Spain)

The Juan Sebastián Elcano Institute of Geography was founded in Madrid in 1940 as part of the Consejo Superior de Investigaciones Científicas. Its directors have been successively Eloy Bullón, Amando Melón, Manuel de Terán, and Antonio López Gómez. From 1944, sections of the institute also existed at Zaragoza (directed by José Manuel Casas Torres) and Barcelona (under Luis Solé Sabarís). In the years 1940 to 1970 the institute played a fundamental role in the training of research geographers, a good number of whom subsequently occupied university chairs of geography and thus contributed importantly to the development of present-day Spanish geography. The work of this center is diffused through monographs and the journal *Estudios Geográficos* (from 1940, four numbers a year), which in fact is the principal journal of Spanish geography. At Zaragoza there was also published the review *Geographica* (from 1954). At the end of the 1960s, with the transfer of J. M. Casas Torres to the chair in the University of Madrid, there was also created in the CSIC the Institute of Applied Geography, which from 1971 published *Geographica*, second series. In 1988, with the restructuring of the CSIC, the two institutes were merged with those of economics as the Institute of Economics and Applied Geography. This institute continues to publish the journal *Estudios Geográficos*, the present editor being Antonio López Gómez.

• *Horacio Capel Sáez*
(translated by C. Julian Bishko)

Kant, Edgar (1902–1978)

Doctorate, University of Tartu, 1934. University of Tartu, 1928–44; University of Lund, 1945–67. *Tartu* (1926); (ed.) *Eesti Atlas* (1938–40); *Suburbanization, Urban Sprawl, and Communication* (1957); *Zur Frage der inneren Gliederung der Stadt* (1962). A. Buttimer in *GBS*, 11 (1987), 71–82; obit. by T. Hägerstrand in *Svensk Geografisk Årsbok*, 54 (1978), 96–101.

Kates, Robert William (b. 1929)

Ph.D., University of Chicago, 1962. Clark University, 1962–86; Brown University, 1986 to the present. *The Environment as Hazard* (with I. Burton and G. White) (1978); (ed.) *Climate Impact Assessment* (with J. Ausabel and M. Berberian) (1985); *Perilous Progress* (with C. Hohenemser) (1985); (ed.) *Hunger in History* (with L. Newman, W. Crossgrove, R. Matthews, and S. Millman) (1989). *WWA*.

Keltie, John Scott (1840–1927)

Student at the University of Edinburgh, 1860–67 *passim*. Royal Geographical Society, 1884–1917. *Geographical Education* (1886); *Applied Geography* (1890); *The Partition of Africa* (1893); *History of Geography* (with O. Howarth) (1913). L. Jay in *GBS*, 10 (1986), 93–98; obits. by G. Chisholm in *SGM*, 43 (1927), 102–105; and H. Mill and D. Freshfield in *GJ*, 69 (1927), 281–287.

Kenya. *See* East Africa.

Keuning, Hendrik Jacob (1904–1987)

Doctorate, University of Utrecht, 1933. University of Groningen, 1948–74. *De Groninger Veenkoloniën: Een sociaal-geografische studie* (1933); *Mozaïek der functies: Proeve van een regionale landbeschrijving van Nederland* (1955); "Activities in the Field of Economic Regionalization in the Netherlands" (with A.C. de Vooys), *Geographica Polonica*, 4 (1964), 107–116; "Standort der Sozialgeographie," pp. 91–98 in *Zum Standort der Sozialgeographie: Wolfgang Hartke zum 60. Geburtstag* (1968). W. W. de Jong, "Zeventig Jaren H. J. K. [Hendrik Jacob Keuning]," pp. 7–22 in *Te Keur voor Keuning* (1974); M. de Smidt, "Dutch Economic Geography in Retrospect," *TESG*, 74 (1983), 344–357; H. van Ginkel, "Nederlandse Geografie na 1950," pp. 255–283 in *Algemene Sociale Geografie: Ontwikkelingslijnen en Standpunten*, ed. by A. G. J. Dietvorst, J. A. van Ginkel, *et al.* (1984).

King, Leslie John (b. 1934)

Ph.D., University of Iowa, 1960. University of Canterbury, 1960-62; McGill University, 1962-64; Ohio State University, 1964-70; McMaster University, 1970 to the present. *Readings in Economic Geography* (with R. H. T. Smith and E. J. Taaffe) (1968); *Statistical Analysis in Geography* (1969); *Cities, Space, and Behavior* (with R. Golledge) (1978); *Central Place Theory* (1984). *WWA*; *CWW*; *GOF* (1973).

Kirchhoff, Alfred (1838-1907)

Doctorate, University of Bonn, 1861. University of Halle, 1873-1904. (ed.) *Landeskunde von Europa* (3 vols., 1886-93); *Allgemeine Erdkunde* (with others) (3 vols., 1896-99); *Mensch und Erde* (1901). E. Meynen in *GBS*, 4 (1980), 69-76; H. Steffen, "Alfred Kirchhoff," *GZ*, 25 (1919), 289-302; M. Schmidt, "Alfred Kirchhoff als akademischer Lehrer," *Geographischer Anzeiger*, 39 (1938), 217-224; obits. by W. Ule in *GZ*, 13 (1907), 537-552; and A. Supan in *PGM*, 53 (1907), 47-48.

Kish, George (1914-1989)

Ph.D., University of Michigan, 1945. University of Michigan, 1940-85. *Economic Atlas of the Soviet Union* (1960, 1972); *History of Cartography* (1973); *A Source Book in Geography* (1978); *La carte: image des civilisations* (1980). *WWA*; *GOF* (1976).

Kliewe, Heinz (b. 1918)

Ph.D., University of Greifswald, 1951; (habil.) University of Greifswald, 1958. University of Jena, 1960-69; University of Greifswald, 1949-60, 1969-81. *Die Insel Usedom* (1960); "Ein Jahrzehnt Quartärforschung am Jenaer Lehrstuhl für Physische Geographie," *Wissenschaftliche Zeitschrift der Universität Jena, Mathematisch-Naturwissenschaftliche Reihe*, 19 (1970), 885-896; "Spätglaziale Marginalzonen der Insel Rügen," *PGM*, 119 (1975), 261-269; "Die Ostseeküste zwischen Boltenhagen und Ahlbeck," *Geographische Bausteine*, 30 (1987).

Kniffen, Fred Bowerman (b. 1900)

Ph.D., University of California, 1930. Louisiana State University, 1929-70. "Louisiana House Types," *AAAG*, 26 (1936), 179-193; *Culture Worlds* (with R. Russell) (1951); "Folk Housing: Key to Diffusion," *AAAG*, 55 (1965), 549-577; *Louisiana, Its Land and People* (1968). *WWA*; *GOF* (1976); *Pioneer America* (July 1971 [Kniffen Festschrift], January 1973, and January 1975); *Man and Cultural Heritage* (Kniffen Festschrift) (ed. by H. Walker and W. Haag) (1974).

Kollmorgen, Walter Martin (b. 1907)

Ph.D., Columbia University, 1940. United States Department of Agriculture, 1938-41, 1943-46; University of Kansas, 1946-76. *The German-Swiss in Franklin County, Tennessee* (1940); *The Old Order Amish of Lancaster County, Pennsylvania* (1942); "Grazing Operations in the Flint Hills-Bluestem Pastures of Chase County, Kansas" (with D. Simonett), *AAAG*, 55 (1965), 260-290; "The Woodman's Assaults on the Domain of the Cattleman," *AAAG*, 59 (1969), 215-239. *GOF* (1970); W. Kollmorgen, "Kollmorgen as a Bureaucrat," *AAAG*, 69 (1979), 77-89; T. Smith (ed.), "Man and Environment: Walter Kollmorgen, the First Forty Years," University of Kansas, Department of Geography, *Occasional Paper*, 1 (1977).

Korea (South)

Modern geography in Korea began with a Silhak scholar named Yi Chung-Hwan (1690-1756) in the Yi Dynasty. He was influenced by Western science and technology by way of China. Yi wrote *Taekni-chi*, which dealt with systematic descriptions and regional accounts of each province of Korea. He made extensive field observations throughout Korea to provide the basic materials for his book.

During the Yi Dynasty, geography was not a part of Confucian scholarship, but was included in the civil service examination. In 1894, the civil service examination was abolished and a modern school system was introduced in Korea.

The geography of Korea and world geography began to be taught in primary and secondary schools, and to a limited extent in higher education. World geography, in particular, was an important subject in the school curriculum because of its presumed ability to broaden the world view of students. During the Japanese occupation (1910–45) geography was taught as a compulsory subject in primary and secondary schools and in the institutions for the training of secondary school teachers.

Geography departments at the university level appeared only after World War II. The first geography department was established in the College of Education in Seoul National University in 1946. In 1958 geography departments were established in the Liberal Arts and Science College of Seoul National University and in Kyonghee University. Since that time the number of geography departments has increased rapidly, from ten in 1970 to twenty-three in 1980 and to thirty in 1990. There are now eight departments offering Ph.D. work in geography and twenty-two departments of geographic education. These departments are all of small size, however; most of them have only three or four members of staff, and none has more than ten full-time members.

Because demand for geographers in teaching is high, but very limited for geographers in non-teaching positions, the curriculum is oriented toward teaching rather than research. Among the leading geographers in Korea since 1960 have been Sang-Ho Kim and Hyuck-Jae Kwon in physical geography, Chi-Ho Lee in population geography, Chan Lee in cultural and historical geography, and Kie-Joo Hyong and Chan-Ki Suh in economic geography. The departments of geography in Seoul National University and Kyungpook National University are the two most influential centers of geographic activity in Korea, because of their higher standards and large number of graduates. More than two-thirds of the geographers in higher education are graduates of those universities, Seoul National University alone supplying more than half of the total.

The Korean Geographical Society is the national professional organization for geographers and has published a journal, *Chiri Hak*, twice a year since 1963. The society has about 500 members, of whom one-third are college and university teachers and members of research institutions. Geography is still strong at the elementary and secondary levels. However, the demand for geography teachers at those levels is gradually decreasing, due in part to decreased enrollments resulting from the rapid drop in the national birth rate. The geography departments in colleges and universities have increased in number but not in size.

According to a survey made by the Korean Geographical Society in 1985, Korean geographers have the following major interests (in descending order of popularity): urban and population geography, economic geography (including industrial location and planning), physical geography, and the cultural and historical geography of Korea. Geographical research is rather a new phenomenon in Korea. The recent economic growth of the country has necessitated geographical research for national and regional planning of resources, land use, industrial location, land reclamation, and environmental problems. Computerized research is becoming common, and quantitative analysis is popular. Cultural and historical geography are growing gradually as a part of Korean studies (*i.e.*, studies of the homeland).

Bibliography

Chiri Hak, no. 13 (1976) (publication of the Korean Geographical Society symposium on thirty years of geography in Korea, 1945–1975).

Kim, Y.-S., "Hanguk Taehak Chirige Hakkwa Ui Songchang," *Chirihak Yongu* (The Geographical Journal of Korea), no. 14 (1989), 31–50.

Kwon, H.-J., "Chiri Hak," *Hanguk Hyondae Munhwasa Taege* (1976), 191–235.

Lee, C., "Hanguk Chirihaksa," *Hanguk Munhwasa Taege* (1968), 681–734.

——— and D.H. Gordon, "Geography in the Republic of Korea," pp. 231–259 in *Geography in Asian Universities*, ed. by R. J. Fuchs and J. M. Street (1976).

Lee, K.-S., *A Geographical Bibliography for Korean Studies* (1982, 1988).

• *Chan Lee*

Kraus, Theodor (1894–1973)

Dr.phil., University of Cologne, 1924. University of Würzburg, 1948–50; University of Cologne, 1924-48, 1950-62. *Die Eisenbahnen in dem Grenzgebieten von Mittel- und Osteuropa* (1924); *Das Siegerland* (1931); *Individuelle Länderkunde und räumliche Ordnung* (1960). E. Meynen in *GBS*, 11 (1987), 83-87; obit. by E. Otremba in *GZ*, 62 (1974), 1-11.

Krebs, Norbert (1876-1947)

Dr.phil., University of Vienna, 1903; (hab.) University of Vienna, 1907. University of Vienna, 1909-17; University of Würzburg, 1917-18; University of Frankfurt, 1918-20; University of Freiburg, 1920-27; University of Berlin, 1927-43. *Länderkunde der österreichischen Alpen* (1913); *Süddeutschland* (1923); *Die Ostalpen und das heutige Österreich* (1928); *Vergleichende Länderkunde* (1951). *Österreichisches Biographisches Lexikon 1815-1950*, no. 18 (1968), 240-242; obits. by H. Hassinger in *Erdkunde*, 2 (1948), 200-202; and H. Slanar in *MOGG*, 92 (1950), 81-85.

Krümmel, Johann Gottfried Otto (1854-1912)

Dr.phil., University of Göttingen, 1876. University of Göttingen, 1878-83; University of Kiel, 1883-1911. *Europäische Staatenkunde* (1880); *Der Ozean* (1886); *Handbuch der Ozeanographie* (1907). J. Ulrich in *GBS*, 10 (1986), 99-104; obits. by M. Eckert in *GZ*, 19 (1913), 545-554; and W. Meinardus in *PGM*, 58 (1912), 281.

Landscape

During the 1980s "landscape" became a key word in Anglo-American human geography, probably more so than at any time since the 1930s. Now, not only are there more empirical studies of landscape, both on the ground and as represented in visual and textual media, but also, and often implicated in such studies, there is a focus on the word itself, its history and ideology. This interest may be seen as part of a new cultural turn in geography. It is manifest not only in the renewal of geography's links with the humanities but also with the broadening of notions of economic and social process, which incorporate a fuller and more complicated sense of the lived world and its representation. In Britain the quickening of interest in landscape is felt across a range of academic disciplines, professions and pressure groups and, if the number of magazine titles and television programs is a reliable guide, among the laity too. The pressing political issues of heritage, nationalism and conservation are frequently articulated in terms of landscape. So were they in the inter-war years when geographers were active in debating them, and in earlier periods before geography was constituted as a discipline, especially the period of the Napoleonic wars. Thus it is not for antiquarian reasons that geographers now interrogate the discourse of landscape in these periods. Then, as now, landscape proved a highly charged, highly contested concept, one that promised to give shape and substance to an array of social, psychological, moral and aesthetic issues, and that threatened to reduce them to a set of pleasing but deceitful images.

We may now look back, perhaps regretfully, at the writing that for so long framed discussions of landscape in professional geography, the chapter on "'Landschaft' and 'Landscape'" in Richard Hartshorne's *The Nature of Geography* (1939). Hartshorne dissected the "multitudinous meanings" of landscape in the writings of German and American geographers, especially of the inter-war years, to demolish their claim for landscape as the definitive concept of the discipline. Out of what he saw as the confusion of the concept Hartshorne discriminated two definitions of landscape: (a) a narrow, particular, pictorial definition, the "field of vision" of the spectator, which was "of little importance in geography" other than perhaps as a preliminary to study proper, and (b) a more extensive, abstract definition, less serviceable to geography as a causal science than the idea of region, for it only considered the visible surface of the earth. "Geography existed long before it took over the term 'landscape,' and might get along very well without it."

It is instructive to go back to the programmatic statement on landscape that Hartshorne had particularly in his sights, Carl Sauer's *The Morphology of Landscape* (1925). Despite Hartshorne's attack, and Sauer's reluctance to develop its arguments, this essay has continued to sustain, if only totemically, the integrity of landscape among geographers, especially American geographers. Hartshorne reduced Sauer's concept of landscape to "area," and indeed Sauer himself makes such an equivalence early in the essay. But the concept of landscape shifts throughout the essay. Rather than offering an enclosed, categorical definition of landscape, developed through example and deduction, the essay conceptualizes landscape in a more emergent way, allowing it to cross-fertilize with an array of geographical concepts and procedures. The form of the essay, so apparently disjointed, empowers the concept of landscape by allowing it a good deal of discursive space, having it frame this, then that, form of knowledge.

Whatever the potential of Sauer's conceptualization of landscape for understanding the dramatic transformations that mark the development of twentieth century capitalism, Sauer and his followers actually developed it in a way that set it apart from such transformations, even in opposition to them. So the "aesthetic morphology of landscape" is for Sauer epitomized by "the rural scenes wherever simple folk have designed and placed their habitations," and the values of the "marketplace" are held to be those that disrupt the integrity of landscape. This tension between money and morality has been written into the discourse of landscape since its inception (in England, as far back as the seventeenth century), and it has been customary to resolve it conservatively, to construct landscape as essentially nostalgic, rustic and static. But there is an alternative resolution (sometimes lurking under the ideological cover of the landscape of nostalgia or written out by nostalgic historians) that constructs landscape in more progressive, dynamic terms. There is an explicit tradition of American writing and painting that does so. A modern representative of this tradition who has been claimed by professional geography is J. B. Jackson, and he is of interest here because he takes Sauer's themes of landscape and learning not to challenge or contain the processes of capitalist transformation but to articulate and celebrate them. In a series of essays in *Landscape*, the journal he founded, Jackson has focused on themes of energy, mobility, contingency and transiency on topics such as trailer parks, hot-rodding, and the highway strip, as well as on apparently more stable landscapes like small towns and farms.

We can return this discussion of the integrity of landscape to its anti-modern axis by comparing J. B. Jackson's conception of landscape with that of the English historian W. G. Hoskins (a comparison first made so fruitfully by D. W. Meinig). Hoskins has proved as fertile an influence on landscape studies in English geography since the 1950s as Jackson has on landscape studies in American geography. Learning through observation is central to both their conceptions of landscape, which is perhaps why both readily reach for the metaphor of "reading" the landscape to denote a knowledge that cannot be picked up at a glance. Like Jackson, Hoskins has always commanded a lay audience, in his case the army of English local historians. Indeed laity and locality are defining terms of Hoskins's conception of landscape, which has a clear ancestry in two centuries of English writing and painting. The narrative of his *The Making of the English Landscape* (1955) positions the integrity of England at its height around 1500, in a comely, detailed, balanced landscape of peasant proprietors and small gentry, before the grandiloquence of Georgian aristocratic power, or that of Victorian industrialists.

In Britain the association of landscape with aristocratic power has provoked a counter-tradition of writing on country life intent on seeing through the allure of landscape, exposing its ideology. Many recent works on the ideology of landscape have taken their cue from two Marxist cultural critics, Raymond Williams and John Berger, especially Williams's *The Country and the City* (1973) on English literature and Berger's *Ways of Seeing* (1972) on European painting. Williams and Berger have influenced a series of writings on the ideology of landscape in English literature, art, photography and design. In geography, the ideology of landscape thesis is most fully sustained in Denis Cosgrove's *Social Formation and Symbolic Landscape* (1984). The origins of landscape are here located in Renaissance Italian city states, and landscape defined as a "way of seeing" that, with a basis in linear perspective, had structural affinities with the "basic techniques of capitalist life" developed at the time, including bookkeeping, surveying and mapping. As a way of seeing, landscape might offer a complex, comprehensive, realistic view of the world, but this is ultimately an exclusive, even illusory view. Landscape is a "visual ideology" that, especially when deployed in rural areas, obscures not only the forces and relations of production but also more plebeian, apparently less pictorial, experiences of the world. "Landscapes can be deceptive," runs the epigraph to the book, a quotation from John Berger. "Sometimes a landscape seems to be less a setting for the life of its inhabitants than a curtain behind which their struggles, achievements and accidents take place."

BIBLIOGRAPHY

Cosgrove, D., and S. Daniels, eds., *The Iconography of Landscape* (1988).

Daniels, S., *Landscape and National Identity* (1990).

———, "Marxism, Culture and the Duplicity of Landscape," pp. 196-219 in *New Models in Geography* (ed. by R. Peet and N. Thrift) (1989).

Jackson, J. B., *Discovering the Vernacular Landscape* (1984).

Meinig, D. W., ed., *The Interpretation of Ordinary Landscapes* (1979).

Mikesell, M. W., "Landscape," pp. 575-580 of *IESS*, vol. 8 (1968).

• *Stephen J. Daniels*

Language, Geography of

The geography of language concerns the distribution of forms of speech or script that contiguous human populations must (if monolingual), or can, employ in communication. The geography of language may either simply ascertain the locational patterns of such linguistically distinct populations and map them, or also investigate the processes responsible for such patterns. Geographers tend to concentrate on larger continua of related usage, such as established literary vehicles and somewhat variable but widespread colloquial idiom, rather than the minute differentiation of particular forms studied by linguists who work on dialectology.

The study of processes of linguistic distribution overlaps broadly with that of migration and the diffusion of innovations, and of course more generally with historical geography. It has succeeded in demonstrating frequent traces of former political and ecclesiastical territories, or earlier economic connections, in surviving linguistic distributions. However, with a few notable exceptions, such discoveries stem mainly from dialectologists, or even sociolinguists, rather than geographers.

The social and political regionalization of human populations commonly reflects linguistic, and often religious, distinctions rather closely. Hence, the geography of language, especially as practiced under the rubric of "geolinguistics," converges with that of ethnic groups and, consequently, pertains importantly to political geography. Conversely, not only economic but also educational and informational policies of modern states exert a major influence on language use that, along with related questions of actual language policy, have attracted interest from geographers.

Since Albert Dauzat's pioneering work (1922), geographers have written a good deal on the topics mentioned, yet have apparently failed so far to develop fully their own distinctively spatial or environmental theories to account for the changing map of languages. And they still have to confront the topic of profounder effects that the language faculty in its many aspects, or linguistic diversity itself, may have on human comprehension of environments and on the human agency on earth. *See also* Toponymy.

BIBLIOGRAPHY

Dauzat, A., *La géographie linguistique* (1922).

Dominian, L., *Frontiers of Nationality and Language in Europe* (1917).

Trudgill, P., *On Dialect: Social and Geographical Perspectives* (1983).

Wagner, P. L., "Remarks on the Geography of Language," *GR*, 48 (1958), 86-97.

———, "The Geographical Significance of Language," pp. 51-61 in *Geolinguistic Perspectives*, ed. by J. Levitt, L. R. N. Ashley, and K. H. Rogers (1987).

Williams, C. H., ed., *Language in Geographic Context* (1988).

• *Philip L. Wagner*

Lautensach, Hermann (1886-1971)

Dr.phil., University of Berlin, 1910; (hab.) University of Giessen, 1928. University of

Giessen, 1928-34; University of Braunschweig, 1934-36; University of Greifswald, 1936-45; University of Stuttgart, 1945-54. *Allgemeine Geographie* (1926), *Portugal* (2 vols., 1932-37), *Korea* (1945), *Iberische Halbinsel* (1964). P. Tilley in *GBS*, 4 (1980), 91-101; H. Beck, "Hermann Lautensach—führender Geograph in zwei Epochen. Ein Weg zur Länderkunde," *Stuttgarter Geographische Studien*, 87 (1974), 1-30 (followed by Lautensach bibliog. on pp. 31-42); obits. by C. Troll in *Erdkunde*, 25 (1971), 161-163; and E. Plewe in *GZ*, 60 (1972), 1-7.

Lefèvre, Marguerite Alice (1894-1967)

D. de l'Univ., University of Paris, 1926. University of Louvain, 1929-64. *L'habitat rural en Belgique* (1926); *La Basse-Meuse* (1935); *Principes et problèmes de géographie humaine* (1946). J. Denis in *GBS*, 10 (1986), 105-110; L. G. Polspoel, "La carrière et l'activité scientifique de Mademoiselle M.A. Lefèvre, professeur à l'Université de Louvain," pp. 5-21 in *Volume Jubilaire offert à M. A. Lefèvre (Acta Geographica Lovaniensia*, vol. 3, 1964); obits. by G. Chabot in *AG*, 78 (1969), 472-473; and L. Polspoel in *Bulletin de la Société belge d'études géographiques*, 37 (1968), 27-31.

Lehmann, Edgar (b. 1905)

Dr.phil., University of Leipzig, 1933; (habil.) University of Leipzig, 1952. Bibliographical Institute of Leipzig, 1930-52; University of Leipzig, 1952-70. *Die Staaten der Erde und ihre Wirtschaft* (Atlas) (1952, 9th ed. 1969); "Carl Ritters kartographische Leistung," *Die Erde*, 90 (1959), 184-222; "Zur theoretischen Grundlegung des Begriffs Region," *Geographische Berichte*, 18 (1973), 41-48.

Leighly, John Barger (1895-1986)

Ph.D., University of California, Berkeley, 1927. University of California, Berkeley, 1927-60. "Some Comments on Contemporary Geographic Method," *AAAG*, 27 (1937), 125-141; "What Has Happened to Physical Geography?," *AAAG*, 45 (1955), 309-315; (ed.) *Land and Life: A Selection from the Writings of Carl Ortwin Sauer* (1963); (ed.) *The Physical Geography of the Sea* (by M. Maury) (1963). J. Parsons in *GBS*, 12 (1988), 113-119; *GOF* (1970); *60 Years of Berkeley Geography* (1983); J. Leighly, "Memory as Mirror," pp. 80-89 in *The Practice of Geography* (ed. by A. Buttimer) (1983); obit. by D. Miller in *AAAG*, 78 (1988), 347-357.

Le Lannou, Maurice (b. 1906)

D.-ès-L., University of Paris, 1942. University of Rennes, 1937-46; University of Lyon, 1947-68; Collège de France, 1969-76. *La géographie humaine* (1949); *Géographie de la Bretagne* (2 vols., 1950-52); *Le déménagement du territoire* (1967); *Europe, terre promise* (1977). *WWF.*

Leszczycki, Stanisław Marian (b. 1907)

Doctorate, University of Cracow, 1932; (hab.) University of Cracow, 1945. University of Cracow, 1928-39, 1945-47; University of Warsaw, 1948-70; Director of Institute of Geography, Polish Academy of Sciences, 1953-78; president of IGU, 1968-72. *Region Podhala* (1938); (ed.) *Narodowy Atlas Polski* (1970); *Geografia jako nauka i wiedza stosowana* (1975); *Nad mapą Polski* (1980). *International Who's Who.*

Lewis, Peirce Fee (b. 1927)

Ph.D., University of Michigan, 1958. Pennsylvania State University, 1958 to the present. "Linear Topography in the Southwestern Palouse, Washington-Oregon," *AAAG*, 50 (1960), 98-111; "Small Town in Pennsylvania," *AAAG*, 62 (1972), 323-351; "Common House, Cultural Spoor," *Landscape*, 19 (no. 2, 1975), 1-22; *New Orleans: The Making of an Urban Landscape* (1976). *GOF* (1976, 1985).

Library of Congress, Geography and Map Division (Washington, D.C.)

The *de facto* national library of the United States, analogous to the British Library in London and the Bibliothèque Nationale in Paris, is called the Library of Congress because it was planned to serve the legislative branch (the Congress) of the federal government. A Hall of Maps and Charts was one of several new units established when the main building of the Library of Congress was opened in 1897. Long known as the Map Division, it was renamed the Geography and Map Division in 1965. The division has custody of the world's largest collection of maps and related materials, including approximately 4,000,000 maps and charts, 52,000 atlases, 350 globes and globe gores, more than 2,000 relief models, and indexes to several million aerial photographs and remote sensing imagery (Landsat 1-4). The chiefs (administrative heads) of the division have been Philip Lee Phillips (1897-1924), Lawrence Martin (1924-46), Robert Platt (1946), Burton Adkinson (1947-49), Arch Gerlach (1950-67), Walter Ristow (1968-78), and John Wolter (1978 to the present).

BIBLIOGRAPHY

Wolter, J., A. Modelski, R. Stephenson, and D. Carrington, "A Brief History of the Library of Congress Geography and Map Division, 1897-1978," pp. 47-90 in *The Map Librarian in the Modern World: Essays in Honour of Walter W. Ristow* (1979).

• *John A. Wolter*

Lichtenberger, Elisabeth (b. 1925)

Dr.phil., University of Vienna, 1949; (hab.) University of Vienna, 1965. University of Vienna, 1972 to the present. *Wien: Bauliche Gestalt und Entwicklung* (with H. Bobek) (1966, 2nd ed. 1978); *The Eastern Alps* (1975); *Die Wiener Altstadt* (1977); *Stadtgeographie* (1985). M. Seger, "Elisabeth Lichtenberger—Forscherpersönlichkeit und wissenschaftlicher Lebensweg," *MOGG*, 127 (1985), 270-283.

Linton, David Leslie (1906-1971)

M.Sc., King's College London, 1930. University of Edinburgh, 1929-40; University of Sheffield, 1945-58; University of Birmingham, 1958-71. *Structure, Surface and Drainage in South East England* (with S. Wooldridge) (1939); *Discovery, Education and Research* (1946); *Air Photographs in Geographical Research* (1947); *The Tropical World* (1962). J. Gold, M. Haigh, and G. Warwick in *GBS*, 7 (1983), 75-83; obits. by R. Waters in *GJ*, 137 (1971), 432-433; and C. Embleton in *TIBG*, no. 55 (1971), 171-178.

Lisbon Geographical Society (Sociedade de Geografia de Lisboa)

The Lisbon Geographical Society was founded in 1875. Its main aims were developing research and spreading the teaching of the several branches of the geographical sciences, especially Portuguese colonial geography. Portuguese interests in Africa as well as in Asia and Oceania were promoted by the society through the organization of scientific expeditions, congresses, lectures, and the publication of a wide range of studies. A great deal of that work was carried out by the first two general secretaries, Luciano Cordeiro, who served from 1875 to 1900, and Ernesto de Vasconcelos (1900-30), as well as by various committees. Since then these committees have collated the efforts of the society's members: historians, anthropologists, biologists, and sociologists, as well as geographers. The *Boletim da Sociedade de Geografia de Lisboa* has been published since 1876. It is one of the means used by the society to maintain exchanges with scientific institutions in other countries. Also noteworthy are the society's large library and its remarkable museum, where valuable anthropological and ethnographic collections are to be found. They demonstrate that the society is one of the most important centers for the study of Portuguese communities worldwide.

BIBLIOGRAPHY

Catálogo-Amostra evocativo do primeiro centenário da Sociedade de Geografia de Lisboa (1977).

Moreno, M., *Sociedade de Geografia de Lisboa—75 anos de actividades ao serviço da ciência e da Nação* (1950).

d'A. Torres, R., "Sociedade de Geografia de Lisboa," in *Dicionário de História de Portugal*, ed. by Joel Serrão, vol. 4 (1971).

• *João Carlos Garcia*

Lobeck, Armin Kohl (1886-1958)

Ph.D., Columbia University, 1917. University of Wisconsin, 1919-29; Columbia University, 1929-54. *Block Diagrams* (1924); *Geomorphology* (1939); *Military Maps and Photographs* (with W. Tellington) (1944); *Things Maps Don't Tell Us* (1956). *WWWA*; obit. by H. Sharp in *GR*, 48 (1958), 584-585.

Location Theory

Location theory is the central repository of concept in economic geography. It serves two functions. As a tool of historical analysis, it seeks to explain why economic activities have succeeded in certain locations and not in others, producing distinctive patterns of world economic geography. In the service of business and government, it is a tool for predicting the most desirable location for enterprises that have yet to be established.

The theory has evolved alongside the development of modern economies from domination by agriculture through the period of industrialization to contemporary service orientation and "high tech" growth. Thus, in 1826, Johan Heinrich von Thünen, a German farmer, developed a theory to explain the location of agricultural cropping patterns with respect to market towns. He showed that quite independently of variations in such resource factors as climate and soils, different agricultural activities would form concentric zones around towns as a result of the interaction of transport costs and land rents. His theory was one of the influences of rent-paying ability on locational choice.

The tradition that Thünen established in seeking the economic factors that determine agricultural locations was complemented at the end of the nineteenth century by another German, economist Alfred Weber, who formulated a theory of industrial location. Assuming a competitive environment, the formula for success was, he argued, to minimize production costs. His theory centered on the components of production costs: transport of raw materials to factory, labor costs in converting raw materials to finished products, and transport of products to market. Each component of cost produced a distinctive locational pattern for the economic activities most influenced by that component.

Next, in the 1930s and 1940s, two more Germans, geographer Walter Christaller and economist August Lösch, developed a theory of retail location, while yet other scholars picked up on Weber's suggestions that there were two kinds of cost advantages arising out of industrial agglomeration: localization economies resulting from the clustering of specialists in the same industry, as in the New York garment industry; and urbanization economies that emerge when diverse activities located in large cities support larger-scale and higher-quality services. Swedish economist Gunnar Myrdal added a theory of regional economic structure, positing a process of "circular and cumulative causation" whereby industrial heartlands have grown more rapidly than raw-material-producing hinterlands.

Since World War II, the most active arenas of theoretical development have been those dealing with less-than-competitive situations where demand and prices may be managed, with problems of uncertainty and risk, and with the growth of the "quaternary" sectors of the economy, involving knowledge-based production and services. In addition, the introduction of modern computers has enabled the optimization techniques of operations research to be applied to the least-cost and profit-maximizing location problems. The result is a research field that retains its dynamism as innovation propels economic development in uncharted directions. In the 1950s and 1960s, economist Walter Isard used location theory as the cornerstone of his attempt to develop a new discipline of Regional Science. His initiatives helped propel economic geography from a largely descriptive field into its contemporary theory-rich scientific form. *See also* Central-Place Theory.

BIBLIOGRAPHY

Christaller, W., *Die zentralen Orte in Süddeutschland* (1933) (trans. by C.W. Baskin as *Central Places in Southern Germany*, 1966).

Isard, W., *Location and Space-Economy* (1956).

Lösch, A., *Die räumliche Ordnung der Wirtschaft* (1944) (trans. by W. Woglom and W. P. Stolper as *The Economics of Location*, 1954).

Ponsard, C., *Histoire des théories économiques spatiales* (1958) (trans. and rev. ed., *History of Spatial Economic Theory*, 1983).

von Thünen, J. H., *Der isolierte Staat* (1826) (trans. by C. M. Wartenberg and ed. by P. Hall as *Isolated State*, 1966).

Weber, A., *Uber den Standort der Industrien* (1909) (trans. by C. J. Friedrich as *Alfred Weber's Theory of the Location of Industries*, 1929).

• *Brian J. L. Berry*

Lowenthal, David (b. 1923)

Ph.D., University of Wisconsin, 1953. Vassar College, 1953–56; American Geographical Society, 1956–72; University College London, 1972–85. *George Perkins Marsh: Versatile Vermonter* (1958); "Geography, Experience, and Imagination," *AAAG*, 51 (1961), 241–260; *West Indian Societies* (1972); *The Past Is a Foreign Country* (1985). *GOF* (1983).

Mabogunje, Akinlawon Lapido (b. 1931)

Ph.D., University of London, 1961. University of Ibadan, 1958-81; President of IGU, 1980-84. *Urbanization in Nigeria (1968); Regional Mobility and Resource Development in West Africa* (1972); (ed.) *Regional Planning and National Development in Tropical Africa* (with A. Faniran) (1977); *The Development Process: A Spatial Perspective* (1981, 2nd ed. 1989). *Who's Who in the World; Africa Who's Who.*

Machatschek, Fritz (1876-1957)

Doctorate, University of Vienna, 1899; (hab.) University of Vienna, 1906. German University of Prague, 1914-24; Swiss Federal Institute of Technology, 1924-28; University of Vienna, 1928-34; University of Munich, 1934-46; University of Tucumán, 1949-51. *Gletscherkunde* (1902); *Die Alpen* (1907); *Geomorphologie* (1919, 6th ed. 1954); *Das Relief der Erde* (2 vols., 1938-40, 2nd ed. 1955). Obit. by G. Fochler-Hauke in *PGM*, 102 (1958), 1-5.

Mackay, John Ross (b. 1915)

Ph.D., University of Montreal, 1949. McGill University, 1946-49; University of British Columbia, 1949-81. *The Mackenzie Delta Area, N.W.T.* (1963); "Ice-Wedge Cracks, Garry Island, Northwest Territories," *Canadian Journal of Earth Sciences*, 11 (1974), 1366-1383; "Pingos of the Tuktoyaktuk Peninsula Area, Northwest Territories," *Géographie physique et quaternaire*, 23 (1979), 3-61; "Pingo Collapse and Paleoclimatic Reconstruction," *Canadian Journal of Earth Sciences*, 25 (1988), 495-511. *GOF* (1987); *CWW*; *Geological Society of America Bulletin*, 93 (1982), 361; M. Church and O. Slaymaker, *Field and Theory: Lectures in Geocryology* (Mackay Festschrift) (1985).

Mackinder, Halford John (1861-1947)

B.A., University of Oxford, 1883, 1884. University of Oxford, 1887-1905; University of Reading, 1892-1903; London School of Economics, 1895-1925 (Director, 1903-1908). *Britain and the British Seas* (1902); *The Rhine* (1908); *Democratic Ideals and Reality* (1919). G. Kearns in *GBS*, 9 (1985), 71-86; W. Parker, *Mackinder: Geography as an Aid to Statecraft* (1982); B. Blouet, *Halford Mackinder, A Biography* (1987); obits. by E. Gilbert in *GJ*, 110 (1947), 94-99; and H. Ormsby in *Geography*, 32 (1947), 136-137.

Macrogeography

Macrogeography is that part of social physics related to spatial and temporal variations of potentials of population (V) and income (U), together with variables representing measures of social intensity and interaction that are closely related to V or U by power function relationships. For mathematical definitions and a related bibliography, see the entry on *Social Physics*.

The potential of population (or income potential) at a place measures the influence or accessibility at that place of all populations (or per capita income-weighted populations). A potential is a field quantity and varies continuously over space. The first maps of V were created by John Q. Stewart with the assistance of Kirk and others (Stewart 1945). Contours of V were interpolated from spot values of V_c at the center of each state, c.

Stewart (1947) recognized that rural population densities predicted by a given population potential varied, depending on the region of the country. The southeastern states had more rural people while the far west had fewer than would be expected in the main sequence states at the same value of V. This regionalization of the country was detected by Stewart (1948) in several other distributions related to V, including farmland values per acre, densities of railroad tracks and death rates. He suggested that the regional differences were due to differences in regional incomes. After William Warntz created income potential U by weighting V by per capita income, Stewart and Warntz (1958) were able to account for the regional differences by substituting U_c for V_c. $U_c = \sum_{i=1}^{n} I_i r_{ic}^{-1}$ where U_c is the income potential for state c, I_i is the income in state i and r_{ic} is the distance between states i and c.

The epitome of this work was Warntz's *Macrogeography and Income Fronts* (1965). Warntz drew an analogy between the income regions and air masses separated by fronts, thus adding notions from the gas laws and meteorology to macrogeography and social physics. Income density varied continuously with income potential across the income fronts (as would air pressure across a meteorological front) while there were major discontinuities of per capita income (analogous to temperature) at the fronts. In a historical survey of census data for the contiguous United States, Warntz demonstrated the generation and development of per capita income fronts and predicted their dissipation by the year 2005 (approximately).

A similar prediction was made by Geoffrey Dutton (1973) based on different criteria. He found that the population-weighted mean of the state population potentials, divided by the urban fraction, equals 562,000 people/mile. The figure has been constant from 1790 to 1970 for the country as a whole. Regional differences have always existed but have declined over time and should disappear by 2010.

The empirical constants, exponents and trends discovered by Stewart and Warntz indicate a self-organized spatial system. These regularities and trends thus act to constrain the collective will and behavior, although some individuals may not perceive or need to obey these constraints at any given time. *See also* Social Physics.

Bibliography

Dutton, G., "Criteria of Growth in Urban Systems," *Ekistics*, 36 (1973), 298-306.

Stewart, J. Q., *Coasts, Waves and Weather* (1945).

———, "Empirical Mathematical Rules Concerning the Distribution and Equilibrium of Population," *GR*, 37 (1947), 451-485.

———, "Demographic Gravitation: Evidence and Applications," *Sociometry*, 11 (1948), 31-58.

Stewart, J.Q., and W. Warntz, "Macrogeography and Social Science," *GR*, 48 (1958), 167-184.

Warntz, W., *Macrogeography and Income Fronts* (1965).

• *Michael J. Woldenberg*

Madrid Geographical Society. *See*

Royal Geographical Society of Madrid.

Malaysia

Geography occupies an important place in the curriculum of the Malaysian educational system. Throughout the secondary school years geography has a regional approach, which in the matriculation years incorporates systematics and elementary cartographic techniques.

At the tertiary level, the trend of geography is influenced by developments in the West. Since the late 1960s, there has been a major shift in emphasis from regional geography, save that on Southeast Asia, to systematic geography, and the curriculum today encompasses spatial, areal and man-land environmental studies, and the earth sciences. Students are additionally equipped with an understanding of cartography, aerial photo interpretation and statistics. The 1980s have seen the incorporation of the computer as a tool in geographical teaching and research. Field work is an integral part of the curriculum, while remote sensing techniques represent a new frontier in research.

Geographical inquiry has evolved from a descriptive to an analytical approach, with an increasing use of quantitative methods. Students and staff are encouraged to present seminars. All courses are conducted in Bahasa Malaysia, the national language.

Apart from the teaching profession, the practice of geography also extends to development and land use planning, research and consultancy work. Administrators in the civil service and statutory boards find geography of value in decisionmaking.

Research is encouraged through the publication of journals such as the *Malaysian Journal of Tropical Geography, Malaysian Geographers, Geographica*, and *Ilmu Alam*.

The major institutions of geographical teaching and research are the University of Malaya, Kuala Lumpur; the National University of Malaysia, Bangi, Selangor; and the University Science Malaysia, Penang. Apart from these, teacher training colleges are also involved in the teaching of geography, while the National Geographical Association of Malaysia aims to maintain and promote the status of geography in the country.

The University of Malaya offers a three-year B.A. degree and four-year B.Sc. degree in geography, while the other two universities mentioned offer a four-year B.A. in geography. An off-campus course of five years (minimum), of which the final year must be spent on campus, is available at the University Science Malaysia, Penang. Postgraduate exercises in all the three universities are in the form of advanced research and thesis writing.

Among the leading geographers in Malaysia today are Professor Tunku Shamsul Bahrin, University of Malaya, land development studies; Professor Voon Phin Keong, University of Malaya, land use studies; Datin Associate Professor Zaharah Mahmud, University of Malaya, historical geography; Professor Sham Sani, National University of Malaysia, air pollution studies; Associate Professor Abdul Samad Hadi, National University of Malaysia, urban studies; Professor Zakaria Awang Soh, National University of Malaysia, geomorphology; and Professor Mahinder Santokh Singh, University Science Malaysia, economic geography.

Bibliography

Bahrin, Tunku Shamsul and P. D. A. Perera, *FELDA, 21 Years of Land Development* (1977).

Bahrin, Tunku Shamsul, P. D. A. Perera, and H. K. Lim, *Land Development and Resettlement in Malaysia* (1979).

Bahrin, Tunku Shamsul and B. T. Lee, *FELDA, Three Decades of Evolution* (1988).

Sham, Sani, *Aspects of Air Pollution Climatology in a Tropical City: A Case of Kuala Lumpur—Petaling Jaya Malaysia* (1979).

Voon, P. K., *Western Rubber Planting Enterprise in Southeast Asia, 1876-1921* (1976).

Zaharah, Mahmud, "A Historical Geographical Account of the Malay Peninsula: An Exercise in Theory and Methodology," *Malaysian Journal of Tropical Geography*, 12 (1985), 37-44.

• *Tunku Shamsul Bahrin*

Marbut, Curtis Fletcher (1863-1935)

M.A., Harvard University, 1894. University of Missouri, 1895-1910; United States Department of Agriculture, 1910-35. *Soils of the Ozark Region* (1910); (trans.) *The Great Soil Groups of the World and Their Development* by K. Glinka (1927); *The Vegetation and Soils of Africa* (with H. Shantz) (1923); *Soils of the United States* (in USDA *Atlas of American Agriculture*) (1935). Obits. by H. Shantz in *AAAG*, 26 (1936), 113-123; and anon. in *GR*, 25 (1935), 688.

Marinelli, Giovanni (1846-1900)

Studied law at the University of Padua in the 1860s but did not complete the Laurea degree. University of Padua, 1879-81; University of Florence, 1882-1900. *Scritti minori* (2 vols., 1908), containing reprints of the following major papers of GM: "Della geografia scientifica e di alcuni suoi nessi collo sviluppo degli studi astronomici e geologici," I, 1-62; "Carlo Roberto Darwin e la geografia," I, 99-141; "Concetto e limiti della geografia" (1893), I, 143-179; "Se e come l'università italiana possa provvedere al fine de preparare insegnanti di geografia per le scuole secondarie," II, 229-245. R. Almagià, "Giovanni Marinelli e la sua scuola," *BSGI*, ser. 7, vol. 11 (1946), 232-233; R. Biasutti, "Giovanni Marinelli (nel centenario della nascità)," *RGI*, 53 (1946), 57-59; obits. by G. Dalla Vedova in *BSGI*, ser. 4, vol. 1 (1900), 629-654; and G. Pennesi in *RGI*, 7 (1900), 305-334.

Marinelli, Olinto (1874-1926)

Laurea, University of Florence, 1896. University of Florence, 1902-26. "Del moderno sviluppo della geografia fisica e della morfologia terrestre," *BSGI*, ser. 4, 9 (1908), 226-248; "Delle corrente litorale del Mediterraneo con particolare riguardo alla costa oriente della Sicilia" (with G. Platania); *Memorie Geografiche* (Florence), vol. 2, no. 5 (1908), 71-230; "I ghiacciai delle Alpe Venete," *Memorie Geografiche*, vol. 4, no. 11 (1910), 5-289; *Atlante dei tipi geografici* (1922). R. Riccardi, "Olinto Marinelli nel centenario della nascità," *BSGI*, 3 (1974), 31-43; A. Sestini, "La figura e l'opera di Olinto Marinelli," *RGI*, 81 (1974), 523-544 (plus OM's bibliography on pp. 617-683); obits. by anon. in *RGI*, 33 (1926), 97-102; and R. Biasutti in *RGI*, 34 (1927), 8-20.

Martin, Lawrence (1880-1955)

Ph.D., Cornell University, 1913. University of Wisconsin, 1906-19; U.S. Department of State, 1920-24; Library of Congress Map Division, 1924-46. *Alaskan Glacier Studies* (1914); *The Physical Geography of Wisconsin* (1916, 1932); (ed.) *The George Washington Atlas* (1932). Obit. by F. Williams in *AAAG*, 46 (1956), 357-364.

Martonne, Emmanuel de (1873-1955)

D.-ès-L., University of Paris, 1902; D.-ès-Sc., Paris, 1907. University of Rennes, 1899-1905; University of Lyon, 1905-09; University of Paris, 1909-44. *La Valachie* (1902), *Traité de géographie physique* (1909), *Les régions géographiques de la France* (1921), *Géographie universelle* (Vol. 4, 2 parts, 1930-31, and Vol. 6, part 1, 1947). J. Dresch in *GBS*, 12 (1988), 73-81; J. Dresch in *LGF* (1975), 35-48; J. Dresch in *Geographisches Taschenbuch 1970/72* (1970), 280-291; R. Beckinsale in *DSB*, 9 (1974), 149-151; O. Ribeiro *et al.* in *Finisterra*, 8 (1973); obits. by A. Cholley in *AG*, 65 (1956), 1-14; and M. Sorre in *IGU Newsletter*, 7 (1956), 3-7.

McBride, George McCutcheon (1876-1971)

Ph.D., Yale University, 1921. American Geographical Society, 1917-22; University of California, Los Angeles, 1922-47. *Agrarian Indian Communities of Highland Bolivia* (1921); *The Land Systems of Mexico* (1923); *Chile: Land and Society* (1936). Obits. by J. Spencer in *GR*, 62 (1972), 428-430; and H. Aschmann in *AAAG*, 62 (1972), 685-688.

McCarty, Harold Hull (1901-1987)

Ph.D., University of Iowa, 1929. University of Iowa, 1923-69. *The Geographic Basis of American Economic Life* (1940); *American Social Life* (1949); "An Approach to a Theory of Economic Geography," *EG*, 30 (1954), 95-101; *A Preface to Economic Geography* (with J. Lindberg) (1966). *GOF* (1971); H. McCarty, "The Cornbelt Connection—Geography at Iowa," *AAAG*, 69 (1979), 121-124; obit. by L. King in *AAAG*, 78 (1988), 551-555.

Mead, William Richard (b. 1915)

Ph.D., London School of Economics, 1946; D.Sc., London, 1967. University of Liverpool, 1946-49; University College London, 1949-81. *Farming in Finland* (1953); *Economic Geography of the Scandinavian States and Finland* (1958); *Finland* (1968); *An Historical Geography of Scandi-

navia (1981). *WW*; *GOF* (1982); W. Mead, "Autobiographical Reflections in a Geographical Context," pp. 44-61 in *The Practice of Geography* (ed. by A. Buttimer) (1983).

Medical Geography

Medical geography is the application of geography to the understanding of health and disease in human populations. This includes the geography of health and medical services. Medical geography was an important field of medicine until the early part of the twentieth century, when it gave way to modern scientific medicine based on microbiology and biochemistry. From 1900 to 1960, interest in the geographical aspects of disease was maintained by a small number of geographers and biomedical scientists mostly in Europe and North America. Notable among these were: Huntington, Jusatz, McKinley, Rodenwaldt, Sorre, Shoshin, and Stamp (geographers), and Ackerknecht, Audy, Doll, Pavlovsky, and Stocks (biomedical).

After 1960, interest in medical geography intensified as the search for causes of ill health moved beyond specific infectious and toxic agents to multiple factors in the environments and behaviors of populations at risk. It also saw new application in health and medical-services planning as increasing demand for services, concern for equitable access, and rising costs required more attention to location and other geographical factors.

The ecology of specific diseases continues to be the main concern of medical geographers. For example, Akhtar, Fonaroff, and Learmonth have contributed to the geography of malaria, Learmonth and Hunter to other infectious diseases, and McGlashan to cancer and other non-infectious diseases. The relations of disease to social and economic development are represented in the works of Jones and Meade; and to climate and seasonal variations by Sakamoto-Momiyama. Mathematical modelling of the spread or diffusion of infectious diseases by Brownlea, Cliff, Girt, Haggett, Pyle and others has become an especially exciting subfield. Studies in health and medical services, both modern and traditional, are represented in the works of Dear, Dever, Gesler, Good, and Shannon. Several general texts have appeared in recent years, *e.g.*, by Howe, Jones, and Meade. Medical geography is a rapidly growing branch of the discipline, contributing both to the understanding of the nature of health and disease itself, and to the nature of the pattern of health and disease in particular places.

Bibliography

Ackerknecht, E. H., *History and Geography of the Most Important Diseases* (1965).

Cliff, A. D., P. Haggett, and J. K. Ord, *Spatial Aspects of Influenza Epidemics* (1986).

Gesler, W. M., *Health Care in Developing Countries* (AAG Resource Publications in Geography) (1984).

Jones, K., and G. Moon, *Health, Disease and Society: A Critical Medical Geography* (1987).

Learmonth, A., *Disease Ecology: An Introduction* (1988).

Meade, M. S., J. W. Florin, and W. M. Gesler, *Medical Geography* (1988).

• *R. W. Armstrong*

Mehedinţi, Simion (1868-1962)

Doctorate, University of Leipzig, 1899. University of Bucharest, 1901-39. *Le pays et le peuple roumain* (1927); *La Roumanie* (1929); *Terra* (2 vols., 1930); *Opere complete* (2 vols., 1943-44). V. Mihailescu in *GBS*, 1 (1977), 65-72; I. Matley, "Simion Mehedinţi and Modern Romanian Geography," *PG*, 37 (1985), 452-458.

Meinig, Donald William (b. 1924)

Ph.D., University of Washington, 1953. University of Utah, 1950-59; Syracuse University, 1959 to the present. *On the Margins of the Good Earth* (1962); *The Great Columbia Plain* (1968); (ed.) *The Interpretation of Ordinary Landscapes* (1979); *The Shaping of America* (Vol. 1, 1986). *WWA*; *GOF* (1971).

Meitzen, August (1822-1910)

Dr.phil., University of Berlin, 1848. Privy Councillor, Prussian and German governments, from 1868; honorary lecturer, 1875-92; honorary professor, University of Berlin, from 1892. *Urkunden schlesischer Dörfer* (1863), *Das deutsche Haus* (1882), *Siedelung und Agrarwesen . . .* (3 vols. plus atlas, 1895). *Biographisches Wörterbuch zur deutschen Geschichte*, 2 (1974), column 1868; *Wer ist's? Unsere Zeitgenossen* (3rd ed., 1908), 893; obits. by anon. in *PGM*, 56 (1910), 86; and A. Fitzau in *GZ*, 16 (1910), 109–110.

Mental Maps

Apart from J. B. Priestley's use of the phrase in a novel of the 1920s, "mental maps" appeared in the early 1960s as a general term for cartographic representations of collective images of residential desirability. Simplifying somewhat, composite representations were created from people's responses to the question, "Where would you like to live?" Mapped as a contoured surface, with hills of desirability and valleys of disdain, the mental maps displayed extraordinary consistency from one location to another. It was postulated that a "national surface," convoluted with a "local dome of desirability," could recreate the shared views of any group of people. While generally confirmed in the United States, the United Kingdom, Sweden, Nigeria and Ghana, exceptions to the "rule" led to a deeper understanding of the role of language, culture, age, political inclination, religious belief and economic status upon the formation of such shared images. Information, and the way it was filtered, became an important issue, and led to the investigation of information and ignorance surfaces. Mental maps also appear to be highly stable over time: small, but gradually accumulating changes reflect gradual shifts in economic opportunities at the regional and national scales.

Some collective forms of human behavior could be predicted from mental maps and their major variants, in particular the migration of young adults, and the evaluation of distances. Both mental maps and migration patterns were predictable in varying degrees by gravity model expressions.

At the local scale (neighborhood to city), a concern for mental maps led to questions of environmental familiarity and wayfinding, particularly by handicapped children and adults. Research focused on the clues in the natural or built environment used by handicapped people to move about the local area. It also provided thought-provoking cartographic representations of the way in which increasingly handicapped people used mental maps that were less and less congruent with conventional maps constructed for wayfinding purposes.

"Mental maps" has become increasingly a generic term, and underpins a concern for such things as geographic images in regional planning and the way in which children learn about basic geographic concepts. A parallel research tradition evaluates the physical environment for human use. *See also* Environmental Perception.

BIBLIOGRAPHY

Golledge, R., and G. Rushton, eds., *Spatial Choice and Spatial Behavior* (1976).

Gould, P., and R. White, *Mental Maps* (1974).

• *Peter Gould*

Mercantile Model of Settlement

First proposed in 1970 by James Vance in his book, *The Merchant's World*, this model seeks to explain the location of settlement based on exogenic forces such as trade and transportation. It was called "mercantile" because mercantilist economic thought was found to be fundamental to the spread of urbanization in most parts of the world, starting with the spread of European trade and settlement plantation commencing in the sixteenth century. Two contrasting forms were found: the creation of more narrowly confined *traders' factories* on the Guinea Coast of Africa, and in India, Indonesia, China, and Japan, where advanced trading partners were already in existence; and *planted settlements* such as those in North America, parts of southern Africa and South America, and in Australia and New Zealand, which have an elaborated and geographically extensive settlement pattern based on the exogenic force of regional or world trade. Applications of the model to most of the New Lands of the world have shown its appropriateness. In contrast is Walter Christaller's central-place theory, founded on endogenic forces and apparently requiring a medieval administrative background for full development. Recent work by European historians has lent increasing support to the notion of a long-distance "mercantile" origin of many cities, even in the European Middle Ages. The largest cities of the world and major regions thereof—Shanghai, Bombay, São Paulo, New York, Chicago, and Sydney merely as examples—were mercantile in origin. In countries founded after the onset of mercantile economic development, such as the United States and Canada, the fundamental settlement pattern at virtually all levels seems to be best explained by this mercantile model with its historically contingent growth stages. A related concept has been that the countryside in many parts of the world grew out of pre-existing cities (mercantile model) rather than the cities growing out of a pre-existing developed countryside (central-place theory).

BIBLIOGRAPHY

Vance, J. E., *The Merchant's World, The Geography of Wholesaling* (1970).

• *James E. Vance Jr.*

Mexican Society for Geography and Statistics (Sociedad Mexicana de Geografía y Estadística)

The first geographical societies were established in the nineteenth century with two principal aims: to promote exploration of poorly known regions, so as to incorporate them into the civilized world, and to diffuse geographical knowledge in all its different aspects.

The Mexican Society for Geography and Statistics was founded on 18 April 1833 by the vice-president of the Republic, Valentín Gómez Farías, and forty intellectuals, among whom corresponding member Alexander von Humboldt stands out.

By the date of its foundation the Mexican Society must be ranked fourth among those of the world, after Paris (1821), Berlin (1828) and London (1830). It is important to note that it has endured down to our own day and now numbers 1,204 members.

The importance of this society, in particular for the country's intellectual and scientific life, has been highly praised by outstanding men and women of the nation.

At the present time, the society's basic objectives are to further and carry out scientific and cultural researches of all kinds, especially those related to national problems. To these ends it is organized into twenty-one sections of different subject matter related to the country's interests. Research work appears in the society's *Boletín*, which is one of the oldest and best-established geographical journals anywhere in the world. Its first number was published in 1839, and since then 140 volumes have appeared, containing important studies and valuable maps that constitute an important contribution to the country's science and culture.

The society represents Mexico in the International Geographical Union, and it is also the sponsor of national geographical congresses, of which there have been eleven.

The society possesses a library of more than 450,000 volumes, mainly works of a geographical and statistical nature, and a map collection that contains numerous maps of great value (including a portolan chart), the majority—dating from the nineteenth century—being originals and unpublished pieces. It also has an archival collection of manuscripts. All these holdings of the society suffered years of indifference and neglect until the period 1985-87, when, thanks to the concern of Professor Dolores Riquelme Rejón, the SMGE's first woman president, they were rescued and restored. The library and map collection are once again accessible as an extremely rich resource for Mexican bibliography and cartography.

The society is located in the Centro Histórico of Mexico City.

BIBLIOGRAPHY

"Edición conmemorativa del Centenario del Boletín de la Sociedad Mexicana de Geografía y Estadística," *Boletín de la Sociedad Mexicana de Geografía y Estadística*, 49 (1939), 9-346.

"Indice general que comprende desde el Tomo I hasta el Tomo LXIII, 1839-1947," *Ibid.*, 64 (1947), 1-27.

Primer Centenario de la Sociedad Mexicana de Geografía y Estadística, 1833-1933, 2 vols. (1933).

• *María Teresa Gutiérrez de MacGregor (translated by C. Julian Bishko)*

Mexico

In Mexico, in the nineteenth century, geography was regarded as one of the most important formative and informative subjects for knowledge of the country, so much so that the profession of geographer was established in 1843, a date early enough to rank it among the first three of its kind anywhere in the world. Some years later, in 1867, the president of the Republic issued a law making geography a required course at various stages of education. It is generally held that systematic teaching of geography on the higher level began in 1943, with the setting-up of a program of study designed to teach the modern concepts of the subject.

In the 1970s, while becoming universal in middle and elementary instruction, the teaching of geography underwent something of a setback. Physical geography came to be absorbed into the natural sciences, and human geography into the social sciences, with the result that knowledge of the field was broken up and reduced to the teaching of unrelated facts.

Although in 1839 publication had begun of the oldest geographical journal in the Americas, the *Boletín de la Sociedad Mexicana de Geografía y Estadística* (for many decades an important journal), it must be said that it was not until the 1960s that the research work of the geographers of this period came to be put together in what are now the country's two most important geographical serials; the *Anuario de Geografía* and the *Boletín del Instituto de Geografía*, both sponsored by the National Autonomous University of Mexico.

The most outstanding geographers of the past fifty years are as follows: Angel Bassols Batalla introduced into Mexico the viewpoint of Marxist geography, and in his many publications underscored its interest for the country's economic regionalization. Enriqueta García has given strong impetus to the study of climatology in Mexico. Jorge L. Tamayo devoted his chief efforts to publishing a series of books that came to fill an important gap in knowledge of the country. Jorge A. Vivó, whose preoccupation was with the teaching of geography, dedicated much effort to reorganizing study programs for the training of geographers along modern lines. This resulted in the formation of several generations of geographers who at the present time represent geography in the country. He also published numerous works on Mexico and initiated geographical research by stimulating his students to publish in the geographical journals under his editorship.

The most important institutions in the teaching of geography are the College of Geography of the National Autonomous University of Mexico, which currently has 121 instructors and 1,128 students; the School of Geography of the Autonomous University of the State of Mexico; and the Faculty of Geography of the University of Guadalajara. In research, there is the Institute of Geography of the National Autonomous University of Mexico, with a staff of 73, and at which most of the country's research work is carried on.

Bibliography

Bassols Batalla, A., *Bibliografía geográfica de México* (1955).

———, *México: Formación de regiones económicas* (2nd ed., 1983).

García Amaro, E., *Precipitación y probabilidad de la lluvia en la República Mexicana y su evaluación* (19 vols., 1973–79).

Tamayo, Jorge L., *Geografía general de México* (4 vols. and an atlas, 1949).

———, *Atlas del agua de la República Mexicana* (1976).

Vivó, Jorge A., *Geografía de México* (4th ed., 1958).

• *María Teresa Gutiérrez de MacGregor (translated by C. Julian Bishko)*

Mikesell, Marvin Wray (b. 1929)

Ph.D., University of California, Berkeley, 1959. University of Chicago, 1958 to the present. *Northern Morocco* (1961); (ed.) *Readings in Cultural Geography* (with P. Wagner) (1962); (ed.) *Geographers Abroad* (1973); (ed.) *Perspectives on Environment* (with I. Manners) (1974). *WWA*; *GOF* (1971).

Military Geography

Geographical knowledge and methods employed to prescribe, describe or analyze the deployment of armed force are the elements of military geography. The geographer's skills and preoccupation bring to bear on the study of military matters a concern for the significance of distance, habitat and spatial scale.

The application of geography to inform military decision has a long history. Much geographical exploration was done for this purpose.

Topographic mapmaking emerged to serve military needs. The map remains the premier instrument of intelligence, decision and command. Intelligence on geographical conditions is fundamental to tactical or strategic success. Field Marshal Montgomery credited victory in battle to "transportation, administration and geography." In a more academic vein, geographers have described the foundations and disposition of military might and its impact on society and the landscape.

Beyond this there is the study of warfare, the analysis of the effect of geographical circumstances on the outcome of violent conflict. Such work may be particular and historical or it may involve attempts at generalization. Whatever the purpose, the circumstances of war need to be examined at different levels of geographical resolution in order to understand fully the effect of location and environmental variety on its outcome.

There are three sets of variables that describe the relevant geographical conditions of war. These are: positional, environmental and topographic. Each set needs to be viewed at a different focal distance. Thus, we have three scales of geographical analysis that we can roughly equate with geopolitical, strategic and tactical levels of decisionmaking. These are not, however, independent realms. What is feasible in grand strategy is constrained by tactical capabilities. Appropriate action at the tactical level may be dictated not by local conditions but by the need to satisfy a geopolitical objective.

Whether such analyses are an attempt to explain past events or predict things to come, if they add to the evidence of the futility of war they will not have been in vain.

Bibliography

Kidron, M., and D. Smith, *The War Atlas* (1983).

O'Sullivan, P., and J. W. Miller, *The Geography of Warfare* (1983).

Peltier, L. C., and G. E. Pearcy, *Military Geography* (1966).

• *Patrick O'Sullivan*

Mill, Hugh Robert (1861–1950)

D.Sc., University of Edinburgh, 1886. Royal Geographical Society, 1892–1900; British Rainfall Organization, 1901–19. *The Realm of Nature* (1891); (ed.) *International Geography* (1899); *The Record of the Royal Geographical Society* (1930). T. W. Freeman in *GBS*, 1 (1977), 73–78; *Hugh Robert Mill: An Autobiography* (1951); obits. by G. R. Crone in *GJ*, 115 (1950), 266–267; and R. N. R. Brown (with I. Bartholomew and A. G. Ogilvie) in *SGM*, 66 (1950), 1–2.

Morocco. *See* North Africa.

Morrill, Richard Leland (b. 1934)

Ph.D., University of Washington, 1959. University of Washington, 1961 to the present. *Geography of Poverty* (1970); *Spatial Organization of Society* (1973); *Political Redistricting and Geographic Theory* (1981); *Spatial Diffusion* (1988). *WWA*; *GOF* (1971, 1986).

Murphey, Rhoads (b. 1919)

Ph.D., Harvard University, 1950. University of Washington, 1952–64; University of Michigan, 1964 to the present. *Shanghai, Key to Modern China* (1953); *An Introduction to Geography* (1961, 4th ed. 1978); *The Outsiders: Westerners in India and China* (1977); *The Fading of the Maoist Vision* (1980). *WWA*; *GOF* (1975).

Murphy, Raymond Edward (1898–1986)

Ph.D., University of Wisconsin, 1930. Pennsylvania State University, 1931–45; University of Hawaii, 1945–46; Clark University, 1946–68. *Pennsylvania, A Regional Geography* (with M. Murphy) (1937); *The American City: An Urban Geography* (1966); *The Central Business District* (1972). A. Bailey, "Professor Murphy and the Blackstone Valley," Chap. 4 of *Through the Great City* (1967); L. J. Johnson, "Raymond E. Murphy: An Appreciation," *EG*, 47 (1971), 97–100 ("Festschrift for Raymond E. Murphy," ed. by C. J. Ryan).

National Council for Geographic Education

The National Council for Geographic Education, known as the National Council of Geography Teachers before 1957, is a U.S. organization dedicated to improving the quality of the teaching of geography at all levels, with special emphasis on the primary and secondary levels. It was conceived in the mind of George Miller in 1914 and was born in December 1915 at the annual meeting of the Association of American Geographers. The *Journal of Geography*, which originated in 1902 in a merger of the *Journal of School Geography* (1897–1901) and the *Bulletin of the American Bureau of Geography* (1900–01), became the official organ of the National Council in 1916.

BIBLIOGRAPHY

Barton, T. F., "Leadership in the Early Years of the National Council of Geography Teachers, 1916–1935," *JG*, 63 (1964), 345–355.

Whittemore, K. T., "Celebrating Seventy-Five Years of *The Journal of Geography*," *Ibid.*, 71 (1972), 7–18.

National Geographic Society (Washington, D.C.)

The National Geographic Society began in 1888 with 165 charter members as a "society for the increase and diffusion of geographical knowledge," with the broadest possible interpretation of geography. Its *Magazine*, which also dates from 1888, became the popular journal that we know today after Gilbert H. Grosvenor (1875–1966) became the editor in 1899 (President, 1920–54) and expanded the society's "membership" (subscription list) enormously by promoting the use of photographs and colorful, but shallow and non-polemical, writing. It was then that academic geography began to turn its back on the society. In 1902 William Morris Davis declined to serve as associate editor of the *National Geographic Magazine* because he "couldn't approve general policy of popularizing at expense of science." He resigned from the society's Board of Managers in November 1904 and founded the Association of American Geographers a month later. A rapprochement between the society and academic geography has come about in recent years, as geographers have been drawn into the society's expanded research program and into its imaginative new schemes for the promotion of school geography. A new quarterly journal, *National Geographic Research*, begun in 1985, includes the work of academic geographers but still reflects the nineteenth-century view of geography as natural history more than the late twentieth-century conception of the field. The *National Geographic Magazine* has attracted imitators, such as the

Geographical Magazine (London, begun in 1935), *Canadian Geographic* (1930), and *Alaska Geographic* (1972), but none can hope to approach its popular success, with more than 10 million subscribers.

BIBLIOGRAPHY

Bryan, C. D. B., *The National Geographic Society: 100 Years of Adventure and Discovery* (1987).

Grosvenor, G. M., "A Hundred Years of the National Geographic Society," *GJ*, 154 (1988), 87-92.

Pauly, P. J., "The World and All That Is in It: The National Geographic Society, 1888-1918," *American Quarterly*, 31 (1979), 517-532.

Neef, Ernst (1908-1984)

Dr.phil., University of Heidelberg, 1932; (habil.) Technical University of Dresden, 1935. Technical University of Danzig, 1936-45; University of Leipzig, 1949-59; Technical University of Dresden, 1959-73. "Topologische und chorologische Arbeitsweisen in der Landschaftsforschung," *PGM*, 107 (1963), 249-259; "Elementaranalyse und Komplexanalyse in der Geographie," *MOGG*, 107 (1965), 177-189; "Der Verlust der Anschaulichkeit in der Geographie, das Beispiel Kulturlandschaft," *Sitzungsberichte der Sächsischen Akademie der Wissenschaften*, 115 (1981), 15-28; "Ausgewählte Schriften," ed. by H. Barthel, *PGM Ergänzungsheft* 283 (1983).

Nelson, Helge (1882-1966)

Doctorate, University of Uppsala, 1910. University of Lund, 1916-47. *Om randdetten och randåsar i mellersta och södra Sverige* (1910); *Canada* (1922); *The Swedes and the Swedish Settlements in North America* (2 vols., 1943). K. Bergsten in *GBS*, 8 (1984), 69-75; T. Hägerstrand, "Proclamations about Geography from the Pioneering Years in Sweden," *Geografiska Annaler*, 68B (1982), 119-125; obit. by K. E. Bergsten in *Svensk Geografisk Årsbok*, 42 (1966), 109-120.

Nepal. *See* South Asia.

The Netherlands

The practice of modern geography in the Netherlands had multiple origins: in trade and commerce, land registry, civil and hydraulic engineering, law, administration (also in the colonies), and the military. The gradual formation of a unified discipline of geography with specialized branches out of a loose and extremely varied set of disciplinary roots came about after the introduction of geography in the secondary schools (1863) and the universities (1874). The introduction of geography in universities as a major subject for study came rather late. When it did come about in 1921, the argument of educating secondary school teachers was less important than in neighboring countries. Two geography majors were established: one in physical geography (natuurkundige aardrijkskunde) and one in human geography (sociale aardrijkskunde). In both majors it was compulsory to take the other subject as a minor. Physical geography was established in a faculty of mathematics and natural sciences, and human geography was established in a faculty of arts and philosophy. To cope with organizational problems and to enhance the unity of geography, the "United faculties of mathematics and natural sciences and arts and philosophy" were formed, in which the full professors from both faculties teaching courses in the geography programs worked together.

THE BEGINNINGS

Between the introduction of geography in universities (1874) and the initiation of separate majors in physical and human geography in 1921, geography could be studied only as a minor subject in either a faculty of mathematics and natural sciences or in a faculty of arts and philosophy. Physical geography was usually taught by a professor of geology, and human geography by a professor of history. The exception was Dr. C. M. Kan, a former teacher of classical languages with a great interest in geography and history, who was appointed in 1877 at the University of Amsterdam as the first full-time professor of "physical and political [*i.e.*, human] geography and the knowledge of the land and people of the [Netherlands] East- and West-Indies." Not surprisingly for a man of his academic background, it was his sincere conviction that physical and human geography were too distinct to be adequately dealt with by the same person, and that cooperation with neighboring disciplines was necessary to achieve a "higher unity" of knowledge and insight. Only after Kan retired in 1907 did the full development of university geography begin. Full-fledged geographical institutes were then established in Amsterdam, supported in particular by the faculty of arts and philosophy, and in Utrecht, where support came mainly from the natural sciences.

In Amsterdam, Kan was succeeded by M.E.F.Th. Dubois and S. R. Steinmetz. Dubois, the discoverer of *Pithecanthropus erectus* (Java Man), was appointed professor of geology, palaeontology, and crystallography, as well as of physical geography. Whereas physical geography developed rather slowly in the University of Amsterdam and reached its full development only after the appointment of J. P. Bakker in 1938, human geography progressed much faster under Steinmetz, who was "professor of political [human] geography, anthropology and the knowledge of land and people of the East-Indian Archipelago" from 1907 to 1933. He introduced a strong social scientific conception of human geography in Amsterdam and laid the foundation for what became known as the Amsterdam School of Sociography. To understand the strong social scientific orientation of human geography in the Netherlands, one must keep in mind that sociology and anthropology did not become major subjects for study in Dutch universities until after World War II. Before then, human geography and economics were the only social sciences in which it was possible to major. In practice, the output of sociography consisted mainly of a series of village-, town-, and region-descriptions in which a concrete picture was given of the differentiation and integration of the human occupants. The scant attention paid to generalization, abstraction, and theory brought about an exodus of sociographers to sociology after World War II. Of the first fifteen full professors of sociology after 1945, thirteen were trained as sociographers. Some of them, such as F. van Heek and J. A. A. van Doorn, designated the sociography of Steinmetz and Ter Veen as an exceptionalist period in the development of sociology: "the sociographic intermezzo."

Utrecht (1908–c.1950)

At the University of Utrecht, physical geography developed more favorably than in Amsterdam, and the relation between physical and human geography there has remained stronger. In 1908, the government decided that among the three great state universities—Leyden, Groningen, and Utrecht—the last would build a full-fledged, well-equipped geographical institute that operated at a high international standard. German universities would serve as the source of inspiration. Because of the support of the Royal Dutch Geographical Society (KNAG) and the Royal Netherlands Meteorological Institute (KNMI), Utrecht had from the start two ordinary (full-time) chairs and two extra-ordinary (part-time) chairs at its disposal. The ordinary chairs were for full professors of what are now known as physical and human geography. One extra-ordinary chair was reserved for meteorology, climatology, and oceanography, maintaining a strong relationship with the KNMI. The other chair was for ethnography and was meant to contribute to a well-balanced education of professionals, mostly civil servants, for the Netherlands East and West Indies. The chairs of human geography and ethnography were established in the faculty of arts and philosophy, the others in the faculty of mathematics and natural sciences. It was agreed, however, that all four professors would work together in the new geographical institute.

But from the beginning there was a major difference of opinion concerning the applied or fundamental character of geography. In the faculty of science, the importance of fundamental research on an international level was very strongly stressed. Others pointed out that it was of great national interest to educate good teachers and civil servants. The chair of physical geography was filled by the German geomorphologist Karl Oestreich, who is known for his research in the Alps, Balkans, and Himalayas. His work on such mountain areas was soon regarded as being of little relevance to the Netherlands, whatever its scientific importance might be. His student and successor, Jacoba Hol, who held the chair of physical geography from 1946 to 1958, took a broader view of geomorphology and applied it to areas closer to home—the hilly landscapes of Limburg—thus making physical geography of more practical use within the Netherlands itself.

Oestreich, a pure geographer, fit well in the nineteenth-century tradition of academic scholarship. For the chair of "political, economic and general geography," preference was given to an applied geographer, an educator, generalist, and schoolteacher who fit the nineteenth-century tradition of advanced professional training of civil servants and business people as well as schoolteachers. Jan Frederik Niermeyer was chosen for this chair. From the beginning, he opposed the dominant German influence on Dutch geography, particularly on school geography. His major accomplishment was the introduction of the French School of Geography, which strongly influenced human geography at Utrecht until the early 1960s. Whereas Niermeyer dismissed geomorphology as being only a part of geology, he took a strong position against the Amsterdam sociography by stressing the economic aspects in his brand of geography. Not long after the introduction of human and physical geography as two separate majors in 1921, Niermeyer became ill, and died in 1923. At first, his declared adversaries, Steinmetz and Oestreich, agreed to replace Niermeyer only by a university lecturer of human geography and to concentrate the study of sociography (human geography) in Amsterdam and physiography (physical geography) in Utrecht. Strong opposition to these ideas by the Utrecht students eventually brought about the appointment of Louis van Vuuren to the chair of "social, economic, colonial and regional" geography at Utrecht in 1927.

Van Vuuren proved to be an extremely imaginative and inspiring leader, with a clear eye for the importance of applied geography. Van Vuuren and his students produced not only regional studies of the Netherlands that were useful for planning purposes, but also undertook thematic studies that had a more enduring influence. Major examples include A. C. de Vooys' rural-urban migration studies and W. Steigenga's research on agricultural labor market problems. These topical studies, more so than the regional ones, made important contributions to the early discussions in Dutch geography about the real content of the concept of "welfare," decades before this became a topic in the international geographical literature.

After World War II, van Vuuren was briefly succeeded by J. O. M. Broek, who introduced the American style of cultural geography to the Netherlands. Broek left Utrecht for the University of Minnesota in 1948 and was succeeded by A. C. de Vooys, who reinstated the earlier emphasis on empirical and applied research.

Physical Geography Since the Early 1950s

Whereas the Amsterdam School of Sociography flourished during the interwar period, physical geography in Amsterdam had to wait until 1946, when J. P. Bakker was appointed its first full professor. Bakker emphasized the morphogenesis of the Holocene lowlands. He also pioneered in the theoretical-mathematical analysis of hillslopes. Physical geography would no longer rely on fieldwork and grand theories alone but would now also make use of well-designed experiments. Process studies, particularly of erosion, became dominant in physical geography in the Netherlands, not only in the University of Amsterdam but also in Utrecht, Enschede, and Wageningen. As a consequence of the great numbers of students in Utrecht (eighty to 100 new first-year students each year, about two thirds of the national total), an important growth of physical geography took place there under professors J. H. J. Terwindt, P. A. Burrough, and E. A. Koster. In close cooperation with physical geography, a separate interdisciplinary department of environmental studies has been established in Utrecht.

Human Geography Since the Early 1950s

The period from the early 1950s to the early 1970s was very much influenced by the rapid expansion of secondary education as well as by demographic and economic growth. For the first time, the demand for secondary schoolteachers became a strong force in the development of university geography. The growth of the population and economy and the demand for more and better housing and transportation facilities led to expanded opportunities for geographers in urban and regional planning. The independence of Indonesia in 1949, however, brought an end to the strong preoccupation of Dutch geographers with the East Indies.

Between the early 1950s and the early 1970s, the number of first-year geography students in Dutch universities rose from about 100 to about 750, but the ratio between the numbers of human geography students and those in physical geography remained the same (4:1).

After World War II, with the defection of the sociographers to the new department of sociology, human geography at the University of Amsterdam began to resemble more closely the type of geography practiced in other universities, but some important differences remained. In Amsterdam, cultural and political factors were accentuated more and economic factors less than in Utrecht or Groningen. The distance between human and physical geography was much greater in Amsterdam. The introduction of demography and of urban and regional planning as majors in Dutch universities caused much the same outflow of talent from the Utrecht geography department as the introduction of sociology did to Amsterdam.

In the 1970s major personnel changes took place in Amsterdam, Groningen, Nijmegen, and Utrecht. The number of full professors and staff members increased considerably, as did the variety of their specializations. The international orientation of Dutch geography became stronger as well. Since the early 1980s, however, the number of students opting for a major in human geography has decreased. In 1987, only 340 first-year students decided to register for human geography, whereas in the 1970s the yearly average had been close to 600. In that same year, a total of 2,500 students were registered for human geography (1,000 in Utrecht, 600 in Amsterdam, 500 in Nijmegen, and 400 in Groningen), already 500 less than in 1982. In the late 1980s, the influx of new students in the human geography departments stabilized at a level of about 350. Human geography in the Free University of Amsterdam was terminated in 1984 and partly merged with human geography in the University of Amsterdam because of the diminishing number of students and the budget decisions of the national government.

The increasing numbers and the growing differentiation brought an end to the Amsterdam-Utrecht rivalry. In the 1980s, the alignment of human geographers in the Netherlands was much more on the basis of their shared disciplinary specialization. The group working in urban geography, including housing studies, is perhaps strongest: F. M. Dieleman and J. van Weesep (Utrecht), R. van Engelsdorp Gastelaars (Amsterdam), J. Buursink (Nijmegen), and G. A. van der Knaap (Rotterdam). Equally strong is the group of full professors in economic geography: J. G. Lambooy (Amsterdam), M. de Smidt (Utrecht), E. Wever (Nijmegen), and W. J. van den Bremen (Groningen). Since the early 1970s, an interesting group specializing in development studies in Third World countries has developed: J. Hinderink (Utrecht), J. G. M. Kleinpenning (Nijmegen), and G. A. de Bruyne (Amsterdam). Historical geography (G. Borger), population geography (H. van Amersfoort), and political geography (H. van der Wusten) get special attention in Amsterdam, whereas Utrecht contributes to rural geography and the methodology of geography (J. Hauer) as well as to population studies and the philosophy and history of geography (J. A. van Ginkel), to regional geography and education (G. A. Hoekveld), to thematic cartography (F. J. Ormeling), and to historical cartography (G. Schilder).

Human geography in the Netherlands has acquired a kaleidoscopic character. Many relations exist with scientists in adjacent disciplines and abroad, as well as with practitioners working on urban problems, housing, labor markets, infrastructure, and development studies.

Bibliography

Dietvorst, A. G. J. and F. J. P. M. Kwaad, eds., *Geographical Research in the Netherlands 1978-1987* (1988).

van Ginkel, J. A. and M. de Smidt, eds., "75 Years of Human Geography at Utrecht," *TESG*, 74 (1983), 313-406.

Heslinga, M. W. "Probleme und Aufgaben der Geographie in den Niederlanden," pp. 9-29 in *Oldenburg und der Nordwesten* (1971).

van Paassen, C. "Human Geography in the Netherlands," pp. 214-234 in *Geography since the Second World War*, ed. by R. J. Johnston and P. Claval (1984).

de Pater, B., and M. de Smidt, "Dutch Human Geography," *PHG*, 13 (1989), 348-373.

Zonneveld, J. I. S. "Physical Geography in the Netherlands," *Erdkunde*, 31 (1979), 1-10.

• *J. A. van Ginkel*

Newbigin, Marion Isabel (1869-1934)

D.Sc., University of London, 1898. Editor, *Scottish Geographical Magazine*, 1902-34; Extra-mural School of Medicine for Women, Edinburgh. *Frequented Ways* (1922); *The Mediterranean Lands* (1924); *Canada: The Great River, the Lands and the Men* (1927); *Plant and Animal Geography* (1936). T. W. Freeman, "Two Ladies," *Geographical Magazine*, 49 (1976), 208; K. C. Edwards in *British Geography 1918-1945* (ed. by R. W. Steel) (1987), 93-94; obits. by anon. in *SGM*, 50 (1934), 331-333; and by E. G. R. Taylor in *GJ*, 84 (1934), 367.

New Zealand

The claim in the preface to Marshall's *Geography of New Zealand* (1905) that "the fortunate inhabitants of New Zealand should be geographers by instinct" has survived its environmentally deterministic origins. Regardless of the paradigmatic flavor of the decade, subsequent high levels of professional activity, enrollments in schools and universities, diverse job prospects for graduates, and success of New Zealand geographers internationally support the idea that the country encourages geographic inquiry. Regionalism and areal differentiation were actively pursued during the 1950s as the country's diversity was recorded and analyzed. Despite the country being far removed from classical isotropic space some fine quantitative work emerged during the 1960s. The prolific R. J. Johnston cut his eye teeth analyzing the urban and regional processes of New Zealand in the 1970s from a variety of perspectives. Adherents to a political economy viewpoint, although puzzled in earlier periods by New Zealand's brand of dominion capitalism, are finding the deregulated New Zealand of the late 1980s a fertile laboratory as the country's new breed of capitalists seem to "jump out of the pages of the best polemical writings of Marx."

A strong argument can be made for human agency being pre-eminent in the strength and variety of emphases in New Zealand geography. The eminent geologist Charles Cotton (Victoria University, Wellington) stimulated interest in the evolution of one of the most diverse fluvial, coastal, glacial and volcanic geomorphologies in any area of comparable size. George Jobberns, although a geologist "turned geographer by declaration," brought a breadth of perspective, passion for the discipline of geography, and personal warmth that was revered at the University of Canterbury in Christchurch (New Zealand's first department of geography, founded in 1937). Kenneth Cumberland (foundation member [1946] and professor [1949] at the University of Auckland) reinforced the international status of New Zealand geography and espoused a Hartshornian philosophy that had considerable influence on the high school curriculum into the 1970s. He was the first editor of the well-respected *New Zealand Geographer*, which was founded in 1944. Keith Buchanan and Harvey Franklin at Victoria University provided alternative viewpoints of geography that were based on more diverse social theory and an international perspective that is evident from the journal *Pacific Viewpoint*. With departments of geography also established at the University of Otago (1946), Massey University, Palmerston North (1964) and the University of Waikato, Hamilton (1965), there are now six university departments in a nation of 3.4 million people.

New Zealand professional geography was in a strong position to capitalize on the international expansion of the subject after World War II, especially in Australia and the United States. From the 1950s, a steady stream of graduate students enrolled for doctoral degrees in the United States, Canada, the United Kingdom and Australia. A significant number stayed to work in North American universities and now hold eminent positions there. Geographers with their first degrees from New Zealand are even more strongly represented on the faculties of Australian departments.

The strength of New Zealand geography is evident in the high schools. From the late 1940s, when it became firmly established in the curriculum, it has remained one of the five most popular subjects in the fifth form (eleventh year of compulsory education) and has strong enrollments in the final year of high school, where it averages about 40 percent of English, the subject attracting most students. In the 1950s high school teaching was the most popular choice of occupation by graduates, but progressively geographers have found employment in an increasingly wide range of occupations and industries in the public and private sectors. In the 1980s only about one in ten masters graduates entered teaching. Graduates in geography are using their research skills in market and project analysis and evaluation, in environmental and resource issues and in the analysis of earth, biological and atmospheric processes. Significant numbers find employment in planning of various types with local authorities and central government agencies.

New Zealand academic geography of the 1980s cannot be simply typified. An increasingly strong home-grown component was combined with the blend of British and North American traditions. Most university appointments in the 1970s and 80s were New Zealanders returning after completing higher degrees overseas. Currently, all six departments have graduate programs, with Auckland (highest enrollments in Australasia), Canterbury and Otago being the largest. Because the universities are strongly regional in their recruitment of students and orientation they offer broadly similar programs, although some specialization is apparent. All departments (except Waikato, where geography is grouped with social sciences) teach both physical and human geography. Geomorphology and hydrology, with different regional emphases, are prominent in the physical geography curricula and a healthy diversity of approach to the study of society and economy is apparent in human geography. Degrees in regional and resource planning are associated with geography departments at Otago and Massey, and Auckland has developed remote sensing and GIS programs.

BIBLIOGRAPHY

Holland, P. G., and W. B. Johnston, eds., *Southern Approaches: Geography in New Zealand* (1987).

Johnston, W. B., "An Overview of Geography in New Zealand to 1984," *New Zealand Geographer*, 40 (1984), 20-33.

Macaulay, J. U., "New Zealand School Geography 1937-1987: Some Put-Downs, Takeovers and Challenges," *New Zealand Journal of Geography*, 85 (1988), 17-21.

Marshall, P., *The Geography of New Zealand* (1905).

McDermott, P., "The Application of Geography in New Zealand," *New Zealand Geographer*, 40 (1984), 44-53.

Moran, W., "Time, Place and New Zealand Human Geography in the 1980s," in *Geographyfor the 1980s* (Proceedings of the Twelfth New Zealand Geography Conference), ed. by F. Owens, D. Johnston, M. Jaspers, and R. Vaughan (1984).

• *Warren Moran*

Nielsen, Niels (1893-1981)

Doctorate, University of Copenhagen, 1924. University of Copenhagen, 1933-64. *Studier over Jaernproduktionen i Jylland* (1924); *Eine Methode zur exakten Sedimentationsmessung* (1935); *Vatnajokull* (1937). N. Jacobsen in *GBS*, 10 (1986), 117-124; obits. by N. Jacobsen in *Geografisk Tidsskrift*, 81 (1981), viii-ix; and *Svensk Geografisk Årsbok*, 58 (1982), 210-215.

Niermeyer, Jan Frederik (1866-1923)

No university degrees. Qualified (outside the university) as a secondary-school teacher by state examination. University of Utrecht, 1908-23; University of Rotterdam, 1913-23. "Penck over de task der geografie," *Tijdschrift van het Koninklijk Nederlandsch Aardrijkskundig Genootschap*, 24 (1907), 1060-1069; "Penck über die Aufgabe der Geographie," *Ibid.*, 25 (1908), 348-351; "Barriere-riffen en atollen in de Oost-Indiese Archipel," *Ibid.*, 28 (1911), 877-893, 29 (1912), 64-65, 225-227, 623-636; *De Aardrijkskunde van de Oost-Indiese Archipel* (1908). M. W. Heslinga, "Sociografie versus Sociale Geografie," pp. 53-67 in *Toen en thans: de sociale wetenschappen in de jaren dertig en nu*, ed. by K. Bovenkerk *et al.* (1978); M. W. Heslinga, "Between French and German Geography: In search of the origins of the Utrecht School," *TESG*, 74 (1983), 317-334; W. J. van den Bremen, "De Nederlandse geografie van 1850 tot 1950," pp. 160-174 in *Algemene Sociale Geografie: Ontwikkelingslijnen en Standpunten*, ed. by A. G. J. Dietvorst, J. A. van Ginkel, *et al.* (1984).

Nigeria

Geography was one of the pioneer disciplines when the first tertiary institution—University College, Ibadan—was established in

1948. Since then, it has been suggested that the development of the subject in the country shows three distinct phases: the inception phase (1948–64), the indigenization phase (1964–70), and the consolidation phase (1971 to the present). The inception phase could also be characterized as the colonial phase, and much of the development was in the department at University College, Ibadan, which became a full-fledged university in 1962. Before that date, the college had a special relationship with the University of London, and Nigerian geography at the time shared much of the idiographic orientation of that source.

By 1964, not only had the number of university departments of geography increased to four, but Nigerians began to dominate the faculty. It was also a period of greater exposure to the quantitative revolution and to U.S. influences. Geography became more theoretical in orientation, with increasing planning applications. The phase of consolidation shows geography being taught in twenty-three of the twenty-nine universities in the country and providing specialization in many subfields of the subject.

Over the past four decades, geography in Nigeria has matured, especially in the older universities at Ibadan, Lagos, Nsukka, Zaria, and Benin. The discipline has produced some outstanding scholars whose works have commanded international attention. Notable amongst these are Akin Mabogunje, whose publications on *Urbanization in Nigeria* (1968) and *The Development Process: A Spatial Perspective* (1981) are highly regarded across the social sciences generally; Reuben Udo, *Geographical Regions of Nigeria* (1970) and *Migrant Tenant Farmers of Nigeria* (1975); Hodder and Ukwu, *Markets in West Africa* (1969); Ojo, *The Climates of West Africa* (1977); Agboola, *An Agroclimatological Atlas of Nigeria* (1979); Faniran and Areola, *Essentials of Soil Study* (1978); Bola Ayeni, *Concepts and Techniques in Urban Analysis* (1979); Paulina Makinwa, *Internal Migration and Rural Development in Nigeria* (1981); Ayoade, *Introduction to Climatology for the Tropics* (1983) and *Tropical Hydrology and Water Resources* (1988); and Hayward and Oguntoyinbo, *Climatology of West Africa* (1987).

Numerous other publications by Nigerian geographers are journal articles, a substantial proportion of which are carried in the *Nigerian Geographical Journal*, the official organ of the Nigerian Geographical Association, founded in 1957. In spite of the difficult economic situation, this journal continues to be published, though not as regularly as in the past. This notwithstanding, the academic status of geography in Nigeria remains quite high and Nigerian geographers continue to be active internationally. One of them, Akin Mabogunje, served as President of the International Geographical Union in the period 1980–84.

Bibliography

Ikhuoria, I. A., "The Nigerian School of Geography," *GeoJournal*, 12 (1986), 107–110.

Okafor, S. I., "Research Trends in Nigerian Human Geography," *PG*, 41 (1989), 208–214.

Salau, A. T., "Geography in Nigeria," *PG*, 38 (1986), 417–419.

• *Akin L. Mabogunje*

North Africa (Algeria, Morocco, and Tunisia)

Modern geography in North Africa probably began with the spatial inventories of Oscar MacCarthy (1815–1894), wherein statistics and maps followed his years of exploration in Algeria and the Sahara. Only later were the fourteenth-century geographical contributions of Ibn Khaldoun and Ibn Battuta appreciated. France's burgeoning geographical societies of the 1870s were more interested in colonial commercial expansion into and beyond the Sahara by means of a trans-Saharan railway. Active geographical societies emerged in Oran in 1878 and Algiers in 1880, both with regular meetings, periodicals, and exchanges. By 1917 the Algiers society had a Moroccan section in Tangiers, and in 1922 a Casablanca geographical society was founded. Such early academic writing was collated by Elisée Reclus in his 1886 *Géographie universelle* volume on North Africa. More formal academic geography teaching started with G. Hardy and with Augustin Bernard's appointment in Algiers in 1894, to be succeeded by E.-F. Gautier in 1902. Together with H. Schirmer's *Le Sahara*, Bernard and Gautier published numerous geographical studies. Notable were Bernard's *Enquête sur l'habitation rurale des indigènes de l'Algérie* (1921) and his *Atlas d'Algérie et de Tunisie* (1923), while Gautier's travels resulted in *Le Sahara algérien* (1908) on physical geography and *La conquête du Sahara* (1910) on human geography. Gautier's *Structure de l'Algérie* (1922) summarized his physical geography researches; its regional divisions were followed by Bernard in his 1937–39 *Géographie universelle* volumes on North Africa. By 1931 Jean Dresch was teaching in Rabat, Morocco, and researching into its geomorphology, while Jean Poncet's rural geography studies of Tunisia reflected teaching developments in Tunis. Post-1945 French colonial geography flourished with North Africanists Dresch at Strasbourg and Hildebert Isnard at Aix-en-Provence. Scientific Saharan research continued with the *Travaux de l'Institut Saharien*, and general geographies were produced by Isnard with his *Le Maghreb* (1966) and by Jean Despois and René Raynal with their regional *L'Afrique du Nord* (1967). Political independence prompted an increase in local Maghrebi geographers, initially French trained, together with new national journals, the *Revue de Géographie du Maroc*, the *Revue Tunisienne de Géographie*, and the now-defunct *Annales Algériennes de Géographie*. Arabization affected university geography teaching and additional geography institutes were established in Oran, Constantine, and Fes. Six colloquia of Maghreb geographers have met, the first in Tunis in 1967, the latest in Nouakchott, Mauritania, in 1987. National geographical societies now flourish in Morocco and Tunisia, but Algeria's single-party clampdown on organizations has prevented a society there. Constitutional changes in 1989 should now permit one. Close connections with French geography remain, partly channelled through the Centre URBAMA (Urbanisation dans le monde arabe) at Tours. Its annual surveys of geographical theses reflect active French-Maghreb research cooperation. URBAMA's collected study, edited by J. F. Troin, on *Le Maghreb, Hommes et Espaces* (1985, 2nd ed. 1987), together with Marc Côte's *l'Algérie ou l'Espace Retourné* (1988), encompass much current research, as will a third *Géographie universelle* North Africa volume currently being edited by Georges Mutin.

BIBLIOGRAPHY

Côte, M., *L'Algérie ou l'espace retourné* (1988).

Findlay, Allan M., Anne M. Findlay, and R. I. Lawless, compilers, *Tunisia* (World Bibliographical Series, vol. 33) (1982).

Findlay, Anne, Allan Findlay, and R. I. Lawless, compilers, *Morocco* (*Ibid.*, vol. 47) (1984).

Lawless, R. I., compiler, *Algeria* (*Ibid.*, vol. 19) (1980).

Sutton, K., and R. I. Lawless, "Progress in the Human Geography of the Maghreb," *PHG*, 11 (1987), 60–105.

Troin, J. F., *et al.*, *Le Maghreb. Hommes et espaces* (1985, 2nd ed. 1987).

• *Keith Sutton*

Norway

In the first forty-five years of its existence as a university discipline, from 1871, geography was taught only as a subsidiary of history and natural science. Geography was, and is, a separate discipline in the school system, but today only at the high school level. But it is a minor discipline, as history is the dominant social science. Before the 1960s the main job market for geographers was schoolteaching, which explains the slow institutionalization of geography and the protracted development of a scientific community of geographers.

Norway's first department of geography was established at the University of Oslo in 1917. The longtime head of the department was Werner Werenskiold (1883–1961). His interests were broad, but his main field was geomorphology, which up to the 1960s attracted the most students. Fridtjov Isachsen (1906–1979), who took the first M.A. in human geography in Norway, was appointed associate professor in human geography in 1931 and full professor in 1947. He was the architect of the change that took place in 1958 and that formed the basis for fast growth in student numbers in the 1960s. Before 1958 students had to choose between physical and human geography from the very first university lesson in the discipline. In 1958 the basic courses were united in geography "grunnfag," which contains half physical, half human geography. In Oslo half of the graduate students (M.A. and Ph.D.) are in physical geography, whereas the graduate students at the younger universities, Bergen and Trondheim, are mainly in human geography. Geography was established in Bergen in 1963, in Trondheim in

1974, and in the fourth university, Tromsø, only at the undergraduate level, in 1982.

In Bergen, Axel Sømme (b. 1899), who had studied in France and completed his doctorate at Oslo in 1931, was appointed associate professor at the Norwegian School of Economics and Business Administration in 1937, and full professor in 1948. But as geography was only offered as an undergraduate course, Sømme unfortunately never obtained a central position as a supervisor for graduate students. Through graduate research programs, however, a couple of later full professors were educated in Bergen: Asbjørn Aase from 1974 at the University of Trondheim and Tore Ouren from 1969 in Bergen. In 1963 Tore Sund (1914-1965) was appointed to the first chair of geography at Bergen University, and the department was established as a joint department with the School of Economics.

The period 1960-77 was one of fast growth in numbers of university posts and geography students. In Oslo and Bergen staff expansion more or less ended in 1973, but some opportunities were opened in the new department at Trondheim. From 1977 to 1984 there was a decline both relatively and in absolute numbers of geography students, but in the past few years there has been a rise in student numbers in geography from 1.4 percent to 1.7 percent of the total number of university students. There are at present forty graduate students per million inhabitants in the country. These figures indicate that the status of geography is worse than in Finland but better than in Denmark and Sweden.

Most research concerns themes from the home country, but since 1970 an increasing number of theses have been based on problems of the developing countries. Settlement change, in connection with the study of regional development of industries in Norway, has been an important area of research. Urban and rural social geography, environmental and recreational geography, cultural landscape study and planning are attracting many new graduate students.

The school system is a declining job market for new graduates, nearly half of whom now find jobs in local and regional planning and administration. Few graduates are unemployed.

BIBLIOGRAPHY

Asheim, B., "A Critical Evaluation of Postwar Developments in Human Geography in Scandinavia," *PHG*, 11 (1987), 333-354.

Hägerstrand, T., and A. Buttimer, eds., *Geographers of Norden* (1988).

Holt-Jensen, A., "The Status of Geography in Norway," *PG*, 37 (1985), 90-93.

• *Arild Holt-Jensen*

Norwegian Geographical Society (Det Norske Geografiske Selskab)

The Norwegian Geographical Society was founded in Kristiania (as Oslo was then called) on June 1, 1889, when an initial twenty-five gentlemen with geographical interests were elected by acclamation and an annual subscription of four kroner was agreed upon. There was overlapping membership with the Norwegian Geological Society, which was established a year later. The society's annual publication, *Det norske geografiske selskabs arbog*, appeared in 1891, the first article covering Nansen's crossing of Greenland. From the outset, Nansen took an interest in the society, which retains a direct connection with Polhøjda, Nansen's Oslo home, which today houses a number of research foundations. A quarterly journal, *Norsk Geografisk Tidsskrift*, replaced the *Arbog* in 1926/27, with assistance from the Fridtjof Nansen Foundation. Membership includes both academic geographers and the lay public. The society also administers modest research funds.

• *W. R. Mead*

Nutritional Geography

The specialization of nutritional geography encompasses research on human food intake and nutritional patterns exhibited by communities, groups, nations, or regions. Such research examines physical, cultural, and economic factors that contribute to adequate nutritional status as well as to malnutrition. Nutritional geographers usually conduct their research using paradigms of cultural ecology, culture history, and functionalism. They set their work within descriptive and clinical analytical frameworks to determine (a) how and why food intakes change, and (b) the nutritional consequences of food-related behavior.

Nutritional geography is a hybrid specialization, one not easily nested within the subfields of geography it touches, *i.e.* cultural, economic, historical, medical, and political geography. Practitioners usually exhibit training and grounding in the fundamentals of geographical analysis, but are trained to appreciate interrelationships between geography and the social science disciplines of anthropology, economics, history, psychology, and sociology. Likewise, nutritional geographers usually demonstrate a sound scientific understanding of biology, chemistry, physiology, and nutrition. Most nutritional geographers have trained in traditional departments of geography, with concomitant education in the biological or nutritional sciences. The title of nutritional geographer is also attached to some biologists, nutritionists or physicians with training and formal educational degrees in epidemiology, geography, medicine, or public health.

Historically, nutritional geography originated with food habit and dietary pattern studies conducted by ancient Greek, Roman, Byzantine Christian, and medieval Jewish and Moslem geographers. During the modern era, the field blossomed in the 1920s, initially in Germany, subsequently in France, then in North America. Between 1950 and 1980 most research contributions to nutritional geography were authored by Americans, Canadians, and Europeans. Important research by African and Asian geographers on nutrition-related themes was a hallmark of the 1980s.

The literature encompassing nutritional geography is broad and vast. Ancient accounts and publications of the past 500 years can be grouped into eight thematic clusters: 1) general studies on malnutrition conducted on a national or regional basis; 2) theoretical and methodological problems of data collection and analysis; 3) archaeological and historical perspectives on diet and nutritional change; 4) influence of religious injunctions and dietary laws on food selection; 5) famine and seasonal hunger; 6) genetic-physiological factors influencing food intake; 7) toxic foods, their selection, human use, and dietary role; and 8) specific foods: their culture-history, dispersal, and dietary-nutritional uses.

Bibliography

Dando, W. A., *The Geography of Famine* (1980).

Grivetti, L. E., "Cultural Nutrition: Anthropological and Geographical Themes," *Annual Review of Nutrition*, 1 (1981), 47-68.

May, J. M., *The Ecology of Malnutrition in the Far and Near East: Food Resources, Habits, and Deficiencies* (1961) (plus numerous other titles in this series).

Newman, J. L., "Some Considerations in the Field Measurement of Diet," *PG*, 29 (1977), 171-176.

Simoons, F. J., "Geography and Genetics as Factors in the Psychobiology of Food Selection," pp. 205-224 in *The Psychobiology of Human Food Selection*, ed. by L. M. Barker (1982).

Vermeer, D. E., and R. E. Ferrell, "Nigerian Geophagial Clay, a Traditional Antidiarrheal Pharmaceutical," *Science*, 227 (1985), 634-636.

• *Louis E. Grivetti*

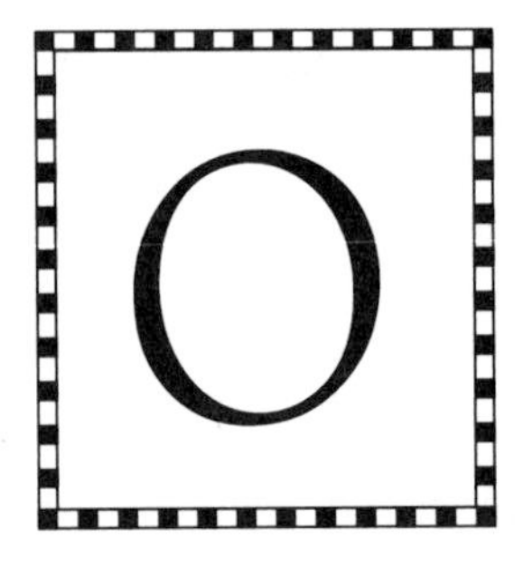

Oberhummer, Eugen (1859-1944)

Dr.phil., University of Munich, 1882. University of Munich, 1886-1903; University of Vienna, 1903-31. *Die Insel Cypern* (1903); *Die Türken und das Osmanische Reich* (1917); (ed.) *Politische Geographie* by F. Ratzel (1923). S. Zimmermann and J. Dörflinger in *GBS*, 7 (1983), 93-100; obit. by H. Hassinger in *PGM*, 90 (1944), 202-203.

Ogawa, Takuji (1870-1941)

D.Sc., Kyoto University, 1909. Kyoto University, 1908-30. *Jinbun Chiri Kenkyu* (1928); *Shina Rekishichiri Kenkyu* (2 vols., 1928-29); *Chishitsu Gensho no Shinkaishaku* (1930); *Nihon Gunto* (1944). U. Tsujita in *GBS*, 6 (1982), 71-76.

Ogilvie, Alan Grant (1887-1954)

B.Sc., University of Oxford, 1915. American Geographical Society, 1920-23; University of Edinburgh, 1923-54. *Some Aspects of Boundary Settlement at the Peace Conference* (1922); *The Geography of the Central Andes* (1922); "The Time-Element in Geography," *TIBG*, 18 (1953), 1-15; *Europe and Its Borderlands* (1957). T. W. Freeman, "Alan Grant Ogilvie," pp. 168-186 in *The Geographer's Craft* (1967); J. N. L. Baker, "A. G. Ogilvie and His Place in British Geography," pp. 1-6 in *Geographical Essays in Memory of Alan G. Ogilvie* (ed. by R. Miller and J. W. Watson) (1959); obit. by J. Bartholomew and D. Linton in *SGM*, 70 (1954), 1-5.

Olsson, Gunnar (b. 1935)

Fil.Dr., University of Uppsala, 1968. University of Michigan, 1966-77; Nordic Institute for Studies in Urban and Regional Planning (Nordplan), Stockholm, 1977 to the present. *Distance and Human Interaction* (1965); *Philosophy in Geography* (with S. Gale) (1979); *Birds in Egg/Eggs in Bird* (1980); *A Search for Common Ground* (with P. Gould) (1982). *Who's Who in the World.*

Ontography

In "An Inductive Study of the Content of Geography," William Morris Davis proposed as geographic the relationship between some "inorganic element of the earth . . . acting as a control . . . and some element of the existence or growth or behavior or distribution of the earth's organic inhabitants, serving as a response." This organic part of geography Davis referred to as ontography. The word *ontography* was apparently minted by Davis in 1902, though ontogeny (the history or science of the development of the individual being) and ontology (the science or study of being) preceded Davis's ontography. Davis's term did not win popularity. But his urging the study of organic response to a physical environment was the geography that dominated the courses of the earliest U.S. geography departments.

Bibliography

Davis, W. M., "An Inductive Study of the Content of Geography," *Bulletin of the American Geographical Society*, 38 (1906), 67-84.

Martin, G. J., "Ontography and Davisian Physiography," pp. 279-289 in *The Origins of Academic Geography in the United States*, ed. by B. W. Blouet (1981).

• *Geoffrey J. Martin*

Ooi Jin Bee (b. 1931)

D.phil., University of Oxford, 1956. National University of Singapore, 1957 to the present. *Peninsular Malaysia* (1976); *The Petroleum Resources of Indonesia* (1982); (ed.) *Natural Resources in Tropical Countries* (1983); *Depletion of the Forest Resources in the Philippines* (1987).

Paassen, Christiaan van (b. 1917)

Doctorate, University of Utrecht, 1957. University of Utrecht, 1947–73; University of Amsterdam, 1973–82. *The Classical Tradition of Geography* (1957); "Human Geography in Terms of Existential Anthropology," *TESG*, 67 (1976), 321–341; "The Philosophy of Geography, from Vidal to Hägerstrand," pp. 17–29 in *Space and Time in Geography*, ed. by A. Pred and G. Törnqvist (1981); "Sociography, Before and After? A Great Antinomy Revisited," pp. 14–40 in *Theoretische Perspectieven in de Sociale Geografie*, ed. by L. van der Laan and B. C. de Pater (1987). *Geografisch Tijdschrift*, vol. 16, no. 4 (1982) (special issue dedicated to van Paassen); *TESG*, vol. 73, no. 6 (1982) (special issue dedicated to van Paassen); H. van Ginkel, "Mirror of a Changing Society: The Development of Residential Geography at Utrecht," *TESG*, 74 (1983), 358–366 (van Paassen on pp. 362–365).

Pakistan. *See* South Asia.

Paris Geographical Society (Société de géographie de Paris)

The Société géographie de Paris, the world's oldest geographical society in continuous existence, was founded in 1821 by a group of specialists in ancient geography such as Jean-Denis Barbié de Bocage, scholars devoted to the Middle East such as Louis Langlès, former members of Napoleon's Egyptian expedition, including Jomard, and successful writers in the field, including Conrad Malte-Brun. Alexander von Humboldt, then in residence in Paris, was among its first foreign members.

Throughout the nineteenth century the society was active in promoting explorations, but because of its inability to raise sufficient funds, it could not finance them directly; instead, the society collected and edited travel accounts—*e.g.*, the account of René Caillié's visit to Timbuktu in 1829. The society's influence relied mainly on its collections, its role as a meeting point, its annual medals, and its journal, the *Bulletin de la Société de géographie de Paris*.

At the time of colonial expansion, from about 1860 to 1885, Charles Maunoir succeeded in making the society prosperous. During the twentieth century, however, it suffered greatly from the inflation generated by the two world wars, which destroyed the society's assets. The launching of the journal *La Géographie* in 1900 was an expression of the will to enlarge the society's audience and to become attuned to the new conceptions of the discipline that were being developed by Vidal de la Blache. But the society attracted more civil servants than scholars. *La Geographie* ceased to be a scientific publication around 1930 and disappeared during World War II.

After 1946 the society was directed by academics for the first time, but its new journal, *Acta Geographica*, did not circulate widely enough to become influential.

Throughout the society's history, it was mainly through its library and through its original maps and travel accounts that the society played a significant role in the development of geography. Its collections are now deposited in the Bibliothèque Nationale.

BIBLIOGRAPHY

Fierro, A., *La Société de géographie 1821-1946* (1983).

Lejeune, D., "La Société de géographie de Paris: un aspect de l'histoire sociale française," *Revue d'histoire moderne et contemporaine*, 29 (1982), 141-163.

Perpillou, A., "La Société de géographie de Paris depuis son centenaire, 1921-1971," *Acta Geographica*, 20 (1975), 3-12.

• *Paul Claval*

Parsons, James Jerome (b. 1915)

Ph.D., University of California, Berkeley, 1948. University of California, Berkeley, 1946-86. *Antioqueño Colonization in Western Colombia* (1949, rev. 1969); *San Andrés and Providencia* (1956); *The Green Turtle and Man* (1962); *Hispanic Lands and Peoples: Selected Writings of James J. Parsons* (ed. by W. Denevan) (1989). *WWA*; *GOF* (1973); J. Parsons, "The Later Sauer Years," *AAAG*, 69 (1979), 9-15; X. Sanclimens, "L'obra de James J. Parsons sobre Espanya," *Documents d'Anàlisi Geogràfica*, 7 (1985), 177-191; essays by R. West and K. Mathewson in *Hispanic Lands and Peoples.*

Partsch, Joseph Franz Maria (1851-1925)

Ph.D., University of Breslau, 1874. University of Breslau, 1876-1905; University of Leipzig, 1905-22. *Die Gletscher der Vorzeit in den Karpathen und in den Mittelgebirgen Deutschlands* (1882); *Schlesien* (4 parts, 1896-1911); *Central Europe* (1903); *Die Geographie des Welthandels* (1927). G. Schwarz in *GBS*, 10 (1986), 125-133; H. Waldbaur, "Zur Erinnerung an Joseph Partsch . . . ," *Die Erde*, 3 (1951), 60-64; obits by E. Brückner in *PGM*, 71 (1925), 179-181; and F. Lehmann in *GZ*, 31 (1925), 321-329.

Passarge, Siegfried (1866-1958)

Doctorates in medicine and geology, University of Jena, 1891-92; (hab.) University of Berlin, 1903. University of Breslau, 1905-08; Colonial Institute in Hamburg, 1908-19; University of Hamburg, 1919-35. *Die Grundlagen der Landschaftskunde* (3 vols., 1919-20); *Vergleichende Landschaftskunde* (5 parts, 1921-30); *Die Erde und ihr Wirtschaftsleben* (1927); *Morphologie der Erdoberfläche* (1929). H. Kanter, "Siegfried Passarges Gedanken zur Geographie," *Die Erde*, 91 (1960), 41-51.

Penck, Albrecht (1858-1945)

Dr.phil., University of Leipzig, 1878; (habil.) University of Munich, 1882. University of Munich, 1883-85; University of Vienna, 1885-1906; University of Berlin, 1906-26. *Die Vergletscherung der Deutschen Alpen* (1882); *Das Deutsches Reich* (1887); *Morphologie der Erdoberfläche* (2 vols., 1894); *Die Alpen im Eiszeitalter* (with E. Brückner) (3 vols., 1901-09). E. Meynen in *GBS*, 7 (1983), 101-108; H. Beck, *Grosse Geographen* (1982), 191-212; H. Louis, "Albrecht Penck und sein Einfluss auf Geographie und Eiszeitforschung," *Die Erde*, 89 (1958), 161-182; G. Engelmann, "Bibliographie Albrecht Penck," *Wissenschaftliche Veröffentlichungen des Deutschen Instituts für Länderkunde*, 17/18 (1960), 331-447; obits. by W. Behrmann in *PGM*, 92 (1948), 190-193; and J. Sölch in *MOGG*, 89 (1946), 7-12, 88-122.

Perception. *See* Behavioral Geography; Environmental Perception.

Personality, Geographical

The concept of geographical personality—the ascription of personality to places as well as to people—became a central theme in the first half of the twentieth century but subsequently lost ground when human geographers began to curb their literary flourishes in order to join the ranks of the social scientists. The first prominent use of the term "personality" was by the French geographer Paul Vidal de la Blache in his *Tableau de la géographie de la France* (1903), the first section of which bears the title, "Personnalité géographique de la France." Vidal borrowed both "tableau" and "personality" from the historian Jules Michelet. Vidal believed that geographical personality grows through time, until a landscape becomes "a medal struck in the likeness of a people." "The word personality," he said, "belongs to the domain and to the vocabulary of human geography." After 1903 the term was made a focal concept in regional descriptions by geographers such as Albert Demangeon, Jean Gottmann, Carl Sauer, and Estyn Evans, and by archaeologists such as Cyril Fox and Glyn Daniel. The Oxford geographer E. W. Gilbert defined geography as "the art of recognizing, describing and interpreting the personalities of regions" (1960). As near-synonyms geographers have employed such terms as "individuality," "physiognomy," and "genius loci" (cf. similar terms in other languages, such as German "Wesen" and "Eigenart").

Bibliography

Dunbar, G. S., "Geographical Personality," pp. 25–33 in *Man and Cultural Heritage: Papers in Honor of Fred B. Kniffen*, ed. by H. J. Walker and W. G. Haag (1974).

Evans, E. E., *The Personality of Ireland* (2nd ed., 1981).

Gilbert, E. W., "The Idea of the Region," *Geography*, 45 (1960), 157–175.

Peru

The modern discipline of geography in Perú, like modern society in that country, has roots that can be traced back to both the colonial and the republican periods. Spanish "Visitas Generales" and "Relaciones Geográficas" from the sixteenth and seventeenth centuries provide a tradition of micro-regional studies, whereas the reports of numerous nineteenth century European scholars (*e.g.*, the German von Humboldt and the Italian Raimondi) began a pattern in which much of Peruvian geographical research has been done by foreigners. Academic geography, on the other hand, is a relatively young discipline in Perú; only one university in the country (San Marcos in Lima) has had a successful geography program for more than a decade. Other strong departments (especially that of the Catholic University in Lima) were established only recently.

The most important geographical organization in Perú has long been the Sociedad Geográfica de Lima, founded in 1888. It has regularly published a *Bulletin* (currently inactive) and sponsored geographical lectures and research, as well as maintained the most important collection of maps in the country. The society's small library has been useful for visiting scholars.

Perú's leading geographer of this century has certainly been Javier Pulgar Vidal, professor of geography at San Marcos for many years and author of almost a dozen books on Peruvian and Latin American geography. His *Geografía del Perú: las Ocho Regiones Naturales* has long been the standard text on Peruvian geography. The encyclopedic eight-volume work, *Gran Geografía del Perú*, provides a more detailed examination of the country's geography. Emilio Romero's *Geografía Económica del Perú* is an older study (first published in 1930) that still provides much information about the economic geography of the country. Important English language texts include classics by Isaiah Bowman (*The Andes of Southern Peru*, 1916) and George Johnson (*Peru from the Air*, 1930) and, in recent years, the numerous articles by William Denevan and Daniel Gade.

Bibliography

Dourojeanni, M. J., *Gran Geografía del Perú* (1986).

Peñaherrera del Aguila, C., *Geografía General del Perú* (1969).

Pulgar Vidal, J., *Geografía del Perú: las Ocho Regiones Naturales* (9th ed., 1987).

Yacher, L., "Geography in Peru," *PG*, 41 (1989), 220–223.

• *James S. Kus*

Pfeifer, Gottfried (1901–1985)

Dr.phil., University of Kiel, 1927; (hab.) University of Bonn, 1936. University of Hamburg, 1947-49; University of Heidelberg, 1949-69. *Die räumliche Gliederung der Landwirtschaft im nördlichen Kalifornien* (1936); *Sinaloa und Sonora* (1939); *Vereinigte Staaten von Amerika* (1973); *Beiträge zur Kulturgeographie der Neuen Welt* (1981). Obits. by G. Sandner in *GZ*, 74 (1986), 1-2; and G. Kohlhepp in *Geographisches Taschenbuch 1987/1988* (1987), 133-157.

Philippson, Alfred (1864-1953)

Dr.phil., University of Leipzig, 1886; (hab.) University of Bonn, 1897. University of Bonn, 1909-29. *Das Mittelmeergebiet* (1904, 4th ed. 1922); *Grundzüge der Allgemeinen Geographie* (3 vols., 1921-24); *Die Griechischen Landschaften* (4 vols., 1950-58), *Die Stadt Bonn* (1947, 2nd ed. 1951). H. Lehmann, "Alfred Philippsons Lebenswerk," *Colloquium Geographicum*, no. 5 (1956), 9-14; H. Lehmann, "Alfred Philippson . . . ," *GZ*, 52 (1964), 1-6; H. Lehmann and C. Troll, "Alfred Philippson, 1864-1953," pp. 205-214 in *Bonner Gelehrte* (1970).

Physical Geography

Mary Somerville's *Physical Geography* (1848) was one of the first and most influential of textbooks in physical geography and gave a clear definition of the field: "Physical geography is a description of the earth, the sea, and the air, with their inhabitants animal and vegetable, of the distribution of these organized beings, and the causes of that distribution . . . man himself is viewed but as a fellow-inhabitant of the globe with other created things, yet influencing them to a certain extent by his actions, and influenced in return. The effects of his intellectual superiority on the inferior animals, and even on his own condition, by the subjection of some of the most powerful agents in nature to his will, together with the other causes which have had the greatest influence on his physical and moral state, are among the most important subjects of this science" (quotation from 4th ed., 1858).

Somerville's view of physical geography had certain similarities with that of Arnold Guyot (*Earth and Man*, 1850) who thought that physical geography should be more than "mere description": "It should not only describe, it should compare, it should interpret, it should rise to the *how* and the *wherefore* of the phenomena which it describes." Guyot regarded physical geography as "the science of the general phenomena of the present life of the globe, in reference to their connection and their mutual dependence." As part of his concern with "connection" and "mutual dependence" he included a consideration of racial differences and performance in relation to environmental controls, a theme that was to be developed by the North American school of environmental determinists over the next six to seven decades.

Thus both Guyot and Somerville saw a human dimension in physical geography and believed that it was more than an incoherent catalogue. This was a view shared by Thomas Huxley, who used the term "physiography" in preference to "physical geography." Similarly, in France, Emmanuel de Martonne, in addition to covering what he regarded as the four main components of physical geography—climatology, hydrography, geomorphology and biogeography—also included a consideration of environmental influences.

Physical geographers played a major role in the establishment of geography in American universities, and made up a large proportion of the founding members of the Association of American Geographers when it was established in 1904, not least because of the energy and influence of W. M. Davis. However, from the 1920s physical geography suffered an eclipse in the USA, and even in Britain it was taught between the two world wars by relatively few. Of the few, most were geomorphologists, and Stoddart (1987) remarks that at that time, "In research, if not in teaching, 'physical geography' meant geomorphology: for while some attention was given to meteorology, climatology, and to some extent pedology and biogeography, it was on the level of elementary service courses for students rather than as a contribution to new knowledge."

In contrast to the works of Guyot, Somerville and de Martonne, physical geography tomes tended increasingly to ignore human and environmental influences. So, for example, Pierre Birot saw physical geography as the study of "the visible surface of natural landscapes as they would appear to the naked eye of an observer travelling over the globe before the interaction of mankind."

Physical geography possibly reached its nadir within geography in the late 1960s and early 1970s when spatial modellers, particularly of urban systems, saw little room for it in the discipline, and even some physical geographers doubted the role of the subdiscipline in a world where regional differences were seen to be declining and where many people were becoming progressively divorced from the reality of their immediate physical surroundings. Since that time, however, because of a concern with the role of natural hazards and, more importantly, because of a concern with human-induced modifications of the environment and their implications for human welfare, environmental concerns have assumed a much more important position in the discipline. Many physical geographers have also sought to explore the applications of their work and have sought to use systems as an integrating framework. Kenneth Gregory (1985) has provided an overview of these and other developments. *See also* Geomorphology.

Bibliography

Chorley, R. J., "The Role and Relations of Physical Geography," *Progress in Geography*, 3 (1971), 87-109.

Cooke, R. U., and J. C. Doornkamp, *Geomorphology in Environmental Management* (1974).

Goudie, A. S., *The Human Impact on the Environment* (1986).

Gregory, K. J., *The Nature of Physical Geography* (1985).

Steers, J. A., "Physical Geography in the Universities, 1918-1945," pp. 138-155 in *British Geography 1918-1945*, ed. by R. W. Steel (1987).

Stoddart, D. R., "Geographers and Geomorphology in Britain between the Wars," pp. 156-176 in *Ibid.*

• *A. S. Goudie*

Pinchemel, Philippe Amand Raymond (b. 1923)

D.-ès-L., University of Paris, 1952. University of Lille, 1946-48, 1954-65; University of Besançon, 1952-53; University of Paris, 1948-51, 1965 to the present. *Géographie de la France* (1964); *Deux siècles de géographie française* (with M. Robic and J. Tissier) (1984); *France, a Geographical, Social and Economic Survey* (1987); *La face de la terre: éléments de géographie* (with G. Pinchemel) (1988). *WWF*; *GOF* (1979).

Place

Place presents itself to us as a condition of human experience. We live our lives in place and have a sense of being part of place, but we also view place as something separate and external to us. Thus we experience place as both a center of meaning and the external context of our actions.

The tension between a relatively subjective, existential view of place and a relatively objective, naturalistic view of place represents a fundamental polarity in our understanding, and helps explain why place has been such an intriguing, yet troublesome, concept in modern geography. From the relatively detached view of the theoretical scientist place becomes an object of the world. From our perspective as subjects place is something that is a part of us and contributes to our sense of who we are, both as individuals and as members of groups.

In modern geography the concept of place has different meanings depending on the perspective taken. For the theoretical geographer place becomes synonymous with location, a coordinate on a map or a point in space. For the regional geographer, place consists of the ensemble of heterogeneous phenomena that give character to an area. For the humanistic-cultural geographer place becomes a context for human action, a context rich in human significance and meaning. This variety and ambiguity are essential features of the concept as it has been employed by geographers.

Bibliography

Berdoulay, V., *Des mots et des lieux: la dynamique du discours géographique* (1988).

Entrikin, J. N., *The Betweenness of Place: Towards a Geography of Modernity* (1990).

Tuan, Y.-F., *Space and Place: The Perspective of Experience* (1977).

• *J. Nicholas Entrikin*

Plant Geography. *See* Biogeography.

Platt, Robert Swanton (1891-1964)

Ph.D., University of Chicago, 1920. University of Chicago, 1920-57. *Latin America: Countrysides and United Regions* (1942); *Field Study in American Geography* (1959). R. Thoman in *GBS*, 3 (1979), 107-116; obits. by R. Hartshorne in *AAAG*, 54 (1964), 630-637; and C. Harris in *GR*, 54 (1964), 444-445.

Poland

The main stages in the development of geography in Poland indicate their dependence on its political conditions.

Stage I (1890-1918). Polish geography (regional geography in particular) tried to develop patriotic feelings through the educational system. The unity of the nation, and that of "the historical Poland" in its boundaries from before the partitions, were the principal ideas strengthening the national identity. Other forms of geographical works served the same purpose. For example, the Polish statistical atlas by E. Romer (1916) played an important role when the boundaries of the reborn state were discussed (1918-21). Since there existed but one autonomous Polish university, in Cracow, Polish geographers studying at the universities in Austria, Prussia, and Russia followed the research patterns of the European universities. Polish scientific institutions, such as the Polish Academy of Sciences and Letters, functioned in accordance with the general aspirations to regain national independence.

Stage II (1918-1939). In the reborn state, geographers in the five university centers—Cracow, Lwow, Poznań, Warsaw, and Wilno—were primarily involved in physical and economic geographical studies of the country. As a rule, each university accepted the nearest region as its research field. Complex studies of the whole country were rather rare, whereas geographical treatments of other countries of the world were even less frequent. In economic and physical geography Polish scholars accepted the paradigms that were then current in the world—for example, the erosion cycles of W. M. Davis.

Stage III (since 1945). Geographers of the above-mentioned universities and of the new ones (Toruń and Wrołcaw) carried on studies related to the physical and economic geographical problems of their respective regions or of the whole country. Since 1950 a leading role in those studies has been played by the Institute of Geography of the Polish Academy of Sciences.

In Poland, whose political existence has been dependent on the general history, three periods in the transformation of Polish geography may be differentiated. They developed in the conditions of international academic influences and of the social, economic, and political changes within the country. Inspired by Thomas Kuhn's theory, Antoni Kuklinski analyzed postwar geography in Poland and indicated the following periods and the prevailing paradigms therein.

Period I (from the end of the nineteenth century to 1950). Geographical work was largely descriptive, rather than analytical, and regional geography was dominant.

Period II (1950-76). This was a period of intensive contacts with Western science, documented by international symposia publications and by fifty volumes of *Geographica Polonica*. The "quantitative revolution" was introduced, and some branches of geography underwent increased mathematization.

Period III (since 1976). The earlier ruling paradigms have declined, and Polish geographers are now concerned with questions related to the current economic, political, and ecological crises.

Bibliography

Babicz, J., "L'école géographique polonaise d'Eugenіusz Romer," pp. 45-48 in *Les écoles géographiques*, ed. by J. Babicz (1980).

———, "Trois étapes de recherches sur l'histoire de la géographie en Pologne," pp. 24–36 in *Studia z dziejów geografii i kartografii*, ed. by J. Babicz (1973).

Kondracki, J., "The West Slav Geographers, Part II.—The Development of Geography in Poland," pp. 122–127 in *Geography in the Twentieth Century*, ed. by G. Taylor (3rd ed., 1957).

Kuklinski, A., "Poland," pp. 132–155 in *Geography since the Second World War*, ed. by R. J. Johnston and P. Claval (1984).

Rykiel, Z., "The Functioning and the Development of Polish Human Geography," *PHG*, 12 (1988), 391–408.

• *Józef Babicz*

Political Geography

Geography in general and political geography in particular emerged as an academic discipline as a result of the imperialist scramble for territories at the end of the nineteenth century. The Franco-Prussian War of 1870–71 was notable for the involvement of students of geographer Carl Ritter as mapmakers on the German side. As a reaction to defeat, France also encouraged the development of geographical institutes and the training of young officers in the methods of military geography. A double relationship developed between academic geography and governmental and business interests, engaged in the mapping and identification of colonial resources. Nationalist ideals were promoted by the respective geographical societies, and the strong popular base of geography in the European states was formed.

The first academic political geography text was Friedrich Ratzel's *Politische Geographie* (1897). In it the German geographer developed a pseudo-scientific argument that could be used to justify the territorial expansion of states. By his "law of the spatial growth of states," he described the expansion of a state through war as a natural progressive tendency. His ideas were drawn from Social Darwinism; just as species compete, so too will large states grow at the expense of smaller neighbors. At the same time, the other founding father of political geography, Halford Mackinder (UK), was considering the changing world-order and attempting to develop a geostrategy for his home state that would help perpetuate British hegemony into the twentieth century. He developed the most famous political geographic model, that of a Eurasian continental bloc as the "geographical pivot of history." To counter the natural strategic advantages of a great power sheltered in the center of Eurasia, yet able to reach to all parts of the world-island (Asia, Europe, and Africa), Mackinder advocated the control of the Eurasian littoral by the British (allied to the U.S., a junior partner). This geographic juxtaposition of Heartland (landpower) and Rimland (seapower) became the most debated and policy-relevant political geographic concept after World War II, when the American policy of containment of the USSR was implemented. The strategic rationale for the policy was derived from Mackinder's work, though it should be noted that Mackinder thought that Germany would be the landpower and that Great Britain would be the seapower. Additionally, the natural protection afforded by the great spaces of the Eurasian landmass was removed by the deployment of U.S. intercontinental ballistic missiles (ICBMs) in the 1950s. Nevertheless, the Mackinder worldview still pervades geostrategic analyses by contemporary U.S. policy analysts.

The Treaty of Versailles (1919) was a geopolitical act with far-reaching consequences, not the least of which was the stimulus given to political geography in the major states. The United States, now a global power, had Isaiah Bowman as Chief Territorial Specialist at the peace conference. His *New World* (1921) advocated an activist globalist foreign policy for the U.S.; this book, too, contains a lot of Social Darwinism in support of territorial extension of U.S. interests. On the French side, Emmanuel de Martonne advised the government on the placement of the new European national boundaries. Mackinder was active in British politics as a member of Parliament and saw service as a foreign emissary of his government on numerous occasions. But nowhere did political geography penetrate the national discussion and halls of power as in Nazi Germany. General-Major Karl Haushofer created "Geopolitik" (geopolitics) as a counter to Anglo-Saxon worldviews. After becoming professor of geography at the University of Munich, Haushofer founded the

Zeitschrift für Geopolitik as a journal to give voice to his ideas and that of other geopoliticians. The key concept was "Lebensraum" (living space) borrowed from Ratzel, and German dominance of Eastern Europe was taken as an essential ingredient in the reestablishment of German power on the continent. Haushofer drew up a global model of panregions, which he conceived as political-economic unions of nation-states, of sufficient size to discourage attack and to permit economic autarchy. Each panregion would have a developed, dominant core and a resource periphery. The four areas would be PanAsia, with Japan as the core; EurAfrica, with a German core; PanAmerica, with a U.S. core; and PanRussia, consisting essentially of the Soviet Union. Haushofer held out the hope that panregions would evolve through accommodation between global powers in recognition of geopolitical realities and mutual interests, but he did not exclude the possible use of force to achieve this global equilibrium. Interestingly, this panregional world has evolved as trade-regions in the late twentieth century. Although Hitler and the Nazis used the ideas of the geopoliticians as justification for their expansionist policies, recent research has revealed that the personal links between Haushofer and Hitler were vastly overstated by U.S. wartime commentators.

Contemporary research in political geography can be subdivided by scale of interest, though an important stream of work has evolved that examines links between global-level trends and state and local interests. For example, various countries have undertaken policies that try to ameliorate the negative effects on the old industrial localities of the shift in manufacturing to the Newly Industrializing Countries. At the international scale, the global shifts engendered by the relative decline of the U.S. and the stresses caused by the clash of alternative worldviews, from Washington, Moscow and numerous Third World centers, remain the subject of much research. An important topic is the intervention of outside states in local conflicts. The cause of most of these conflicts is the attempt to control the same territory by competing nationalisms (for example, in Palestine), and so the international scale intersects with the national territorial divisions. Other important national-level topics are the distribution of government revenues and taxes (which regions benefit and which lose) and the electoral geography of states. Here, both the system of translating votes into seats, especially the delimitation of voting districts, and the bases of party support, are major research topics. Finally, at the local (usually metropolitan) scale, the geography of access remains the dominant theme, though there is increasing interest in the "neighborhood" effect, whereby residents of particular milieux develop a local identity and use this identity to choose candidates and maintain political activism.

The best general definition of political geography is "who gets what, where and how." It is equally applicable to the three scales of inquiry in the discipline. The checkered history of the discipline reminds us that there is a close connection between the concerns of governments and the dominant research themes of political geography. More than most research topics, political geography is a creature of its sociopolitical environment.

Bibliography

Girot, P., and E. Kofman, eds., *International Geopolitical Analysis* (1987).

Johnston, R. J., and P. J. Taylor, eds., *A World in Crisis: Geographical Perspectives* (2nd ed., 1989).

Parker, G., *Western Geopolitical Thought in the Twentieth Century* (1985).

Political Geography Quarterly (1982-).

Taylor, P. J., *Political Geography: World-Economy, Nation-State and Locality* (2nd ed., 1989).

Taylor, P. J., and J. W. House, eds., *Political Geography: Recent Advances and Future Directions* (1984).

• *John O'Loughlin*

Population Geography

Population geography is an important but late-blooming specialty within the larger discipline, a tardiness explained by past inadequacies in census data and registers of vital events, but perhaps even more by the fact that human beings are relatively inconspicuous elements in the visible landscape. But since the 1950s, with the pioneering work of such scholars as Glenn Trewartha, V. V. Pokshishevskii, William William-Olsson, Pierre George, and Jacqueline Beaujeu-Garnier, demographic topics have come into their own in the world of geography—as is also true for other relatively intangible social and psychological facets of the human scene. Several textbooks and collections of readings have appeared, along with various atlases; many universities now offer courses in population geography; and a formal interest group has materialized within the U.S. profession. Geographers are also active within national groups such as the Population Association of America and the International Union for the Scientific Study of Population. Work of geographic interest appears in such leading journals as the U.S. *Demography*, the British *Population Studies*, and the French *Population*. At the international scale, the Population Commission of the International Geographical Union has performed yeoman work in stimulating research, staging conferences, and publishing symposial volumes.

The effective mapping of population numbers, change, and other characteristics at various scales was a major preoccupation of the subfield during its formative years, and much progress, both theoretical and substantive, has occurred, in tandem with advances in general statistical cartography and computerized mapping. The nature of the subject is such that most population geographers have worked with, or shared interests with, sociological and economic demographers. Within the geographic fold, there is substantial overlap with students specializing in urban, social, medical, and various types of applied geography, as well as with those specializing in the problems of the less-developed countries.

In principle, population geographers can and should involve themselves with all the standard demographic topics. Nevertheless, although a few analysts have investigated the geography of fertility, of mortality, of educational, marital, and household attributes, and of population projections, the major investments of energy have been in two sectors relatively neglected by demographers: the spatial patterning of human populations and temporal changes therein, and migration and spatial mobility in all their dimensions. Indeed it is in the latter area that geographers have made their most meaningful contributions to population studies, and can be expected to continue doing so.

Bibliography

Beaujeu-Garnier, J., *Geography of Population* (2nd ed., 1978).

Demko, G. J., H. M. Rose, and G. A. Schnell, *Population Geography: A Reader* (1981).

Jones, H. R., *A Population Geography* (1981).

Kosinski, L. A., and R. M. Prothero, eds., *People on the Move: Studies on Internal Migration* (1975).

Trewartha, G. T., *A Geography of Population: World Patterns* (1969).

Woods, R., *Theoretical Population Geography* (1982).

• *Wilbur Zelinsky*

Portugal

The Sociedade de Geografia de Lisboa (Geographical Society of Lisbon) was founded in 1875. It brought together scholars from various areas, and had features similar to those of its European counterparts. The society developed an extensive research program on the Portuguese Empire (cartographic surveys, organization of scientific expeditions, etc.), and from its founding has published its *Boletim* and, between 1965 and 1973, the journal *Geographica*.

The institutionalization of geography as an autonomous science, however, came about with the introduction of its teaching in the universities, through Professors Silva Telles at Lisbon, A. Ferraz de Carvalho at Coimbra, and A. Mendes Corrêa at Oporto.

At Coimbra geography gained great momentum with the creation of the Centro de Estudos Geográficos, which between 1950 and 1967 published an important *Boletim*. Amorim Girão, its founder, also developed in that city a notable teaching program, while publishing numerous studies. This work has been continued by A. Fernandes Martins and, more recently, in the reorganization of the Instituto de Estudos Geográficos and the publication of the *Cadernos de Geografia* (since 1983).

In 1942, the justly best-known Portuguese geographer, Orlando Ribeiro, settled at Lisbon. A disciple of Emmanuel de Martonne and Albert Demangeon, Ribeiro founded in 1943 the Centro de Estudos Geográficos de Lisboa, and in

1949 organized the Sixteenth International Geographical Congress, and has since directed an important group of research scholars. Since 1966 the Centro has published the journal *Finisterra* and various collections of studies.

Meanwhile, two new departments of geography arose, one at the University of Oporto (1973), the other at the Universidade Nova of Lisbon (1979), each also with a research center. Oporto since 1985 has also put out an annual journal.

The teaching of geography also became established alongside other sciences and in other universities at Sá da Bandeira (Angola), Lourenço Marques (Mozambique), Ponta Delgada (Azores), Aveiro and Évora. The impact made itself felt in regional planning, cartography, etc.

A final reference must be made to the studies on tropical geography carried out since the 1940s by Portuguese geographers in Africa, India and Oceania. The present Centro de Geografia of the Instituto de Investigaçao Cientifica Tropical is the heir to all this work, and publishes the geographical series of the journal *Garcia de Orta*. *See also* Lisbon Geographical Society.

BIBLIOGRAPHY

Alcoforado, M. J., *O clima da região de Lisboa* (1989).

Bibliografia Geográfica de Portugal (1948, 1982).

Gaspar, J., *A área de influência de Évora* (1972).

Girão, A., *Atlas de Portugal* (1941).

Ribeiro, O., *Portugal, o Mediterrâneo e o Atlântico* (1945).

Ribeiro, O., H. Lautensach, and S. Daveau, *Geografia de Portugal* (1987-).

• *João Carlos Garcia*
(translated by C. Julian Bishko)

Possibilism

This interpretation of man-environment relationships, developed by Vidal de la Blache and others during the formative phase of French geography, is usually contraposed to environmental determinism. In essence it postulates that the environment, whether natural or culturally modified, determines neither human history nor present conditions, presenting rather a range of "possibilities" that are selectively developed by different societies according to their respective "genres de vie," or modes of living.

Occasionally possibilism has been viewed as an assertion of free will coupled with the denial of environmental influence, but this is to misinterpret its thrust. It was an accepted principle that the earth's surface was differentiated into areas unequally fit for human occupance; indeed, a hierarchical ordering of potentialities was involved. Furthermore, different genres de vie both (a) assimilated the experience of former milieux into their fabric and (b) consolidated into social complexes that channeled further environmental relationships. But it was characteristically insisted that the initiative lay with man rather than with nature, that deterministic laws did not govern historical developments and regional patterns, and that selection within the varying range of potentialities was contingent—it was not predetermined, necessary, or synchronous. In Febvre's simile, possibilities were like the keys of a piano: selection of notes varies with society and time. See also *Genre de Vie*.

BIBLIOGRAPHY

Febvre, L., *La terre et l'évolution humaine* (1922).

Lukermann, F., "The 'Calcul des Probabilités' and the Ecole Française de Géographie," *CG*, 9 (1965), 128-137.

Vidal de la Blache, P., "Les genres de vie dans la géographie humaine," *AG*, 20 (1911), 193-212, 289-304.

• *Gordon R. Lewthwaite*

Pred, Allan Richard (b. 1936)

Ph.D., University of Chicago, 1962. University of California, Berkeley, 1962 to the present. *The Spatial Dynamics of U.S. Urban-Industrial Growth, 1800-1914* (1966); *Urban Growth and the Circulation of Information: The United States System of Cities, 1790-1840* (1973); *City-Systems in Advanced Economies* (1977); *Urban Growth and City-Systems in the United States, 1840-1860* (1980). *GOF* (1980); A. Pred, "The Academic

Past through a Time-Geographic Looking Glass," *AAAG*, 69 (1979), 175-180; A. Pred, "From Here and Now to There and Then," pp. 86-103 in *Recollections of a Revolution* (ed. by M. Billinge, D. Gregory, and R. Martin) (1984).

Price, Archibald Grenfell (1892-1977)

D.Litt., University of Adelaide, 1932. University of Adelaide, 1925-57. *The Foundation and Settlement of South Australia* (1924); *White Settlers in the Tropics* (1939); *White Settlers and Native Peoples* (1949); *The Western Invasion of the Pacific and Its Continents* (1963). J. Powell in *GBS*, 6 (1982), 87-92; obit. by anon. in *GJ*, 144 (1978), 180-181.

Quantitative Methods in Geography

Although statistical methods had been used in geographical studies before World War II, the development of a vigorous quantitative geography was a phenomenon of the 1950s and 1960s. Variously called statistical geography, quantitative geography, quantitative methods, and, more broadly, spatial analysis, the new area was concerned with the application of descriptive and inferential statistics to geographical problems and data. While other fields developed their branches of econometrics, sociometrics and psychometrics, geographers seemed unwilling to follow with geometrics if only because of the confusion with geometry.

As others have noted, the development of a statistical approach in the social sciences generally can be viewed in the postwar context of an increased interest in technological innovation and technical approaches to planning and policy. The use of statistical methods expanded dramatically in the mid-1950s. The occasional statistical paper of the prewar years became a large number of papers on a variety of topics that employed statistical techniques from regression analysis to factor analysis. To an older, more classically trained group of geographers the changes were indeed a "quantitative revolution."

During the 1950s there were a number of individuals who established programs and courses in statistical methods and gathered graduate students who became versed in the new techniques. William Garrison at Washington, Torsten Hägerstrand at Lund, Harold McCarty at Iowa and Edward Taaffe at Northwestern were influential users of the new techniques, and Garrison's students (notably Brian Berry, Michael Dacey, Arthur Getis, Duane Marble, Richard Morrill, and Waldo Tobler at Washington) diffused the techniques to major departments of geography throughout the United States. The establishment of the *Papers of the Regional Science Association* in 1954 and the formation of the new journals *Geographical Analysis* and *Environment and Planning A* in 1969 provided an outlet for technical papers that had not always received a welcome in the traditional geography journals. They also legitimized the new focus.

Most of the early research in statistical methods was quite derivative of work in allied fields, especially in economics and sociology, but increasingly the statistical work has achieved an independence from those studies, and the new research in statistical methods plays an important role in the training and preparation of professional geographers. Despite a radical critique of the quantitative approach and a rejection of the search for a value-free approach to geographic inquiry, the humanist-scientist debate has been short-lived, and the 1980s saw the movement away from geography as a spatial science and toward a focus on spatial processes and spatial relations. Much of the recent statistical and/or quantitative geography is concerned with understanding the link between spatial processes and their outcome as spatial patterns. In the most current work the analysis of spatial problems has recognized the limitations of simple techniques of statistical inference and extended the approaches to techniques that elucidate the underlying data structures. The methods of analysis that are increasingly relevant are those that focus on non-random samples, sometimes of the whole population, and deal with situations in which spatial autocorrelation is an essential part of the analysis. Techniques of categorical data analysis are currently among the most important ways of dealing with geographic data. Refocusing on a geographical core of relationships in space emphasizes quantitative geography as a tool and not as a methodological core in itself. Whether spatial analysis will develop new forms of mathematical expression is a question for the future.

BIBLIOGRAPHY

Berry, B. J. L., and D. F. Marble, eds., *Spatial Analysis: A Reader in Statistical Geography* (1968).

Clark, W. A. V., and P. L. Hosking, *Statistical Methods for Geographers* (1986).

Gaile, G. L., and C. J. Willmott, eds., *Spatial Statistics and Models* (1984).

Haggett, P., *Locational Analysis in Human Geography* (1965).

King, L. J., *Statistical Analysis in Geography* (1969).

Wrigley, N., *Statistical Applications in the Spatial Sciences* (1979).

• *W. A. V. Clark*

Quantitative Revolution

"Quantitative Revolution" was the not-quite-accurate label applied to a partial transformation of geographic methodology and subject orientation, beginning in the latter 1950s and extending through the 1960s. In fact, it was part of a wider trend emphasizing theory construction, mathematical modelling, and use of statistical methods in the behavioral and social sciences. Although there were precursors even in the 1930s, the specific appellation followed a coincidence of personalities at the universities of Washington and Iowa in the 1950s. Under the leadership of William Garrison and Edward Ullman at Washington and Harold McCarty at Iowa, a cadre of zealous graduate students (including Brian Berry, Duane Marble, William Bunge, Michael Dacey, Richard Morrill, Edwin Thomas, and Leslie King) embraced a vision of a nomothetic geography devoted to a theory of spatial relations and organization, whose propositions would be verified by appropriate statistical and econometric methods. They were inspired in part by the arguments of Fred Schaefer at Iowa and from the beginnings of Regional Science, and influenced by exposure to theory and statistics in sociology and economics. In several institutions, the new ideas came through cartography, as at Wisconsin (Arthur Robinson). An initial resistance only redoubled our fervor. The "discussion paper" series (1957) spread rapidly from Washington to other schools and, in fact, the subversive activity gained surprisingly quick acceptance from an expanding number of faculty and students. By 1959, the first trained Ph.D. disciples were placed at such major schools as Chicago, Northwestern, and Michigan, and were being quite widely published, despite the perceived barriers. By 1959, too, NSF-sponsored symposia and workshops on "quantitative methods" occurred, spreading the message to Canada, Britain, Australia, New Zealand, and another "generation" of protagonists (such as John Cole, Richard Chorley, Stanley Gregory, Peter Haggett, and then Ronald Johnston, David Harvey, and others). Also, connections were established with an independent center of theoretical work led by Torsten Hägerstrand in Sweden. Dozens of books were published, beginning in the mid-1960s, espousing the new paradigm; Harvey's *Explanation in Geography* was the first broad philosophical exposition of method, William Bunge's *Theoretical Geography* the first major statement of substance.

While the transformation was "quantitative" in the sense of establishing the utility of statistics and mathematics in expressing geographic concepts, and while argument over such methods was clamorous, the truly significant change was qualitative—that is, from a view of geography as regional, descriptive and integrating to a view of geography as spatial, theoretical and topically specialized. Thus in only one decade the revolution was successful in the sense of forcing the legitimacy of new methods and goals. The truest believers would argue that these changes saved university-level geography from extinction, following the decline of its teacher-training mission. In any event, it is also the case that the revolution was but partial or superficial. It has come to complement rather than displace earlier traditions and methodologies. Indeed, in a kind of "counterrevolution" in the 1970s and 1980s, the very orthodoxy of "spatial, theoretical" has been subject to a manifold critique—not so much with respect to the utility of statistics and mathematics, but to the meaningfulness of extant spatial theory.

Bibliography

Billinge, M., D. Gregory, and R. Martin, eds., *Recollections of a Revolution* (1984).

Bunge, W., *Theoretical Geography* (1962, 2nd ed. 1966).

Burton, I., "The Quantitative Revolution and Theoretical Geography," *CG*, 7 (1963), 151–162.

Chorley, R., and P. Haggett, eds., *Models in Geography* (1967).

Cole, J., and C. A. M. King, *Quantitative Geography* (1968).

Harvey, D., *Explanation in Geography* (1969).

• *Richard Morrill*

Radical Geography

Radical geography insists that geographical phenomena (distributions, landscapes) can be understood only in terms of socially divisive and conflict-ridden economic and social processes. Early examples could perhaps be found in some of the writings of Elisée Reclus and Peter Kropotkin. However, the term "radical geography" came into common use only after the social and political turmoil of the late 1960s. There are important French, Italian, British and U.S. "schools" of radical geography. The journals *Hérodote* (Paris) and *Antipode* (Worcester, Massachusetts) have been the most important publication outlets, but today "radical" work is published in all journals rather than being restricted to a few "sectarian" outlets.

Major differences have developed since the late 1960s among those studies (a) that adopt a welfare focus, tracing spatial disparities in wealth, welfare, and social well-being to inequalities emanating from regional, national, or global social structures (*e.g.*, the work of David Smith); (b) that adopt a Marxist viewpoint in tracing welfare inequalities to the class divisions and economic logic of a particular mode of production, which today is capitalism (*e.g.*, the work of David Harvey); and (c) that argue from a radical cultural position in order to identify the social practices and ideas that give rise to and legitimize the social inequalities underlying geographical phenomena (*e.g.*, the work of Derek Gregory and of Denis Cosgrove).

There is today much less explicit political argument than was once the case. Radical geography has to a considerable extent been academicized and institutionalized. Some of its most notable practitioners hold prestigious university chairs. Only the development of a feminist critique over the past ten years has introduced a fresh source of overt political commitment, often to the discomfort of the more established nonfeminist positions.

A continuing source of unity is a shared critique of the view of geography as spatial science, which is seen as isolating geographical phenomena from their socioeconomic contexts. Radical geography continues to provide an important source of lively thinking antithetical to many disciplinary nostrums and reminds *all* geographers that processes of inequality and conflict should be addressed rather than assumed away.

Bibliography

Corbridge, S., *Capitalist World Development: A Critique of Radical Development Geography* (1986).

Harvey, D., *Social Justice and the City* (1973).

Peet, J. R., ed., *Radical Geography* (1977).

"Symposium 'Reconsidering Social Theory: A Debate,'" *Society and Space*, 5 (1987), 367–434.

• *John Agnew*

Raisz, Erwin Josephus (1893-1968)

Ph.D., Columbia University, 1929. Harvard University, 1931-50. *General Cartography* (1938); *Principles of Cartography* (1962); *Atlas of Florida* (with others) (1964). L. Yacher in *GBS*, 6 (1982), 93-97; obit. by A. Robinson in *AAAG*, 60 (1970), 189-193.

Ratzel, Friedrich (1844-1904)

Ph.D., University of Heidelberg, 1868. Technical University of Munich, 1875-86; University of Leipzig, 1886-1904. *Die Vereinigten Staaten* (2 vols., 1878-80), *Anthropogeographie* (2 vols., 1882-91), *Völkerkunde* (3 vols., 1885-88), *Politische Geographie* (1897). M. Bassin in *GBS*, 11 (1987), 123-132; H. Wanklyn, *Friedrich Ratzel, a Biographical Memoir and Bibliography* (1961); G. Buttmann, *Friedrich Ratzel: Leben und Werk eines deutschen Geographen* (1977); obits. by J. Brunhes in *La Géographie*, 10 (1904), 103-108; and K. Hassert in *GZ*, 11 (1905), 305-325, 361-380.

Reclus, Elisée (1830-1905)

University-level education at Protestant seminary in Montauban, France, 1848-49, and the University of Berlin, 1851. Hachette publishing house, Paris, 1858-71; New University of Brussels, 1894-1905. *La terre* (2 vols., 1868-69), *Nouvelle géographie universelle* (19 vols., 1876-94), *L'homme et la terre* (6 vols., 1905-08). B. Giblin in *GBS*, 3 (1979), 125-132; G. Dunbar, *Elisée Reclus, Historian of Nature* (1978); M. Fleming, *The Geography of Freedom: The Odyssey of Elisée Reclus* (1988); obits. by F. Schrader in *La Géographie*, 12 (1905), 81-86; and P. Geddes in *SGM*, 21 (1905), 490-496, 548-555.

Region

Regions are intellectual constructions that divide the earth's surface into areal units. The regional concept has been associated with geography's chorological tradition, which has emphasized the study of areal differentiation. Many types of regions have been discussed by geographers. The most fundamental distinction is between specific and generic regions. Specific regions are areas of the earth's surface that are viewed as possessing an individuality or wholeness. Generic regions are types of regions, either units of areal classification or functional types. The same areas may be seen in both ways. For example, Southern California may be described in terms of those qualities of its cultural and physical environment that give it an individual character, or as a type of region, such as a metropolitan region. A single study will typically employ both viewpoints.

Generic regions have remained important concepts in modern geography in subfields outside of chorology. They have been especially prominent in studies concerning metropolitan spatial organization and economic development. Specific regions have been seen as a vestige of a traditional, descriptive geography, and thus the study of such regions declined in importance during the 1960s with the heightened concern for a more scientific human geography. More recently, however, the specific region has gained in significance with the attempt by geographers to combine their traditional interest in areal differentiation with a concern for the human experience of place and territory.

Bibliography

Gilbert, A., "The New Regional Geography in English and French-Speaking Countries," *PHG*, 12 (1988), 208-228.

Paasi, A., "The Institutionalization of Regions: A Theoretical Framework for Understanding the Emergence of Regions and the Constitution of Regional Identity," *Fennia*, 164 (1986), 105-146.

Whittlesey, D., "The Regional Concept and the Regional Method," pp. 19-68 in *American Geography: Inventory and Prospect*, ed. by P. E. James and C. F. Jones (1954).

• *J. Nicholas Entrikin*

Regional Geography

The study of the areal differentiation of the earth's surface. Regional geographers present narrative-like syntheses that examine heterogeneous phenomena as they are found together in place and that consider the manner in which these phenomenal complexes vary among places. Such studies have their intellectual origins in the writings of classical Greek scholars, but may be seen as rooted in our experience of, and our naive curiosity about, the similarities and the differences among places.

Regional geography enjoyed its greatest prestige in modern geography during the first half of the twentieth century. The influence of the writings of regionalists such as Paul Vidal de la Blache in France, Alfred Hettner in Germany, H. J. Mackinder and A. J. Herbertson in Great Britain, and Carl Sauer and Richard Hartshorne in the United States helped to establish its prominence. Its prestige declined somewhat in the mid-twentieth century in association with the concern to transform geography into a more quantitative, positivistic, spatial science.

Regional geographers offered geography a distinct methodology that appeared to avoid environmental determinism through an emphasis on the concrete relations between particular peoples and their environments rather than on the general laws governing the human adaptation to the environment. Variations of neo-Kantian arguments concerning idiographic and nomothetic concept formation were used to justify the scientific character of studies of the individual case. According to these arguments, the regional geographer sought a causal, yet non-lawful, understanding of the relationship of humans to their environment. This was to be accomplished through an emphasis on the contingent connections among causal series of both natural and human events that contributed to the character of a place, or, at a larger scale, a region.

The role of regional studies in a modern scientific geography has been a matter of continuing controversy. During the 1960s, spatial analysts condemned much of regional geography as overly subjective and unscientific. At the same time they incorporated certain types of regional analysis into their conception of a nomothetic geography. Spatial analysts recast regional geography into a form of areal classification, functional analysis or systems analysis. More recently, some geographers have promoted a "theoretical" regional geography. All such attempts to remake regional geography into a nomothetic and theoretical study result in an emphasis on regions as generic types.

The study of specific or individual regions has been at the center of much of the debate concerning regional geography. For some, such studies represent the highest achievement of the geographer; for others, these studies offer mere descriptions and areal inventories, and hence are of only minor importance for a scientific geography. The controversy surrounding such studies apparently has not diminished their appeal, however, as is illustrated by the continual redemption and rebirth of regional geography during this century. A part of this appeal may be associated with the vantage point from which the regional geographer views places. In moving between the insider's and the outsider's view of place, the regional geographer offers insights into the diversity and the complexity of the relationships between people and their environment. *See also* Areal Differentiation; Chorography/Chorology; Compage.

BIBLIOGRAPHY

Entrikin, J. N., and S. D. Brunn, eds., *Reflections on Richard Hartshorne's The Nature of Geography* (1989).

Gilbert, A., "The New Regional Geography in English and French-Speaking Countries," *PHG*, 12 (1988), 208–228.

Hart, J. F., "The Highest Form of the Geographer's Art," *AAAG*, 72 (1982), 1–29.

Hartshorne, R., *The Nature of Geography* (1939).

Sauer, C. O., *The Morphology of Landscape* (1925).

Thrift, N., "On the Determination of Social Action in Space and Time," *Environment and Planning D: Society and Space*, 1 (1983), 23–57.

• *J. Nicholas Entrikin*

Regional Science

Regional Science is the scientific study of human adaptation of space. Research has shown how use of the earth's surface shows distinct patterns. Understanding in rigorous fashion the spatial characteristics of settlements, and of production, exchange and consumption, constitutes the focus of Regional Science.

The regional scientist emphasizes linkages between various forms of human activities (such as choice of where to live and the journey to work). The style of analysis uses complex models and seeks evidence from large-scale data bases; the skills required are often mathematical and statistical. While elements of probability and chance are often introduced, the underlying premise is that decisions and behavior are rational, and that aggregate patterns can ultimately be explained: sometimes by optimizing models, sometimes by statistical regularity.

Regional Science had its start just after World War II, as an offshoot of applied urban and regional economics. Study of industrial location, agricultural land use, and distribution of urban places soon was expanded to consider urban land values, commuting and other travel behavior, and interactions between the human and the physical environment. As the research agenda expanded, contributions came from theoretical and mathematical geography, demography, operations research, urban and regional planning, ecology, and other disciplines. Work in the field has come to embrace the search for statistical techniques especially appropriate to spatial analysis, the creation of generalized spatial equilibrium constructs, and design of spatial models.

Regional Science investigations have traditionally sought to blend the descriptive with the analytic, seeking to understand *why* given spatial patterns and processes arise (*e.g.*, why the South is poorer than the North). In recent years, more attention has been given to the study of normative considerations, including conflict management. This has led to development of alternate criteria of optimal behavior and states, such as social justice in residential housing patterns and the search for instruments to achieve them.

Regional Science is taught in a number of universities abroad as well as in the United States (e.g., the University of Pennsylvania in Philadelphia, where the first graduate studies program was established, in the late 1950s). Membership in about a dozen societies worldwide, linked in the Regional Science Association, includes several thousand specialists, as well as practitioners of related disciplines including geography. There are many Regional Science research centers. The chief journals are the *Papers* of the Regional Science Association (1954-), *Journal of Regional Science* (1958-), and the *International Regional Science Review* (1975-). Regional scientists also work in government offices as planners and other professionals, in industry and trade (locating branch plants, for example), and as consultants to businesses and local governments.

BIBLIOGRAPHY

Bourne, L., ed., *Internal Structure of the City* (2nd ed., 1982).

Bourne, L., and J. Simmons, eds., *Systems of Cities* (1978).

Isard, W., *Location and Space-Economy* (1956).

———, *Introduction to Regional Science* (1975).

Isard, W., *et al.*, *Methods of Regional Analysis* (1960).

Richardson, H., *Regional Economics* (1979).

• *Thomas A. Reiner*

Religion, Geography of

In contrast to the vast amount of work on the topic by historians and theologians, the original research on religion generated by geographers is slight in volume, incomplete in coverage, and relatively recent in date. Inhibiting rapid progress in this crucial sector of human geography and delaying its entry into the standard teaching curriculum have been related problems of rigorously defining the items to be observed and the inadequacies in, or even lack of, mappable official statistics for much of the world. But if most students have tended to shy away from treating religion, still for certain regions and countries—notably the Middle East and India—the topic is unavoidable, and interesting analyses have been forthcoming. At the world scale, we are fortunate in having three monographs, by Pierre Deffontaines, Martin Schwind, and David Sopher, that at least serve to initiate serious discourse. In the United States, Wilbur Zelinsky and James Shortridge have wrestled with seriously deficient data from private sources to sketch the main spatial features of denominational membership, while historian Edwin Gaustad has produced an admirable atlas of the historical geography of U.S. religion.

Insofar as there have been significant advances in this field, they have been channeled in three directions, which are not entirely unrelated. The impact of religious belief and practice on the visible landscape, *e.g.*, in the Mormon region, has intrigued a number of students. Related to such concerns has been a steady growth in cemetery studies by geographers as well as archaeologists and folklorists. Next, the geography of religion has worked its way into many publications dealing primarily with political and ethnic questions because of the powerful relevance of religion to such matters in so much of the world. Finally, we have a recent upsurge of attention to the geography of pilgrimages in terms not only of the flows of people but also the shaping of the character of their destinations. Conspicuously missing from the arena of active scholarship is any work on the geographic implications of overseas European and North American missionary efforts, despite a fine pioneering essay by Hildegard Binder Johnson.

Bibliography

Deffontaines, P., *Géographie et religions* (1948).

Johnson, H. B., "The Location of Christian Missions in Africa," *GR*, 57 (1967), 168-202.

Schwind, M., ed., *Religionsgeographie* (1975).

Shortridge, J. R., "A New Regionalization of American Religion," *Journal for the Scientific Study of Religion*, 16 (1977), 143-153.

Sopher, D. E., *Geography of Religions* (1967).

Zelinsky, W., "An Approach to the Religious Geography of the United States: Patterns of Church Membership in 1952," *AAAG*, 51 (1961), 139-193.

• *Wilbur Zelinsky*

Remote Sensing

Remote sensing refers to several processes for acquiring data about the attributes of phenomena without being in direct contact with them. The technology of modern remote sensing consists of four stages: 1) designing sensors and platforms to collect digital or photographic data; 2) developing the means for managing the data and the communication links necessary for retrieving and transmitting them from archives, or in real time; 3) designing the hardware and software necessary to process the acquired data into forms that are useful; and 4) developing useful applications for the data. Forms of remote sensing typically include techniques for nondestructive testing (such as in detecting forgeries by ultraviolet photography), or noninvasive sensing (such as in biomedical and dental imaging by X-ray), and earth resource sensing from aircraft and spacecraft. This last form is the one most often referred to in the literature as "remote sensing."

The term "remote sensing" entered the lexicon of science in the early 1960s to distinguish the new equipment and procedures from an already well-established industry focused on aerial photography and photo interpretation. The Environmental Research Institute of Michigan (ERIM), the Department of the Navy's Office of Naval Research (ONR), the National Aeronautics and Space Administration (NASA), the United States Geological Survey (USGS), the then-named American Society of Photogrammetry (ASP) (now renamed the American Society for Photogrammetry and Remote Sensing [ASPRS]), and a few U. S. universities constituted the developing core of the new technology. In 1972, this small research community reached experimental maturity when NASA launched its first polar orbiting satellite (called the Earth Resources Technology Satellite, or ERTS-1) carrying a multispectral scanner. Since then there have been four additional satellites, and they have all been named Landsats (1 to 5). In addition to the multispectral scanner (MSS), which scans four spectral channels and achieves a nominal 80 meter ground resolution, another sensor, called the thematic mapper (TM), has flown aboard Landsats 4 and 5. The TM scans seven spectral channels and achieves a nominal 30 meter ground resolution. The next satellite in this series (Landsat 6) is scheduled for launch in 1991 or early 1992 and, like its predecessors, will carry both MSS and TM sensors. The French "Système probatoire d'observation de la terre" (SPOT-1) was launched in February 1986, ushering in the era of internationalization of remote sensing. In 1990 most of the world's spacefaring nations, including the Soviet Union, India, People's Republic of China, Japan, Brazil, Italy, Canada, Australia and the consortium of countries forming the European Space Agency (ESA), have, or intend to have, earth observation satellites. The Civil Land Remote Sensing Commercialization Act of 1984 and President Reagan's Space Policy of 1988 clearly state that remote sensing technology has graduated from theoretical/experimental status to operational/commercial status.

In the context of earth resources, modern remote sensing is based on detecting, measuring, and recording variations in the spectral responses of terrain objects and features and on repeating these functions in order to discover the rates and directions of change. In addition to the Landsat series, many different sensors and sensor systems are used for these purposes, including specially designed cameras and films, airborne and satellite multispectral scanners, spectrometers, radiometers, thermal and ultraviolet scanners, and radar. Most measurements are made by instruments operating in the wavelength region from ultraviolet through the visible, infrared, and microwave spectra. These instruments record reflected or emitted radiation in specific frequencies and variable bandwidths, depending upon detector sensitivities and applications requirements.

Applications of remote sensing data are numerous in assessing the natural and cultural landscapes of the earth. Agricultural, forestry and rangeland sensing are undertaken to monitor general plant vigor or vegetation condition, disease infestations, yield potential, total production, and changes in these conditions through time. Geological and geophysical applications include mineral and hydrocarbon exploration, fault and lineament mapping, assessment of geological hazards, and general geological surveys for structure and lithology. For water resources, applications focus on surface and groundwater exploration and mapping, water quality assessment, snow melt measurement, assessment of changes in reservoir levels, and the monitoring of soil moisture conditions. Archeologists use remote sensing for noninvasive or nondestructive exploration of known sites (burial mounds, tombs, etc.), locating new cultural sites or features like ancient irrigation systems and structures, and as a general technique for archeological surveys before new land development.

The most recent trends to overtake remote sensing technology are generally referred to as Geographic Information Systems (GIS) and Land Information Systems (LIS). These are spatial data technologies that, when combined with the spectral technologies of remote sensing, provide scientists and resource managers with powerful new tools for detecting, measuring, monitoring, and managing environmental changes in both the cultural and physical landscapes.

Among the most important remote sensing journals are *Photogrammetric Engineering and Remote Sensing*, published monthly by the American Society for Photogrammetry and Remote Sensing; *Remote Sensing of Environment*, published quarterly by Elsevier, New York; *International Journal of Remote Sensing*, a bimonthly published by Taylor and Francis, London; and *Geocarto International: A Multi-Disciplinary Journal of Remote Sensing*, published quarterly by the Geocarto International Center, Hong Kong.

Bibliography

Campbell, J. B., *Introduction to Remote Sensing* (1987).

Chen, H. S., *Space Remote Sensing Systems* (1985).

Colwell, R. N., ed., *Manual of Remote Sensing* (2nd ed. 1983).

Gower, J. F. R., ed., *Oceanography from Space* (1981).

Holz, R. K., ed., *The Surveillant Science: Remote Sensing of the Environment* (1985).

Lillesand, T. M., and R. W. Kiefer, *Remote Sensing and Image Interpretation* (2nd ed. 1987).

• *Stanley A. Morain*

Resource

A resource is an attribute of the Earth that humans perceive to be valuable. Most scholars restrict the term to certain physical and biological materials or processes (*i.e.*, "natural resources"), but people, capital and information are sometimes also included.

During the nineteenth century the concept of resources developed in response to concerns expressed by Thomas Malthus and others about the Earth's capacity to support rapidly growing and industrializing populations. Early analysts used inventories to assess the adequacy of existing supplies of vital materials. The limitations of this approach became clear when it was realized that there is no absolute set of resources. Resource appraisals are strongly affected by prevailing scientific knowledge, available technologies, existing norms, current forms of social organization and other factors. According to Zimmerman's classic statement, "resources are not, they become; they are not static but expand and contract in response to human wants and human actions."

The number of phenomena that are defined as resources has greatly increased throughout the twentieth century. As scientific research reveals that the Earth's life-support systems are dependent on a complex web of interrelationships between nature and society, some scholars have suggested that everything on Earth should now be considered a resource.

BIBLIOGRAPHY

Dasmann, R. F., *Environmental Conservation* (1968).

Firey, W., *Man, Mind and Land: A Theory of Resource Use* (1960).

Zimmerman, E. W., *World Resources and Industries* (1933, rev. 1951).

• *James K. Mitchell*

Resource Geography

Resource geography focuses on the adequacy of natural resources to satisfy human demands. It is a branch of people-environment studies that addresses issues of resource scarcity, sufficiency and quality, and the sustainability of human life on Earth. Resource geographers also pay special attention to disparities between the location of resources and the spatial distribution of demand. Well-developed subfields are concerned with water, energy, agricultural and marine resources, and with natural hazards.

The central problem of resource geography is simply stated: How can a finite Earth meet ever-growing demands for resources? In the first half of the twentieth century, geographers attempted to answer this question via resource inventories and mapping projects. The approach was fieldwork-oriented, empirical and applied. After World War II notable developments in theory began to emerge, although concern for public policy applications remained a distinguishing trait. The best known contribution to theory was made by Gilbert White, who employed concepts of boundedly rational decisionmaking under conditions of uncertainty to illuminate the range of choice that faced resource managers in more developed and less developed societies. Students and co-workers later elaborated this model to take account of behavioral variables that influence the perception of resources and alternative uses.

The dominant philosophical and methodological stance of resource geographers has been utilitarian, pragmatic and managerial, with particular attention devoted to links between individual decisionmaking and contemporary public policy. The idealistic and evolutionary aspects of landscape, which are of concern to cultural geographers, have received less scrutiny. Resource geographers have shown that perceptions of resources are highly variable and that resource use systems involve many interacting factors that are susceptible to human manipulation. Increases in demand have been met by adjusting factor permutations. In the past, society successfully kept pace with changing resource demands by these means, but there are now ominous signs of burgeoning environmental instability and degradation that suggest humanity is pushing against a variety of limits to further growth. A synthesis of resource geography and landscape geography may now be in the making as a consequence of the need to develop broader intellectual paradigms for the study of global environmental change.

BIBLIOGRAPHY

Clark, W. C., and R. E. Munn, eds., *Sustainable Development of the Biosphere* (1986).

Kates, R. W., "The Human Environment: The Road Not Taken, The Road Still Beckoning," *AAAG*, 77 (1987), 525–534.

Kates, R. W., and I. Burton, eds., *Geography, Resources, and Environment*, Vol. I: *Selected Writings of Gilbert F. White*, Vol. II: *Themes from the Work of Gilbert F. White* (1986).

Mitchell, B., *Geography and Resource Analysis* (1979).

Turner, B. L., "The Specialist-Synthesis Approach to the Revival of Geography: The Case of Cultural Ecology," *AAAG*, 79 (1989), 88–100.

• *James K. Mitchell*

Ribeiro, Orlando (b. 1911)

D. Let., University of Lisbon, 1936. University of Coimbra, 1941–42; University of Lisbon, 1943–81. *Portugal, o Mediterrâneo e o Atlantico* (1945); *A ilha do Fogo e as suas erupções* (1954); "Portugal," in *Geografía de España y Portugal* (ed. by M. de Terán) (1955); *Mediterrâneo, ambiente e tradição* (1968). I. do Amaral, *Bibliografia científica de Orlando Ribeiro* (1981); *Livro de Homenagem a Orlando Ribeiro* (1984–88); O. Ribeiro, "Cinquenta anos de vida cientifica e universitaria," *Revista da Faculdade de Letras de Lisboa*, 5th ser., no. 6 (1986), 11–20.

Richter, Eduard (1847-1905)

Doctorate, University of Vienna, 1885. University of Graz, 1886-1905. *Das Gletscherphänomen* (1873); *Die Gletscher der Ostalpen* (1888); *Atlas der österreichischen Alpenseen* (with A. Penck) (1896). J. Goldberger in *GBS*, 10 (1986), 143-148; obits. by G. Leukas in *GZ*, 12 (1906), 121-135, 193-212, 252-277; and R. Marek in *MOGG*, 49 (1906), 161-255.

Richthofen, Ferdinand von (1833-1905)

Doctorate, University of Berlin, 1856; (habil.) University of Vienna, 1860. University of Bonn, 1875-83; University of Leipzig, 1883-86; University of Berlin, 1886-1905. *China* (5 vols. plus 2-vol. Atlas, 1877-1912); *Aufgaben und Methoden der heutigen Geographie* (1883); *Führer für Forschungsreisende* (1886). A. Kolb in *GBS*, 7 (1983), 109-115; R. Beckinsale in *DSB*, 11 (1975), 438-441; E. Plewe and W. Lauer in *Colloquium Geographicum*, 17 (1983), 7-23; B. Zuckermann, "Die Bedeutung Ferdinand von Richthofens als Geologe und Geograph," *Geographische Berichte*, 28 (1983), 1-15.

Robequain, Charles (1897-1963)

D.-ès-L., University of Paris, 1929. University of Poitiers, 1931-33; University of Rennes, 1933-37; University of Paris, 1937-63. *Le Thanh Hoa, étude géographique d'une province annamite* (1929); *L'Indochine française* (1935); *Evolution économique de l'Indochine française* (1939); *Le monde malais* (1946). J. Delvert in *LGF* (1975), 145-151; obit. by P. Gourou in *AG*, 72 (1964), 1-7.

Robinson, Arthur Howard (b. 1915)

Ph.D., Ohio State University, 1947. University of Wisconsin, 1946-80. *The Look of Maps* (1952); *Elements of Cartography* (later editions with others) (1953, 5th ed. 1984); *The Nature of Maps* (1976); *Early Thematic Mapping in the History of Cartography* (1982). *WWA*; *GOF* (1972).

Robinson, John Lewis (b. 1918)

Ph.D., Clark University, 1946. University of British Columbia, 1946-84. *Resources of the Canadian Shield* (1969); *Studies in Canadian Geography: British Columbia* (1972); *British Columbia: One Hundred Years of Geographical Change* (1973); *Concepts and Themes in the Regional Geography of Canada* (1989). *WWA*; *CWW*; *International Who's Who*; *Contemporary Authors*.

Romania

Although there is some significant geographical material attributable to Dimitrie Cantemir (1673-1723), who produced a description of Moldavia at the request of the Berlin Academy, the development of the discipline effectively takes off with the formation of the Geographical Society (Societatea de Ştiinţe Geografice) in 1875, inspired by the international congress held in Paris that year. The society's *Buletin* helped the cause of geography in primary education before the subject was provided for at the universities of Bucharest and Iasi (in 1900 and 1904, respectively). A third center of university geography teaching emerged at Cluj (now Cluj-Napoca) with the enlargement of the state after World War I. There has also been some activity at Cernauti (until the province of Bucovina was lost in 1940) and at Craiova, where a new university offered geography for a number of years in the post-war period. University graduates helped to establish geography in a strong position in the secondary schools, where the subject continues to flourish.

Simion Mehedinţi (1869-1962) was very active before World War I. He studied under Ratzel, Richthofen, and Vidal de la Blache, and attracted much attention through his philosophical paper of 1894 on the place of geography among the sciences ("Locul geografiei între ştiinţe"). But momentum increased during the inter-war years with the completion of a number of outstanding doctorate theses. The French tradition of the regional monograph may be detected, but there was a strong ethnographical bias to bring out enduring relationships with local natural resources and press the politically important point about continuity of occupation from Daco-Roman times. Geography in any country is conditioned by national tasks and in Romania there was a strong response to the pressing political need to cement the new territories of "România Mare" (Greater Romania) into a coherent state to withstand the revisionist policies of some neighboring states. There was a strong interest in Romania shown by some foreign geographers, especially the Frenchman Emmanuel de Martonne, whose half century of dedicated work included such epics as *La Valachie* (1902) and *Recherches sur l'évolution morphologique des Alpes de Transylvanie* (1907).

There was some specialization in geomorphology and, more especially, in the social and historical field, where contributions included the researches of Vinţila Mihăilescu (1890-1978) on urban structure and the classification of rural settlements. It was Mihăilescu who used his contacts with the Ministry of Education of the wartime administration in Bucharest to obtain support for a geography research institute that began its work at Pucioasa in the Subcarpathians before moving to the capital. The institute, which is now subordinated to Bucharest University, publishes a journal in Romanian and another (the geographical series of "Revue Roumaine") using the major world languages. The institute has also produced major studies on the geography of Romania, including a national atlas, and a series of handbooks dealing with the various administrative regions. In its early years the institute had to cope with the communist revolution, which swept aside a number of leading personalities, at least for a time. However, Mihăilescu found his way back in 1958 and maintained a close association through many years of retirement and right up to his death.

Geography in Romania is alive and well, with a steady research output that finds its way into the journals of the institute and of the three universities where the subject is still taught. Meanwhile the Romanian Geographical Society (now known as the Societatea de Ştiinţe Geografice din R. S. România) caters to the teaching profession with a monthly journal (*Terra*) and branch activities throughout the country. However, the research program is somewhat restricted, with a concentration on the geography of Romania and on those topics that are most relevant to the national development plans. There is activity in both physical and human geography, but in the latter case historical and humanistic perspectives are only weakly developed. It is also disappointing to see that Romania's international profile has faded somewhat in recent years with the retirement of the leading figures of the earlier post-war period and the continuing economic crisis that has meant a substantial reduction in resources.

Bibliography

Mihăilescu, V., "L'institut de géographie: 30 années d'existence," *Revue Roumaine: Géographie*, 18 (1974), 257-260.

———, "La société roumaine de géographie à son centenaire," *Revue Roumaine: Géographie*, 19 (1975), 7-24.

Oanceâ, D. I., "Atlasul geografic national al României," *Studii și cercetări: geografie*, 18 (1971), 207-216.

Şandru, I., and V. Cucu, "The Development of Geographical Studies in Romania," *GJ*, 132 (1966), 43-48.

Turnock, D., "Urban Development and Urban Geography in Romania: The Contribution of Vintila Mihailescu," *GeoJournal*, 14 (1987), 181-202.

• *David Turnock*

Romer, Eugeniusz (1871-1954)

Doctorate, University of Lwów, 1894. University of Lwów, 1899-1931. Published numerous articles, maps, and atlases on Polish subjects; see his collected works, *Wybor prac* (4 vols., 1960-64). J. Babicz in *GBS*, 1 (1977), 89-96; J. Babicz in *DSB*, 11 (1975), 524-525; J. Babicz, "L'école géographique polonaise d'Eugeniusz Romer," pp. 45-48 in *Les écoles géographiques* (1980).

Rosberg, Johan Evert (1864-1932)

Lic.Phil., University of Helsinki, 1900. University of Helsinki, 1898-1929. *Bottenvikens finska deltan* (1895); *Finlands geografii* (1911); *Lappi* (1911); *Bygd och obygd* (1919). S. Jaatinen in *GBS*, 9 (1985), 101-108; S. Jaatinen, "Johan Evert Rosberg—Liv och Garning," *Terra*, 100 (1988), 13-31; obit. by J. Granö in *Terra*, 44 (1932), 73-75.

Roxby, Percy Maude (1880-1947)

B. A., University of Oxford, 1903. University of Liverpool, 1904-44. "The Distribution of Population in China," *GR*, 15 (1925), 1-24; "East Anglia," pp. 143-166 in *Great Britain* (ed. by A. Ogilvie) (1928); "The Scope and Aims of Human Geography," *SGM*, 46 (1930), 276-299. T. W. Freeman in *GBS*, 5 (1981), 109-116; Freeman, "Percy Maude Roxby," pp. 156-168 in *The Geographer's Craft* (1967); E. Gilbert, *British Pioneers in Geography* (1972), 211-226; obits. by H. Fleure in *GJ*, 109 (1947), 155-156; and *Geography*, 32 (1947), 36.

Royal Danish Geographical Society (Kongelige Danske Geografiske Selskab)

The Royal Danish Geographical Society was founded on November 18, 1876, and enjoys royal patronage. In 1877, the first volume of *Geografisk Tidsskrift* was published and thereafter has appeared once a year, except in the period 1964-69, when it appeared semiannually. Since 1959, the society has had its headquarters in the Geographical Institute of the University of Copenhagen. In 1940, a monograph series—*Folia geographica danica*—was established. The society also initiated the successive volumes that constitute the *Atlas over Danmark*. Membership has always been drawn from both academic geographers and a broad range of the lay public. Medals awarded for distinguished contributions to geography include the Egede medal for polar research, the Galathea medal for research other than polar, and the Vitus Behring medal.

BIBLIOGRAPHY

Geografisk Tidsskrift, 76 (1977), i-xv.

• *W. R. Mead*

Royal Dutch Geographical Society (Koninklijk Nederlands Aardrijkskundig Genootschap)

The Royal Dutch Geographical Society (KNAG) was founded 3 June 1873. Among the founding fathers were educators, administrators, bankers, traders, and former navy captains. The first president was Dr. P. J. Veth, a famous orientalist who was professor of the geography and anthropology of the East Indies at Leyden University. The major aim of this geographical society was to "stimulate the awakening (public) interest in geography and by doing so to increase the knowledge of which among others the commerce, shipping, industries, colonisation and emigration could take advantage." From the start it was agreed that this society should not become a so-called "learned" or scientific society. The major argument for this was the conviction that "geography itself holds such an extraordinary position between science and practice."

As early as 1874 the KNAG started to publish its own journal, the *Tijdschrift van het Aardrijkskundig Genootschap*, which reflected in its contents the varied interests of its majority of laymen members—land and people, culture and nature, science and practice.

Some major expeditions to "increase the knowledge of the earthglobe" were undertaken, particularly in the first half of the twentieth century. The most important of these were the expeditions to various parts of Indonesia. Since the early 1960s, however, scientific expeditions have not been an important feature of KNAG life.

Because the KNAG chose so overtly to be a society for the interested layman, and was so strongly oriented to exploration, global knowledge, the Netherlands Indies, and the anthropological and natural science parts of the discipline, professional geographers did not regard the KNAG as *their* society.

In 1909 Dr. H. Blink, who was the first professor of economic geography in Rotterdam (1913-23), founded the Vereeniging voor Economische Geographie (VEG—Association for Economic Geography). A year later this association started its own journal (*Tijdschrift voor Eco-*

nomische Geographie). In 1948 the VEG decided to add the word "sociale" to its name and to the title of the journal as well, thus creating a terminological difficulty in the Netherlands, where "sociale geografie" from the beginning was meant to be equivalent to "géographie humaine," or "human geography," and to include economic geography.

The other major society that developed outside the KNAG was the Geografische Vereeniging (GV—Geographical Union). It represented the union in 1948 of several older societies and associations on the regional and national levels that found their membership among professional geographers. The GV's first board included Prof. Dr. J. J. Fahrenfort, Prof. Dr. J. O. M. Broek, and Dr. A. C. de Vooys. In 1948 the GV started publishing its own journal, *Geografisch Tijdschrift*, which was strongly oriented to the teaching of geography at all levels. To reach the teachers in all parts of the country, the GV began a variety of activities on the regional level and won widespread support through its down-to-earth approach.

Since World War II, and in particular after the independence of Indonesia, the character of the KNAG changed profoundly. Applied geography, human and physical, within the Netherlands itself became paramount. In the KNAG laymen began to lose influence and interest as the numbers of professional geographers, in particular those working in urban and regional planning, increased.

In 1967 the inevitable and highly desirable merger of the KNAG, VEG, and GV finally came about and a new start was made. Instrumental in this major event were R. Schrader, then president of the GV, and Prof. Dr. J. P. Bakker, then president of the KNAG. The new organization retains the name of the oldest society, the KNAG; its major aim is to promote the importance of geography for society. The annual membership figure has risen to 3,000.

At present the KNAG publishes three journals—*Tijdschrift voor Economische en Sociale Geografie* (*TESG*), *Geografisch Tijdschrift*, and, since 1977, *De Nieuwe Geografenkrant*. Since 1985 it has issued the *Netherlands Geographical Studies*, a series of monographs and collections of articles devoted to specialized themes. Every two years Dutch-speaking geographers from the Netherlands and Belgium meet in the Nederlandse Geografen Dagen to discuss matters of mutual professional interest. The numbers of participants in these congresses usually range from about 400 to 700.

Bibliography

R. Schrader, "Honderd Jaar Koninklijk Nederlands Aardrijkskundig Genootschap," *Geografisch Tijdschrift*, 8 (1974), 234-402.

• *J. A. van Ginkel*

Royal Geographical Society (London)

The Geographical Society of London, which quickly assumed the further qualification of "Royal," was founded in 1830 as a continuation of the African Association (established 1788) and the Raleigh Travellers' Club (1826). Even more than its contemporaries in France, Germany, and Italy, the RGS placed great emphasis on geographical exploration, possibly because of the greater extent and significance of Britain's colonial empire. Although both empire and exploration have declined, this interest is still manifest in the activities and publications of the society. The renascence of university geography in the period from the 1880s to the 1920s owed much to RGS support, material and moral. A diploma course in surveying was offered at the RGS from 1897 to 1948. The library, map collections, archives, and publications of the society have been very important to geographers and to all others who require geographical information. The quarterly *Geographical Journal*, which began in 1893 as a successor to the Society's *Journal* (1830-80) and *Proceedings* (1855-92), retains its appeal to both professionals and laymen, although it is not regarded as a proper vehicle for the publication of analytical or statistical studies of a high order. British academic geographers started the Institute of British Geographers in 1933 out of dissatisfaction with the conservative publication policies of the RGS, but relations between the two organizations have been cordial and the institute occupies office space in the society's headquarters at No. 1 Kensington Gore in London.

Bibliography

Cameron, I., *To the Farthest Ends of the Earth: The History of the Royal Geographical Society 1830-1980* (1980).

Freeman, T. W., "The Royal Geographical Society and the Development of Geography," pp. 1-99 in *Geography Yesterday and Tomorrow*, ed. by E. H. Brown (1980).

Mill, H. R., *The Record of the Royal Geographical Society 1830-1930* (1930).

Royal Geographical Society of Madrid (Real Sociedad Geográfica de Madrid)

The Real Sociedad Geográfica de Madrid was founded in 1876, at a time when efforts were under way to promote geographical studies in Spain, following the creation in 1870 of the Instituto Geográfico y Estadístico (headed by Carlos Ibañez) and the realization of the Mapa Topográfico Nacional (scale 1:50,000), of which the first sheet was published in 1875. Key figures of the founding era were the geographer Fermín Caballero (1800-1876), the first president, and especially the soldier and cartographer Francisco de Coello (1822-1899). In the same year as the founding a journal was published, the *Boletín de la Sociedad Geográfica de Madrid*, which has continued without interruption to the present day.

As laid down in its statutes, the society had as its objectives: (1) "to promote the advance and diffusion of geographical knowledge in all its branches"; (2) to study "the territory of Spain and its provinces and possessions overseas"; and (3) to study also "those countries with which important relations already exist or where development of such appears opportune." Interest in colonial and mercantile themes was, as in other geographical societies of the time, very great. With the support of the S.G. de M. (which soon received the title of "Royal") and of outstanding figures like Joaquín Costa, a Spanish Congress of Colonial and Mercantile Geography was organized, a Spanish Society of Africanists and Colonialists founded, and a *Revista de Geografía Colonial y Mercantil* established (1897). From 1885 on there also existed in Madrid a Commercial Geographical Society and a *Revista de Geografía Comercial*. The Real Sociedad Geográfica de Madrid, at its founding, had 626 members, including geographers, naturalists, historians, military engineers, and politicians. Among its most outstanding personalities were Eduardo Saavedra, Martín Ferreiro, Federico Botella, Cesareo Fernández Duro, Rafael Torres Campos, Ricardo Beltrán Rozpide, and José Gavira.

With the development of geography in the universities, the society went through a phase of decline that extended over several decades and affected also the caliber of the works published in the *Boletín de la Real Sociedad Geográfica de Madrid*. In recent years, however, greater contact with academic geographers encourages hope for the renewal of this institution.

Bibliography

Asúa, M. de, "Reseña de las tareas de la Corporación en sus primeros cincuenta años de vida," *Boletín de la Real Sociedad Geográfica de Madrid*, 66 (1926), 220-262.

Torroja y Miret, J. M., "La Real Sociedad Geográfica de Madrid en el LXXV anniversario de su fundación, octubre de 1952," *Ibid.*, 89 (1953), 21-32.

Vilá Valentí, J., "Origen y significado de la Sociedad Geográfica de Madrid," *Revista de Geografía* (Barcelona), 11 (1977), 5-21.

• *Horacio Capel Sáez*
(translated by C. Julian Bishko)

Royal Scottish Geographical Society

The Royal Scottish Geographical Society was founded in Edinburgh in 1884. The prime movers in the establishment of the society were the cartographer John George Bartholomew and Mrs. A. L. Bruce, daughter of the African explorer David Livingstone. The Edinburgh society differed from its London counterpart by having a far larger proportion of its members from scientific and academic backgrounds and, conversely, a smaller proportion of the "leisured rich," who formed the backbone of the Royal Geographical Society. Exploration was definitely subordinate to study and education. Almost immediately the Scottish society started a journal, the *Scottish Geographical Magazine*, which began as a monthly but now appears three times a year. The *Magazine* became a world-class journal during the long editorship (1902-34) of Marion Newbigin. The society, and especially J. G. Bartholomew, played a major role in the establishment of geography in the University of Edinburgh. George Chisholm was appointed lecturer in 1908, and a professorship was established in 1931, with A. G. Ogilvie as the first holder of the chair.

BIBLIOGRAPHY

Adams, I., A. Crosbie, and G. Gordon, *The Making of Scottish Geography: 100 Years of the R.S.G.S.* (1984).

Lochhead, E., "The RSGS: The Setting and Sources of Its Success," *SGM*, 100 (1984), 69-80.

Newbigin, M. I., "The Royal Scottish Geographical Society: The First Fifty Years," *SGM*, 50 (1934), 257-269.

Rühl, Alfred (1882-1935)

Doctorate, University of Berlin, 1905; (habil.) University of Marburg, 1909. University of Marburg, 1909-12; University of Berlin, 1912-35. (trans.) *Die erklärende Beschreibung der Landformen* by W. M. Davis (1912); *Vom Wirtschaftsgeist im Orient* (1925); *Vom Wirtschaftsgeist in Amerika* (1927); *Einführung in die allgemeine Wirtschaftsgeographie* (1938). H. Harke in *GBS*, 12 (1988), 139-147; H. Harke, "Alfred Rühls unvergessene Verdienste um die Geographie," *Geographische Berichte*, 27 (1982), 193-199; H. Harke, "Alfred Rühls Anteil bei der Umwandlung der Wirtschaftsgeographie vom Dilettantismus zur Wissenschaft," *PGM*, 129 (1985), 253-261.

Russell, Richard Joel (1895-1971)

Ph.D., University of California, Berkeley, 1926. Louisiana State University, 1928-66. *Physiography of Lower Mississippi River Delta* (1936); *Culture Worlds* (with F. Kniffen) (1951); *River Plains and Sea Coasts* (1967). H. Walker in *GBS*, 4 (1980), 127-138; obits. by F. Kniffen in *AAAG*, 63 (1973), 241-249; and W. McIntire in *GR*, 63 (1973), 276-279.

Russia. *See* Union of Soviet Socialist Republics.

Salisbury, Rollin D (1858-1922)

B.S., Beloit College, 1881. University of Chicago, 1892-1922. *Geology* (with T. Chamberlin) (1904-06), *Physiography* (1907), *Elements of Geography* (with H. Barrows and W. Tower) (1912); *Modern Geography* (with Barrows and Tower) (1913). W. Pattison in *GBS*, 6 (1982), 105-113; W. Pattison, "Rollin Salisbury and the Founding of Geography at the University of Chicago," pp. 151-163 in *The Origins of Academic Geography in the United States* (ed. by B. Blouet) (1981); obit. by T. Chamberlin in *Journal of Geology*, 30 (1922), 480-481.

Sanke, Heinz (b. 1915)

Ph.D., Berlin School of Economics, 1941; (habil.) Humboldt University, 1950. Political School for Trade Unions, Bernau, 1946-48; Humboldt University, 1948-80. *Die Erdölwirtschaft des Imperialismus in ihren geographischen Grundlagen* (1951); (ed.) *Politische und ökonomische Geographie, eine Einführung* (1956, 3rd ed. 1960); "Entwicklungsprobleme der geographischen Wissenschaften," *Sitzungsberichte der Akademie der Wissenschaften zu Berlin*, 9 (1968); "Zur Entwicklung der Geographie an der Humboldt-Universität zu Berlin," *Geographische Berichte*, 19 (1974), 163-175.

Sapper, Karl Theodor (1866-1945)

Dr.phil., University of Munich, 1888; (hab.) University of Leipzig, 1900. University of Tübingen, 1902-10; University of Strassburg, 1910-19; University of Würzburg, 1919-32. *Mittel-Amerika* (1921, 2nd ed. 1927), *Die Tropen* (1923); *Allgemeine Wirtschafts- und Verkehrsgeographie* (1925, 2nd ed. 1930); *Geomorphologie der feuchten Tropen* (1935). F. Termer, *Karl Theodor Sapper* (1966); obit. by F. Termer in *PGM*, 92 (1948), 193-195.

Sauer, Carl Ortwin (1889-1975)

Ph.D., University of Chicago, 1915. University of Michigan, 1915-23; University of California, Berkeley, 1923-57. *Geography of the Ozark Highland of Missouri* (1920); *Agricultural Origins and Dispersals* (1952); *The Early Spanish Main* (1966); *Sixteenth Century North America* (1971). J. Leighly in *GBS*, 2 (1978), 99-108; *GOF* (1970, 1985); J. N. Entrikin, "Carl O. Sauer, Philosopher in Spite of Himself," *GR*, 74 (1984), 387-408; M. Kenzer (ed.), *Carl Sauer: A Tribute* (1987); obits. by J. Leighly in *AAAG*, 66 (1976), 337-348; and J. Parsons in *GR*, 66 (1976), 83-89.

Sauer, Jonathan Deininger (b. 1918)

Ph.D., Washington University (St. Louis, Missouri), 1950. University of Wisconsin, 1950-67; University of California, Los Angeles, 1967-89. *Coastal Plant Geography of Mauritius* (1961); *Plants and Man on the Seychelles Coast* (1967); *Cayman Islands Seashore Vegetation* (1982); *Plant Migration* (1988). *WWA*.

Schlüter, Otto (1872-1959)

Ph.D., University of Halle, 1896; (habil.) University of Berlin, 1906. University of Halle, 1911-51. *Die Siedelungen im nordöstlichen Thüringen* (1903); *Die Ziele der Geographie des Menschen* (1908); *Ferdinand von Richthofen's Vorlesungen über Allgemeine Siedlungs- und Verkehrsgeographie* (1908); *Die Siedlungsräume Mitteleuropas in frühgeschichtlicher Zeit* (2 parts, 1952-58). M. Schick in *GBS*, 6 (1982), 115-122; appreciations by R. Käubler in *Die Erde*, 95 (1964), 5-15; and H. Lautensach in *PGM*, 96 (1952), 219-231; obit. by R. Käubler in *PGM*, 103 (1959), 241-243.

Schmieder, Oskar (1891-1980)

Dr.phil., University of Heidelberg, 1914; (habil.) University of Bonn, 1919. University of Kiel, 1930-44, 1945-56. *Länderkunde Südamerikas* (1932); *Länderkunde Nordamerikas* (1933); *Die Neue Welt* (2 vols., 1962-63); *Die Alte Welt* (2 vols., 1965-69). W. Lauer, *Oskar Schmieder zu seinem 70. Geburtstag: Beiträge zur Geographie der Neuen Welt* (1961); O. Schmieder, *Lebenserinnerungen und Tagebuchblätter eines Geographen* (1972).

Schmitthenner, Heinrich (1887-1957)

Dr.phil., University of Heidelberg, 1911; (hab.) University of Heidelberg, 1919. University of Leipzig, 1928-45; University of Marburg, 1946-55. *Tunesien und Algerien* (1924); *Chinesische Landschaften und Städte* (1925); *China im Profil* (1934), *Lebensräume im Kampf der Kulturen* (1938, 2nd ed. 1951). E. Rosenkranz in *GBS*, 5 (1981), 117-121; E. Plewe, "Heinrich Schmitthenner," *PGM*, 98 (1954), 241-243; obit. by E. Plewe in *Marburger Geographische Schriften*, 7 (1957).

Scotland.

See Royal Scottish Geographical Society; United Kingdom.

Semple, Ellen Churchill (1863-1932)

M.A., Vassar College, 1891. University of Chicago, 1906-24; Clark University, 1921-32. *American History and Its Geographic Conditions* (1903); *Influences of Geographic Environment* (1911); *Geography of the Mediterranean Region* (1931). A. Bushong in *GBS*, 8 (1984), 87-94; J. Wright, "Miss Semple's 'Influences of Geographic Environment': Notes toward a Bibliobiography," *GR*, 52 (1962), 346-361; obits. by C. Colby in *AAAG*, 23 (1933), 229-240; and R. Whitbeck in *GR*, 22 (1932), 500-501.

Sequent Occupance

The term "sequent occupance" was coined by the Harvard geographer Derwent Whittlesey in 1929 to refer to the succession of stages of human occupation of regions. The term had some popularity among U.S. geographers in the 1930s and 1940s but declined to extinction in the 1950s because it failed to live up to its author's promise as an analytical tool in human geography. Marvin Mikesell has said that "sequent occupance probably became popular because it was an 'analogue model', built in emulation of the erosion cycle and the concepts of succession and climax in plant geography." In recent decades historical geographers have tended to follow Andrew Clark's view of their field as the study of "changing geographies" or "the geography of change," and the "snap-shot" approach of sequent occupance no longer holds appeal. *See also* Historical Geography.

Bibliography

Mikesell, M. W., "The Rise and Decline of 'Sequent Occupance': A Chapter in the History of Ameri-

can Geography," pp. 149-169 in *Geographies of the Mind: Essays in Historical Geosophy in Honor of John Kirtland Wright* (ed. by D. Lowenthal and M. J. Bowden) (1976).
Whittlesey, D. S., "Sequent Occupance," *AAAG*, 19 (1929), 162-165.

Sestini, Aldo (1904-1988)

Laurea, University of Florence, 1928. University of Cagliari, 1938-42; University of Milan, 1942-51; University of Florence, 1928-38 *passim*, 1951-74. "Studi geografici sulle città minori della Toscana: I, Arezzo," *RGI*, 45 (1938), 22-65, 89-121; *Introduzione allo studio della geografia* (with G. Nangeroni) (1947); *Il Paesaggio* (1963); *Scritti minori* (1989). P. Innocenti, "Presentazione," pp. 5-11 in *Scritti Geografici in onore di Aldo Sestini*, vol. 1 (1982) (Sestini's bibliography, pp. 1135-1150, of vol. 2).

Settlement Geography

Settlement geography focuses upon study of cultural landscape morphology, especially rural settlement forms, although urban built environments also belong to the subject matter. Some practitioners include settlement *process*, mainly analysis of frontier land occupance, but emphasis upon *form* remains the core of settlement geography. Three landscape elements provide the subject matter: *settlement pattern* types, or the distribution of buildings, as reflected in various clustered and dispersed arrangements, village layouts, and town plans; *cadastral* and *field patterns*, revealing the manner of partitioning land for human use; and types of fixed-site *greater artifacts*, in particular houses, farmsteads, barns, fences and bridges. Traditional types and patterns have received far more attention than modern, popular ones, both because settlement geography is deeply rooted in the humanities and nineteenth century romanticism, and because folk forms possess a demonstrated diagnostic ability, allowing such landscapes to be "read," yielding historical geographical explanations. Field observation provides the basic methodology. The persistent emphases, then, can be summed up in the words *rural, folk* and *landscape*.

The subdiscipline derives largely from a single founder, the German scholar August Meitzen, and his seminal 1895 work, *Siedelung und Agrarwesen* . . . . He was the first to join the three subject components into a comprehensive analysis of rural landscapes and to promise a diagnostic quality in traditional settlement forms, claiming that "we walk . . . in every village among the ruins of antiquity" and that the folk house is "the embodiment of a people's soul." While most of his historical explanations proved to be erroneous, in particular his attempts to link certain settlement patterns to ancient ethnic influences, Meitzen laid the foundation for what came later and deserves most of the credit for introducing the concept of cultural landscape into geography.

Shortly after Meitzen's work appeared, two young German geographers, Robert Gradmann and Otto Schlüter, combined his historical emphasis with a concern for economic function, physical environmental site, and rigorous field work to produce the subdiscipline of *Siedlungsgeographie*. From that time, the study of settlement form played a major role in German geography, with scores of practitioners. Indeed, Schlüter helped found a landscape purist school, which held that only visible elements were suitable for geographical study. Curiously, German geographers largely neglected house types, focusing upon settlement and field patterns.

French geographers, already deeply humanistic in focus, quickly absorbed these German developments, producing their own special *géographie de l'habitat*. Pivotal figures included Paul Vidal de la Blache, who published "De l'habitation sur les plateaux limoneux du Nord de la France" as early as 1899; his student Jean Brunhes, who devoted fully one-fourth of his 1910 classic *La géographie humaine* to settlement geography; and Albert Demangeon, whose abundant works are summarized in *La France* (1946). As in Germany, the later practitioners were so numerous as to defy listing, although the Belgian scholar Marguerite Lefèvre deserves mention because she was one of the few Francophones to write extensively on settlement geography. While the French acknowledged the German influence, their *géographie de l'habitat* exhibited emphases rather different from

Siedlungsgeographie, most notably in their focus on house and farmstead types and rejection of the historical-genetic emphasis.

The British geographers, by contrast, accepted the German focus on settlement pattern and, above all, field systems, within the context of historical geography rather than a separate settlement subdiscipline. In part, too, their intellectual roots were indigenous, derived largely from Frederic Seebohm (1833–1912), whose 1883 classic *English Village Community* also influenced Meitzen. One culmination of this line of work was the monumental 1973 *Studies of Field Systems in the British Isles*, edited by Alan R. H. Baker and Robin A. Butlin. A quite remarkable anomaly in the British school was E. Estyn Evans, whose books on Ireland pieced back together Meitzen's threefold interests and genetic approach, while adding anthropological and ecological sophistication, as well as a fondness for field observation.

North American settlement geography reveals a mixture of German and French influences. Most important is the "Berkeley school," where Carl O. Sauer introduced Meitzen's (and Eduard Hahn's) historical emphasis, Schlüter's landscape purism, and an ethic of field research. Curiously, the French focus upon house types prevails in the Berkeley group, perhaps because U. S. settlement, field, and cadastral patterns are somewhat monotonous. The most influential among Sauer's students has been Fred B. Kniffen, whose seminal article on houses appeared in 1936. Kniffen listed Meitzen, Sauer, Schlüter, and Brunhes as formative influences, in addition to Berkeley anthropologists. Many other geographers, linked directly or indirectly to Berkeley, Sauer, and Kniffen, pursued these interests, most prolifically Terry G. Jordan, Richard Pillsbury, Peter O. Wacker, and Wilbur Zelinsky. Henry Glassie carried Kniffen's influence into U. S. folklore. Independent of the Berkeley group, in the 1940s at Wisconsin Glenn T. Trewartha worked on settlement morphology.

Settlement geography as here defined suffered a decline during the hegemony of economic, theoretical, urban, and applied geography after about 1965. Kniffen's Association of American Geographers presidential address in that year, "Folk Housing: Key to Diffusion," though rich in promise, ironically marked the end of an era. The association directory, which contained fifty-six self-professed settlement geographers in 1961, no longer even lists such a specialty. In Europe, similarly, the convening of the Vadstena Symposium on agrarian landscapes and the almost simultaneous appearance of Gabriele Schwarz's comprehensive settlement text around 1960 improbably preceded a decline of the subdiscipline. Even so, geographical attention to cultural landscape survives on both sides of the Atlantic and seems a legacy likely to continue. *See also* Historical Geography; Human Geography.

Bibliography

Brunhes, J., *La géographie humaine* (1910).

Helmfrid, S., ed., "Morphogenesis of the Agrarian Cultural Landscape" (Vadstena Symposium), *Geografiska Annaler*, 43 (1961), 1–328.

Kniffen, F. B., "Louisiana House Types," *AAAG*, 26 (1936), 179–193.

Meitzen, A., *Siedelung und Agrarwesen der Westgermanen und Ostgermanen, der Kelten, Römer, Finnen und Slawen* (3 vols. plus atlas, 1895).

Sauer, C. O., "Morphology of Landscape," *University of California Publications in Geography*, 2 (1925), 19–54.

Schwarz, G., *Allgemeine Siedlungsgeographie* (1959).

• *Terry G. Jordan*

Shabad, Theodore (1922–1987)

Ph.D., Columbia University, 1976. *New York Times*, 1953–87; Columbia University, 1976–87. *Geography of the USSR* (1951); *China's Changing Map* (1956, 1972); *Basic Industrial Resources of the USSR* (1969); *The Soviet Energy System* (with L. Dienes) (1979). *GOF* (1981); obits. by P. Lydolph in *AAAG*, 78 (1988), 556–561; and by C. Harris, V. Kotljakov, and S. Lavrov in *Soviet Geography*, 28 (1987), 376–387.

Shaler, Nathaniel Southgate (1841–1906)

B.S., Harvard University, 1862. Harvard University, 1864–1906. *Nature and Man in Amer-*

ica (1891); (ed.) *The United States of America* (2 vols., 1894); *Man and the Earth* (1905); *Autobiography* (1909). W. Koelsch in *GBS*, 3 (1979), 133-139; D. N. Livingstone, *Nathaniel Southgate Shaler and the Culture of American Science* (1987).

Shantz, Homer LeRoy (1876-1958)

Ph.D., University of Nebraska, 1905. University of Illinois, 1926-28; University of Arizona (President), 1928-36; U. S. Department of Agriculture, 1910-26, 1936-44 ("annuitant collaborator" from 1945 onward). *The Vegetation and Soils of Africa* (with C. Marbut) (1923); "Agricultural Regions of Africa," *EG*, vols. 16-19 (1940-43); *The Use of Fire as a Tool in the Management of the Brush Ranges of California* (1947); *Photographic Documentation of Vegetational Changes in Africa over a Third of a Century* (with B. Turner) (1958). *WWWA*; obits. by C. Sauer in *GR*, 49 (1959), 278-280; and E. Pruitt in *AAAG*, 51 (1961), 392-394.

Shiga, Shigetaka (1863-1927)

Bachelor of Agriculture, Sapporo Agricultural College, 1884. Waseda University, 1895-1927. *Chirigaku kōgi* (1889); *Nihon Fūkeiron* (1894); *Chiri kōwa* (1906). S. Minamoto in *GBS*, 8 (1984), 95-105; obits. by N. Yamasaki in *Chirigaku hyōron* (Geographical Review of Japan), 3 (1927), 449-452; and Satō in *Chirigaku zasshi* (Journal of Geography of Japan), 39 (1927), 361-362.

Simoons, Frederick John (b. 1922)

Ph.D., University of California, Berkeley, 1956. Ohio State University, 1956-57; University of Wisconsin, 1957-66; Louisiana State University, 1966-67; University of Texas, 1967-69; University of California, Davis, 1969-89. *Northwest Ethiopia* (1960); *Eat Not This Flesh* (1961); *A Ceremonial Ox of India* (with E. Simoons) (1968). *60 Years of Berkeley Geography*.

Singapore

The teaching of geography in Singapore began as early as 1928, when Raffles College, a liberal arts college, was founded by the British colonial administration. Degree courses in geography were given in 1949 when the first University of Malaya (now Malaysia) was established. Singapore was then part of Malaya. The first chair of geography was held by the late Professor E. H. G. Dobby. He and two other full-time members of the staff were responsible for developing the department of geography into a full-fledged teaching and research unit. Changes associated with the political scene in the Singapore/Malaysia region led to the separation of Singapore from Malaysia in 1965. With this division separate universities were established in Malaysia. In Singapore, the original University of Malaya was renamed the University of Singapore. The original department of geography remained in Singapore, although some of the members moved to Kuala Lumpur to staff the department of geography that was established there. In 1980, the two universities in Singapore—University of Singapore and Nanyang University—were merged to form the National University of Singapore. Staff members from both departments of geography were similarly merged to form an enlarged department.

The National University of Singapore is modelled along Commonwealth universities' lines, with undergraduates reading for degrees over three years and graduates obtaining postgraduate degrees up to the doctoral level. Geography is offered to students in the Faculty of Arts and Social Sciences, who may elect to read it as one of the subjects leading to the degree of Bachelor of Arts. Students who qualify are eligible to take geography for a further year when the successful candidates are awarded the Bachelor of Arts Honours degree. The department also offers postgraduate degrees from the Masters to the doctoral levels.

Over the years, the department of geography in Singapore has grown steadily in terms of both staff and students. In 1949, there were only three full-time staff members and 39 students. By 1970, staff members had increased to nine and student numbers to 155. Today, there are sixteen full-time staff and 550 students. There are also four graduates working for their M.A. and one for her Ph.D.

The standards and quality of the degree offered at the National University of Singapore are safeguarded by a system of external examinership. External examiners normally visit the department once in their two-year term of office. Examiners are appointed usually from universities in the U.K., U.S.A. and Australasia. Among those so appointed were Professors Keith Buchanan, R. O. Buchanan, P. R. Crowe, Norton Ginsburg, Peter Haggett, B. W. Hodder, Austin Miller, Mansell Prothero, J. E. Spencer, L. Dudley Stamp, Robert Steel and Paul Wheatley. Currently the external examiner is Professor Peter Scott from the University of Tasmania.

Among the many geographers who have been associated with the department in the period since its inception and who have taught in it at one time or another were Professors J. M. Blaut, W. L. Dale, D. W. Fryer, R. D. Hill, Robert Ho, W. Neville, S. Nieuwolt, J. J. Nossin, Paul Wheatley, R. Wikkramatileke and Yeung Yue-man. In addition, the department has gained academically from geographers who have spent extended periods in Singapore as visiting professors. Among such scholars were Professors J. O. M. Broek, T. R. Leinbach, J. E. Spencer and G. Thomas.

One of the major achievements of the late Professor Dobby was to found the *Malayan Journal of Tropical Geography*, later renamed the *Journal of Tropical Geography* and now split into the *Malaysian Journal of Tropical Geography* and the *Singapore Journal of Tropical Geography*. It was started in 1953 and has an uninterrupted publication record extending to the present day. The department now also publishes the *Tropical Geomorphology Newsletter*, which is circulated to interested geomorphologists throughout the world.

BIBLIOGRAPHY

Ho, R., "Geography at the Universities: Department of Geography, University of Malaya," *Oriental Geographer*, 3 (1959), 105-108.

Ooi Jin Bee, "Geography in Singapore," pp. 400-432 in *Geography in Asian Universities*, ed. by R. J. Fuchs and J. M. Street (1976).

• *Ooi Jin Bee*

Sion, Jules (1880-1940)

D.-ès-L., University of Paris, 1908. University of Montpellier, 1910-40. *Les paysans de la Normandie orientale* (1909); *Géographie universelle*, vol. 7, *Méditerranée, péninsules méditerranéennes* (2 parts, with M. Sorre and Y. Chataigneau, 1934); *Géographie universelle*, vol. 9, *Asie des moussons* (2 parts, 1928-29); *La France méditerranéenne* (1934). W. van Spengen in *GBS*, 12 (1988), 159-165; G. Pinchemel in *Deux siècles de géographie française* (1984); obits. by E. de Martonne in *AG*, 49 (1940), 152-153; and L. Febvre in *Annales d'histoire sociale*, 13 (1941), 81-89.

Smith, Joseph Russell (1874-1966)

Ph.D., University of Pennsylvania, 1903. University of Pennsylvania, 1903-19; Columbia University, 1919-44. *The Organization of Ocean Commerce* (1905); *Industrial and Commercial Geography* (1913, 1925, later editions with M. Phillips and T. Smith); *The World's Food Resources* (1919); *North America* (1925, later eds. with M. Phillips). V. Rowley, *J. Russell Smith: Geographer, Educator, and Conservationist* (1964); obits. by O. P. Starkey in *AAAG*, 57 (1967), 198-202; and by D. J. Orchard in *GR*, 57 (1967), 128-130.

Smith, Robert Henry Tufrey (b. 1935)

Ph.D., Australian National University, 1962. University of Melbourne, 1961-62; University of Wisconsin, 1962-70; Queen's University, 1970-72; Monash University, 1972-75; University of British Columbia (President *pro tem*, 1985), 1975-85; University of Western Australia (Vice Chancellor), 1985-88; National Board of Employment, Education and Training (Chairman), Canberra, 1988 to the present. "Method and Purpose in Functional Town Classification,"

AAAG, 55 (1965), 539–548; "Concepts and Methods in Commodity Flow Analysis," *EG*, 47 (1970), 404–416; "A Theory of the Spatial Structure of Internal Trade in Underdeveloped Countries" (with A. Hay), *Geographical Analysis*, 1 (1969), 121–136; "Periodic Market-Places and Periodic Marketing," *PHG*, 3 (1979), 471–505, vol. 4 (1980), 1–31. *CWW; Who's Who in Australia.*

Social Geography

Social geography celebrates the relative detachment of social life from the physical environment. It is a young subdiscipline whose concerns stretch from measuring the spatial pattern of social structures to interpreting the meaning of social relations in terms of the symbolism of place or territory. Its intellectual roots are usually traced to Robert Park and the Chicago Schools of Human Ecology and Sociology, which flourished during the 1920s and 1930s, and to the ideas of Vidal de la Blache published in the early 1900s. Nevertheless, the self-conscious life of social geography spans less than forty years, and includes such diverse proponents as E. W. Gilbert, Emrys Jones, and Anne Buttimer, whose distinctive approaches are ample testimony to the subdiscipline's eclecticism of content, style, and philosophical orientation.

Social geography gained most of its original impetus in Europe, especially the United Kingdom, where the enthusiastic pursuit of "spatial sociology" during the 1960s and 1970s reflected both the unprecedented popularity of sociology itself and the legacy of a quantitative revolution. In the United States, the subject's early influence was eclipsed by a traditional, Sauerian cultural geography, and did not begin to appear on research agendas and undergraduate curricula until the late 1970s and early 1980s, by which time it had been infused with the more humanistic philosophies.

Today, social geographers' interests continue to range widely, from spatial structure to social process, and from theories of social justice to the principles of social welfare and the practice of social policy, using a spectrum of methodologies ranging from statistical abstraction to detached contemplation and action research. This vibrancy has earned the subdiscipline an annual review in the journal *Progress in Human Geography*. Currently, the subject has perhaps three unifying themes. First, an emphasis on the interdependence of social and spatial organization. This is best expressed in the importance attached to measuring residential segregation, mapping social interaction, and charting the symbolism and meanings of territory. Second, an interest in aspects of social stratification other than those stemming directly from the division of labor. This is reflected in a ready enthusiasm for the social geographies of "race" and racism, gender and patriarchy, ethnicity and nationalism. Third, a growing sensitivity to the relevance of culture, more in the vein of Raymond Williams than that of Carl Sauer, but to an extent sufficient to herald the merger, in 1988, of the social and cultural study groups of the Institute of British Geographers. *See also* Human Geography.

Bibliography

Buttimer, A., "Social Space in Interdisciplinary Perspective," *GR*, 59 (1969), 417–426.

Cater, J., and T. Jones, *Social Geography: An Introduction to Contemporary Issues* (1989).

Eyles, J., ed., *Social Geography in International Perspective* (1986).

Gilbert, E. W., and R. W. Steel, "Social Geography and Its Place in Colonial Studies," *GJ*, 106 (1945), 118–131.

Jackson, P., *Maps of Meaning* (1989).

Jackson, P., and S. Smith, *Exploring Social Geography* (1984).

• *Susan Smith*

Social Physics

The term covers the use of analogs of Newtonian gravity, gravitational energy and LaGrangian potential models to describe and predict, mathematically, the movement and interactions of people based on their location. Substituting population (p) for Newton's mass, we have $F = kp_1p_2r^{-2}$ where F is force, k is a constant (although often neglected it can reflect regional differences) and r is the distance between the centers of two populations. Since energy (E) is force times distance, $E = kp_1p_2r^{-1}$. Potential at a point c (V_c) is the energy acting on one person at c induced by the populations at all places $i = 1$ to n. Thus, $V_c = k \sum_{i=1}^{n} p_i r_{ic}^{-1}$. A correction is made in r for the city population located at c to avoid an infinite potential caused by allowing $r = 0$ (Stewart and Warntz 1958).

The values of F, E and V are calculated from population and distance data. Dependent variables such as movements of people, goods, letters or telephone calls are correlated with F or E, and rural land values, rural population densities, per capita income, income densities and other proxies for interaction are correlated with V. All these relations can be expressed by means of a power function, $y = ax^b$, where y is the observation to be correlated with x, and x is the calculated value of F, E or V. These relations are plotted on double logarithmic graph paper and produce straight lines with remarkably little scatter. The intercept of the line is given by a, and the slope of the line is given by b. The values of a and b can vary regionally and with time. Potential is often mapped on a national scale. The variables that are highly correlated with potential, although rarely mapped, would produce surfaces of similar topology.

The term "social physics" was used by the Belgian statistician L. A. J. Quetelet (1796–1874) and the French philosopher A. Comte (1798–1857) in the mid-nineteenth century and was reintroduced by the U.S. astrophysicist J. Q. Stewart in 1947. The force law was suggested by Carey in 1858 and used by Young in 1924 to predict migration and by Reilly in 1929 to delimit shopping area boundaries. In 1885 Ravenstein used an energy formulation to describe migration flows. LaGrange's use of potential, first devised to measure the attraction of several planets on a point in space and extended later to predict magnetic and electrical fields, was introduced to population studies by Stewart in 1940. The subset of spatial and temporal variations of V and U (income-weighted V or "income potential") together with their correlated dependent variables, comprise "macrogeography."

Later formulations are generically classified as social physics or as "gravity laws," along with the previously described laws for F, E and V. These newer relations use adjustable exponents for distance and population and, in some cases, specify constraints on flows from origins, along routes and to destinations. These refinements may produce a better fit to observed flows. Wilson has derived such relations by analogy to statistical mechanics (Tocalis 1978).

Stewart, Zipf, Lotka, Warntz and other investigators found important regularities that have a strength approaching physical law. Thus, said Stewart (1947), "Meanwhile, let 'social planners' beware! Water must be pumped to flow uphill, and natural tendencies in human relations cannot be combatted and controlled by singing to them. . . . The city or national planner likewise must adapt his studies to natural principles." This advice has been followed by some city and regional planners.

Applications of social physics to theoretical and applied problems in human geography can be found, for instance, in the works of T. Hägerstrand, C. Harris, D. Huff, I. Lowry, J. R. Mackay and A. Wilson. *See also* Macrogeography.

Bibliography

Janelle, D. G. and B., "Patterns of Contact and Influence in the Life of a Spatial Scientist: William Warntz," pp. 189–201 in *Geographical Systems and Systems of Geography: Essays in Honour of William Warntz*, ed. by W. J. Coffey (1988).

Stewart, J. Q., "The Gravity of the Princeton Family," *Princeton Alumni Weekly*, 40 (1940), 409–410.

———, "Empirical Mathematical Rules Concerning the Distribution and Equilibrium of Population," *GR*, 37 (1947), 451–485.

Stewart, J. Q., and W. Warntz, "Physics of Population Distribution," *Journal of Regional Science*, 1 (1958), 99–123.

Tocalis, T. R., "Changing Theoretical Foundations of the Gravity Concept of Human Interaction," pp. 66–124, 154–167 in *The Nature of Change in Geographical Ideas*, ed. by B. J. L. Berry (1978).

Webber, M. J., *Explanation, Prediction and Planning: The Lowry Model* (1984).

• *Michael J. Woldenberg*

Sölch, Johann (1883–1951)

Ph.D., University of Vienna, 1906. University of Innsbruck, 1920–28; University of Heidelberg, 1928–35; University of Vienna, 1935–51. *Die Landformung der Steiermark* (1928), *Die Ostalpen und Österreich* (1930), *Die Landschaften der Britischen Inseln* (2 vols., 1951–52). J. Matznetter in *GBS*, 7 (1983), 117–124; obits. by

H. Bobek in *PGM*, 96 (1952), 110-112; and H. Kinzl in *MOGG*, 96 (1954), 3-31.

Soil, Geography of

The geography of soil involves studying the morphology or profiles of soils, detecting their genesis, and inferring their development. It also involves the mapping of their distribution on the land surface and the assessment of their potential for use by man, in agriculture, forestry, urban development, or waste disposal treatment. It also involves the negative consequences of their use by man—slope erosion, drainage basin siltation, and responses to pollution. Thus the geography of soils interprets soils as natural objects and assesses them as our most fundamental resource.

Most germane to the study is the concept of soil as the result of the interaction of five geographical factors: *parent material*, the geomorphic substrates from and on which soils form; *climate*, the agent of soil formation, mainly through rainfall/evaporation and its seasonal balance; *organisms*—the interaction of the soil with vegetation, especially litter fall and roots, and its own internal biota, forming humus. Then there is the close interdependence of *geomorphology* in soil formation and soil stability, for all soils are related to the landforms on which they are sited, the best example being the catena of soils between dry hilltop and humid valley floor. Finally, soils are not static but develop through *time*, to form a distinct soil profile, appropriate to the environment and past influences.

Developed soils show diagnostic horizons that represent the impact of various processes. The leaching of clay from topsoil (A horizon) to subsoil (B horizon) forms an argillic B horizon, so typical of humid warm temperate areas and the dry season tropics. Highly weathered soils of the perhumid tropics show oxic horizons, typically reddish or yellowish and of great thickness. Soils are marked by morphologic criteria according to *color*; or, according to the dominance of *mineral or organic components*; by *texture*—the dominant particle size; and by *nutrient status* or base exchange capacity (the acid/base ratio) and, joined with this, the contents of available N and P, and the pH, or soil acidity, which largely determine the fertility of soils.

In recent times new concepts of soil have emerged, more suited to the view of soil as a resource and as a fundamental part of man-modified ecosystems. The soil is viewed as an openended system, as an input-output model that implies the interdependence of adjacent soil-sites, a change in one involving and changing others. This concept is best represented in the idea of geoecology or landscape ecology. Other advances are the use of remote sensing and digital mapping of soil properties and the use of light and electron microscopy in examining thin sections of soils to study pores, soil and root interaction, and soil cohesion. Thus the many soils of the world are now more realistically understood, though their ubiquity is not to be interpreted as representing an inexhaustible resource, for many soils are at risk, deteriorating under intensive or ill-advised use.

Bibliography

Bunting, B. T., *The Geography of Soil* (2nd ed., 1967).

Buol, S. W., F. D. Hole, and R. J. McCracken, *Soil Genesis and Classification* (2nd ed., 1980).

Foth, H. D., and J. W. Schafer, *Soil Geography and Land Use* (1980).

Kubiëna, W. L., *Micromorphological Features of Soil Geography* (1970).

Richter, J., *The Soil as a Reactor: Modelling Processes in the Soil* (1987).

U.S. Department of Agriculture, Soil Survey Staff, *Soil Taxonomy*, USDA Agricultural Handbook 236 (1975).

• *Brian T. Bunting*

Sopher, David Edward (1923-1984)

Ph.D., University of California, Berkeley, 1954. California State University, Sacramento, 1956-63; Syracuse University, 1964-84. *The Sea Nomads* (1965); *Geography of Religions* (1967); (ed.) *An Exploration of India* (1980); (ed.) *The City in Cultural Context* (with J. Agnew and J. Mercer) (1983). *60 Years of Berkeley Geography* (1983); eulogy by D. Meinig and biographical sketch by R. Singh in *National Geographical Journal of India*, 33 (1987), iv-xii.

Sorre, Maximilien (1880–1962)

D.-ès-L., University of Paris, 1913. University of Bordeaux, 1919–22; University of Lille, 1922–31; University of Clermont-Ferrand (Rector), 1931–34; University of Aix-Marseilles (Rector), 1934–40; University of Paris, 1941–48. *Les Pyrénées méditerranéennes* (1913); *Géographie universelle*, vol. 14; *Mexique, Amérique centrale* (1928); *Géographie universelle*, vol. 7; *Méditerranée, Péninsules méditerranéennes* (1934); *Les fondements de la géographie* (4 vols., 1943–52). Obit. by P. George in *AG*, 71 (1962), 449–459, reprinted in *LGF* (1975), 185–195.

South Africa, Republic of

South African geography grew from a school subject into a full academic discipline in the early twentieth century, following developments in the United Kingdom and the Netherlands. The needs of school teachers were reflected in the foundation of both the South African Geographical Society (1917) and the Society for Geography (1957), the former based in Johannesburg and the latter in Stellenbosch. The two societies publish the *South African Geographical Journal* and the *South African Geographer*, respectively, for the dissemination of research and techniques.

Geography as a university discipline was founded only after World War I, with the appointment of Dr. P. Serton to the University of Stellenbosch in 1921, followed by J. H. Wellington to the University of the Witwatersrand, Johannesburg. The first generation of academic geographers were recruited almost exclusively from the United Kingdom and the Netherlands and the links remained strong until the 1960s. All the universities established geography departments, reflecting the subject's strength at school. There are thus some twenty-six departments and sub-departments, over half of which were established since 1959 at new universities.

South African university departments are small, and this has been an inhibiting factor in research. The initial pioneering studies were largely physical in orientation, reflecting the diversity of the landscape, and are epitomized by Wellington's major work, *Southern Africa*. Thereafter a greater balance was evident as the quantitative, relevancy and Marxist upheavals exerted their effect. Pioneering work on central place theory was undertaken by K. S. O. Beavon (University of the Witwatersrand); the peculiar nature of the South African city was explored by R. J. Davies (University of Cape Town); T. J. D. Fair (University of the Witwatersrand); traced the broad spatial inequalities present in South Africa; and A. J. Christopher (University of Port Elizabeth) examined the question of frontier land settlement and segregation. Physical geography has not been neglected, as P. D. Tyson (University of the Witwatersrand) has undertaken a major research program directed toward an understanding of the recurrence of drought.

As befits a country with complex problems, professional geographers have played an important part in the analysis of the spatial aspects of rural and urban development. Through the years geographers have contributed their expertise to official commissions, government departments, private organizations and a host of projects aimed at contributing to an understanding and solution of the spatial inequalities of the country.

Bibliography

Beavon, K. S. O., *Central Place Theory: A Reinterpretation* (1977).

Christopher, A. J., *The Crown Lands of British South Africa* (1984).

Fair, T. J. D., *South Africa: Spatial Frameworks for Development* (1982).

Jackson, S. P., "The South African Geographical Society, 1917–1977," *South African Geographical Journal*, 60 (1978), 3–12.

Serton, P., *Suid-Afrika en Brasilie: 'n sosiaal geografiese vergelyking* (1959).

Wellington, J. H., *Southern Africa* (2 vols., 1955).

• *A. J. Christopher*

South Asia (India, Pakistan, Bangladesh, Sri Lanka, and Nepal)

The publication of Elisée Reclus's *L'Inde et l'Indo-Chine* (1883), volume 8 of his *Nouvelle géographie universelle*, provided the first modern synthesis of the geography of South Asia, a remarkable feat for one who had never set foot in the region. Not until the publication of O. H. K. Spate's *India and Pakistan* (1954), however, was there a comprehensive, modern, university-level text. Spate's work, both systematic and regional in its approach, was substantially supplemented by the regional accounts in the weighty, multi-authored *India* (1971), published under the editorship of R. L. Singh.

To Aligarh Muslim University falls the distinction of establishing (in 1930) India's first department of geography. Since that date, additional departments have been founded in virtually every major university of South Asia and in scores of affiliated colleges as well, with a distinct regional bias in favor of northern India. Relatively speaking, Pakistan and Bangladesh lag behind India—if proportional numbers of practitioners, departments, and publications are accepted as guides—while Sri Lanka and Nepal fare well relative to their limited size. Certain departments have developed strong specializations; for example, urban geography at Banaras Hindu University and development studies at the Centre for the Study of Regional Development at Jawaharlal Nehru University in New Delhi.

Along with university departments, geographic societies have also proliferated, many of them publishing—often irregularly—their own journals. The oldest society was the Bombay Geographical Society, which functioned from 1832 to 1873 and commenced publishing its proceedings in 1837. Not until 1926, with the founding of the Madras Geographical Association and the launching of its *Journal*, did South Asia give rise to a geographical society and periodical that would last to the present day. Regrettably, each of South Asia's many societies tended to draw on a limited region for its journal authors and clientele. That situation was substantially ameliorated with the founding in 1978 of the National Association of Geographers, India (NAGI), largely modeled on the Association of American Geographers. NAGI's membership is truly countrywide, and since 1980 it has held annual conventions, each year in a different city, that have brought together hundreds of geographers from academic and other professions. Among other leading societies (with dates of founding) are the Geographical Society of India (1933, called the Calcutta Geographical Society until 1951), the Bombay Geographical Association (1935), the Ceylon Geographical Society (1938), the National Geographical Society of India (1946), the Pakistan Geographical Association (1948), the Bangladesh Geographical Society (1956, originally the East Pakistan Geographical Society), and the Institute of Indian Geographers (1979).

Outside academe, South Asian geographers have found abundant work opportunities in United Nations development activities, in various organs of their respective national governments, and in state, regional, and municipal planning agencies. A geographer, Harka Gurung, was for a time the head of Nepal's Planning Commission. Important employers within India have included the National Atlas Organization, established in 1956 under the directorship of S. P. Chatterjee and subsequently renamed the National Atlas and Thematic Mapping Organization; the Regional Survey/Geography Unit of the renowned Indian Statistical Institute; the National Council of Applied Economic Research; the Office of the Registrar-General, which is responsible for the decennial censuses, for which important national and state census atlases have been prepared since 1961; the Diagnostic Survey of the Damodar Valley Authority (modeled on the Tennessee Valley Authority in the United States); the Calcutta Metropolitan Planning Organization; and the Central Arid Zone Research Institute.

South Asia is exceptionally well mapped. The preindependence Survey of India and its several post-independence successor organizations have prepared topographic maps at a variety of scales, whereas special-use maps have come from a wide variety of domestic and foreign sources. Among the latter we may cite the *Carte Internationale du Tapis Végétal et des Conditions Ecologiques* (1:1,000,000, 1962–) sponsored by the French Institute of Pondicherry; a variety of medium- and small-scale maps of Bangladesh, Nepal, Bhutan, and parts of India made by the World Bank and based largely on remote sensing imagery; and *The Historical Atlas of South Asia* (1978), funded by many U.S. governmental and private agencies and edited by Joseph E. Schwartzberg, which treats a much wider range of subjects than the title suggests and is supplemented by a very extensive text.

Thousands of indigenous scholars with advanced degrees (including well over 500 geography Ph.D.s in India alone) now work on and in South Asia. These are joined by hundreds more from Europe, the United States, and Japan. A number of the more exciting recent publications (e.g., *An Exploration of India*, edited by David Sopher, 1980) have entailed collaboration among authors of diverse national backgrounds. Within Europe, the attention accorded South Asia by British geographers is somewhat less than one might expect, whereas numerous German geographers have entered the field. An exceptionally good textbook, *Südasien* (1977), edited by Jürgen Blenck, Dirk Bronger, and Harald Uhlig, exemplifies this burgeoning interest.

South Asian geographers have been active contributors to the work of the International Geographical Union. India's S. P. Chatterjee served as its president from 1964 until 1968, when the International Geographical Congress was held in New Delhi, and several other South Asians have served as Chairs or Vice-Chairs of I.G.U. Commissions.

Bibliography

Chatterjee, S. P., *Fifty Years of Science in India: Progress of Geography* (1964), plus supplement issued in 1968, *Progress of Geography in India (1964-1968)*.

Schwartzberg, J. E., "The State of South Asian Geography," *PHG*, 7 (1983), 232-253.

Sopher, D. E., "Toward a Rediscovery of India: Thoughts on Some Neglected Geography," pp. 110-133 in *Geographers Abroad*, ed. by M. W. Mikesell (1973).

• *Joseph E. Schwartzberg*

Space

Space serves as one of the basic structures of human experience. Conceptions of space vary, however, with levels of cognitive development and across cultures. Geographical space has traditionally referred to the areal relations among phenomena on the earth's surface. The geographer's spatial perspective is most evident in the cartographic, chorological and spatial-analytic traditions. Cartographers abstract from real spatial relations in the construction of maps; chorologists study patterns of regional differentiation; and spatial analysts seek generalizations about the spatial organization of society.

The philosophy of Immanuel Kant has been one of the primary intellectual sources of the modern scientific geographer's spatial perspective. Kant's arguments about spatial and temporal intuition have been used as a basis for distinguishing between geography and history. The geographer orders phenomena in space, just as the historian orders events in time. Thus, the geographer's concept of place and region may be seen as similar to the historian's concept of period.

This spatial perspective was given an abstracted form during the 1960s when neo-Kantian chorology was successfully challenged by positivistic spatial analysis. The most extreme expression of this view was that of geography as applied geometry. This "spatial separatist" view has been criticized on logical grounds by those who have noted the contradictions inherent in the notion of empirical, spatial laws.

Critics of spatial analysis have sought to "reconnect" space to the world of experience and social relations. For example, humanistic geographers have considered the experience of space and the affective character of human spatial perception, and social geographers have argued for the essential interconnection of the social and the spatial. Geographers thus continue to regard the spatial perspective as being central to their disciplinary worldview, but are less concerned with making space their primary "object" of study.

Bibliography

Gregory, D., and J. Urry, *Social Relations and Spatial Structures* (1985).

Hartshorne, R., "The Concept of Geography as a Science of Space, from Kant and Humboldt to Hettner," *AAAG*, 48 (1958), 97-108.

Sack, R. D., *Conceptions of Space in Social Thought* (1980).

• *J. Nicholas Entrikin*

Spain

The civil war of 1936-39, the exile of a large number of scientists and intellectuals, and the repression of the early years of the Franco dictatorship brought about a serious discontinuity in the development of contemporary Spanish geography.

Down to 1939, geography had several principal developments: (1) the advancement of basic cartography, carried on by the Instituto Geográfico y Catastral and agencies of the armed forces; (2) the work of the Real Sociedad Geográfica de Madrid, founded in 1876 and publisher from that year of the *Boletín de la R.S.G.*; (3) the teaching of geography on the primary and secondary levels, with some excellent teachers and a long tradition in the composition of textbooks (to give an idea of the latter's importance it suffices to cite a checklist, *El Libro de Geográfia en España, 1800–1936* [Capel, Solé and Urteaga, 1988], which identifies some 2,300 geographical works intended for teaching; reprintings aside, these amount to 1,800 titles, with 750 authors who published at least one work in the period cited); (4) instruction in geography for teachers on the primary level, carried on by the normal schools and the Escuela Normal Superior del Magisterio of Madrid (in these, highly important theoretical and methodological advances were made in the 1910s and 1920s, with such outstanding figures as Ricardo Beltrán y Rózpide and Pedro Chico); (5) university-level geography in the Faculties of philosophy and letters, aimed above all at the training of instructors of geography and history in secondary education; (6) the physical geography of the natural scientists, pursued especially in the Science Faculties or scientific institutes, and to which notable contributions were made by Eduardo Hernández Pacheco, Emilio Huguet del Villar, and other geologists, meteorologists, and geobotanists.

After 1939, most of the above-cited trends continued to be important, but geography acquired a new impulse as the result of two factors. The first was the creation at Madrid, within the Consejo Superior de Investigaciones Científicas, of the Instituto Juan Elcano, which since 1941 has published *Estudios Geográficos*, the most important Spanish geographical review; and, somewhat later, the creation of the Instituto de Geografía Aplicada at Zaragoza (subsequently at Madrid), publisher of the journal *Geographica*. The second factor was the creation and endowment of university chairs of geography in Faculties of philosophy and letters, and subsequently of geography and history.

Cartography has been carried on in its basic aspects by the Instituto Geográfico Nacional, which by 1977 employed more than 200 engineers and geographers and about 1,000 technicians. Some of these, such as R. Núñez de las Cuevas and F. Vázquez Maure, have also participated in other geographical enterprises.

In the development of geography at the universities an important role has been played by certain professors: A. Melón and Manuel de Terán at Madrid, J. M. Casas Tomás at Zaragoza, J. Vilà Valentí at Barcelona, A. López Gómez at Valencia, and Joaquín Bosque Maurel at Granada. Around these names some of the most active research centers of the present day have been organized. Since the 1960s, development has been very important, with the creation of university departments and specialized studies in practically all the thirty Spanish public universities. Many university departments have also initiated a publications policy of issuing doctoral dissertations, research monographs, and journals. The majority of the twenty-seven scientific geographical journals have this origin. Outstanding among these are the *Revista de Geografía* of the University of Barcelona (since 1967, edited by J. Vilà Valentí), *Geocrítica* of the University of Barcelona (since 1976, edited by Horacio Capel), and *Anales de la Universidad Complutense* of Madrid (since 1981, edited by Joaquín Bosque Maurel). At the present time more than 400 university instructors offer courses in geography in Spanish universities and also make contributions to geographical research. Spanish geographers are also organized in the Asociación de Geógrafos Españoles (about 700 members), which puts on national congresses and specialized meetings, and in various societies, such as the Real Sociedad Geográfica of Madrid (some 500 members), which continues to publish its *Boletín de la R.S.G.* Spanish geographers actively also participate in scientific and professional organizations such as the Asociación de Ciencia Regional and others.

Within the universities geography is structured around three areas of scientific knowledge: physical geography, human geography, and regional geographical analysis. These are grouped in various ways in the university departments and institutes. In Spanish universities as a whole the number of students pursuing specialized studies in geography fluctuates between 1,500 and 2,000, with a tendency to level off or decline.

Research activity extends to practically all fields of physical and human geography, with outstanding contributions in the areas of geomorphology and climatology on one hand, and in urban, rural, and population geography on the other. Notable also are the contributions to geographical theory and to the history of geographical thought. In the field of regional geography (an old tradition in this country), attention centers above all on the geography of Spain, one of the most prominent shortcomings being the scant attention paid to other areas of the world. *See also* Juan Sebastián Elcano Institute of Geography; Royal Geographical Society of Madrid.

BIBLIOGRAPHY

Capel, H., "La Geografía española actual," in *La Ciencia española*, ed. by J. M. López Piñero (1989).

Capel, H., J. Solé, and L. Urteaga, *El libro de geografía en España* (1988).

López, F., R. Morell, L. Urteaga, and J. Vilagrasa, "La enseñanza de la geografía y el empleo de los geógrafos," *Geo Crítica*, no. 64 (1986).

Suárez de Vivero, J. L., "La Geografía en el desarrollo científico español 1875-1914," *II Coloquio Ibérico de Geografía, Lisboa 1980, Comunicaçoes*, vol. 2 (1983), 235-248.

Vilà Valentí, J., "Origen y significado de la fundación de la Real Sociedad Geográfica," *Revista de Geografía*, 11 (1977), 5-36.

———, "The Iberian Peninsula and Latin America," pp. 264-281 in *Geography since the Second World War*, ed. by R. J. Johnston and P. Claval (1984).

• *Horacio Capel Sáez*
(translated by C. Julian Bishko)

Spate, Oskar Hermann Khristian (b. 1911)

Ph.D., University of Cambridge, 1937. University of Rangoon, 1937-39; London School of Economics, 1945-51; Australian National University, 1951-89. *India and Pakistan* (1954, 3rd ed. with A. Learmonth 1967); *Let Me Enjoy* (1965); *Australia* (1968); *The Pacific since Magellan* (3 vols., 1979-88). T. Perry, "Oskar Spate: A Personal Impression," pp. xiii-xix in *Of Time and Place: Essays in Honour of O. H. K. Spate* (ed. by J. Jennings and G. Linge) (1980).

Spatial Analysis. *See* Quantitative Methods in Geography.

Spencer, Joseph Earl (1907-1984)

Ph.D., University of California, Berkeley, 1936. University of California, Los Angeles, 1940-75. *Land and People in the Philippines* (1952); *Asia, East by South* (1954); *Philippine Island World* (with F. Wernstedt) (1967); *Cultural Geography* (with W. Thomas) (1969). *GOF* (1970, 1987); *60 Years of Berkeley Geography* (1983); obit. by H. Nelson in *AAAG*, 75 (1985), 595-603.

Sri Lanka. *See* South Asia.

Stamp, Laurence Dudley (1898-1966)

D.Sc., University of London, 1921. University of Rangoon, 1923-26; London School of Economics, 1926-58; President of IGU, 1952-56. *The British Isles* (with S. Beaver) (1933); (ed.) *The Land of Britain: The Report of the Land Utilisation Survey of Britain* (92 parts, 1936-46); *The Land of Britain: Its Use and Misuse* (1948); *The Geography of Life and Death* (1964). M. Wise in *GBS*, 12 (1988), 175-187; *Land Use and Resources: Studies in Applied Geography: A Memorial Volume to Sir Dudley Stamp*, Institute of British Geographers (1968); obits. by M. Wise in *GJ*, 132 (1966), 591-594; and S. Beaver in *Geography*, 51 (1966), 388-391.

Stanislawski, Dan (b. 1903)

Ph.D., University of California, Berkeley, 1944. University of California, Berkeley, 1942-45; University of Washington, 1945-47; University of Pennsylvania, 1947-49; University of Texas, 1949-62; University of Arizona, 1963-79. *The Anatomy of Eleven Towns in Michoacán* (1950); *The Individuality of Portugal* (1959); *Portugal's Other Kingdom, the Algarve* (1963); *Landscapes of Bacchus* (1970). *60 Years of Berkeley Geography* (1983).

Statistical Geography. *See* Quantitative Methods in Geography.

Steel, Robert Walter (b. 1915)

M.A., University of Oxford, 1941. University of Oxford, 1939-56; University of Liverpool, 1957-74; University College of Swansea (Principal), 1974-82. (ed.) *Geographers and the Tropics: Liverpool Essays* (with M. Prothero) (1964); (ed.) *Liverpool Essays on Geography: A Jubilee Collection* (with R. Lawton) (1967); *The Institute of British Geographers: The First Fifty Years* (1983); (ed.) *British Geography 1918-1945* (1987). *WW.*

Steers, James Alfred (1899-1987)

B.A., University of Cambridge, 1921. University of Cambridge, 1922-66. *The Unstable Earth* (1932, 5th ed. 1950); *The Coastline of England and Wales* (1946, 2nd ed. 1964); *The Coastline of Scotland* (1973); *Coastal Features of England and Wales* (1981). *WW*; *GOF* (1982); obits. by D. Stoddart in *TIBG*, 12 (1988), 109-115; and anon. in *GJ*, 153 (1987), 436-438.

Steinmetz, Sebald Rudolf (1860-1940)

Ph.D., University of Leyden, 1892. University of Utrecht, 1895-1900; University of Leyden, 1900-08; University of Amsterdam, 1908-33. *De beteekenis van de volkenkunde voor de studie van mensch en maatschappij* (1908); "Die Stellung der Soziographie in der Reihe der Geisteswissenschaften," *Archiv für Rechts- und Wirtschaftsphilosophie* (1912-13), 492-501; *De nationaliteiten van Europe* (1920); "De differentiatie der schoolgeographie en het goed recht der sociographie," *Tijdschrift van het Koninklijk Nederlands Aardrijkskundig Genootschap*, 43 (1926), 674-684. H. N. ter Veen, "Steinmetz' betekenis voor de aardrijkskunde in Nederland," *Men en Maatschappij* (Steinmetz Number), 9 (1933), 19-21; W. H. Vermooten, *De mens in de geografie* (1941); M. W. Heslinga, "Sociografie versus sociale geografie," pp. 53-67 in *Toen en thans: de sociale wetenschappen in de jaren dertig en nu*, ed. by K. Bovenkerk *et al.* (1978); W. J. van den Bremen, "De Nederlandse geografie van 1850 tot 1950," pp. 160-174 in *Algemene Sociale Geografie: Ontwikkelingslijnen en Standpunten*, ed. by A. G. J. Dietvorst, J. A. van Ginkel, *et al.* (1984).

Stoddart, David Ross (b. 1937)

Ph.D., University of Cambridge, 1964. University of Cambridge, 1962-88; University of California, Berkeley, 1987 to the present. (ed.) *Regional Variation in Indian Ocean Coral Reefs* (with C. Yonge) (1971); (ed.) *The Northern Great Barrier Reef* (with C. Yonge) (1978); *Biogeography and Ecology of the Seychelles Islands* (1984); *On Geography and Its History* (1986). *GOF* (1979).

Strahler, Arthur Newell (b. 1918)

Ph.D., Columbia University, 1944. Columbia University, 1945-68. *Physical Geography* (1951, 4th ed. 1975); *The Earth Sciences* (1963, 2nd ed. 1971); *Environmental Geoscience* (with A. H. Strahler) (1973); *Physical Geology* (1981). *WWA.*

Sudan

The teaching of geography in the Sudan started with the introduction of scholastic education by the British administration during the first decade of the twentieth century, and the evolution of the subject has been associated with the development of the educational process and institutions ever since.

Currently, the teaching of geography starts in the third grade (age nine) and remains compulsory all through the general education levels, *i.e.*, up to the twelfth grade. Geography as taught gets wider and deeper as one moves up the educational ladder, mainly on a regional basis but also systematically. The curricula start with the immediate local habitat and end with the study of all continents, with focus on the Arab World and Africa and with further emphasis on the Nile Basin.

The general objectives of teaching geography include the promotion of (1) the ability to identify and interpret natural and social phenomena; (2) the understanding of nature and society and their interrelations; (3) awareness of the interrelatedness of all human communities; (4) national and international citizenship; and (5) positive personal and social attributes such as cooperation, patience, exactness and responsibility.

At the higher level, geography has continued to grow steadily in popularity and has captured the respect of academicians and employers since 1940, when Gordon Memorial College became a center of higher education. There are now separate geography departments at the University of Khartoum, the Islamic University, and Cairo University (Khartoum Campus). In other universities geography is incorporated in other disciplines, *e.g.*, rural development, natural resources, and environmental studies. In the University of Khartoum (which is the oldest, biggest, best-established and most highly regarded university) geography is so popular that geography students constitute 30 to 37 percent of the total student population in the Faculty of Arts, which houses more than ten departments. The department offers forty-three courses covering almost all aspects of geography. Practical geography and fieldwork constitute an important component of the training of students. Degrees awarded are B.A., M.A. and Ph.D.

Geographical societies play an invaluable role in the dissemination of geographical knowledge. These societies are active in most schools and universities. The Sudan Geographical Society has been reactivated.

Sudanese geographers have gained a national reputation for applied research pursued independently or interdisciplinarily, or in collaboration with foreign universities, whether for academic purposes, for government agencies or for international organizations. Most of the research addresses pragmatic issues pertaining to the development of the Sudanese economy or society. These issues include deterioration of the ecosystems, drought and desertification, famine and food security, rural-urban migration and population displacement, regional disparities, sedentarization of the nomads, production relations and problems, urbanization, and refugee problems.

Bibliography

Elbushra, S., "The Sudan," in *Essays on World Urbanization*, ed. by R. Jones (1975).

Elnour, A. H. B., "Teaching and Research in Geography Department, University of Khartoum: An Appraisal," paper presented at the workshop on Teaching and Research in Eastern and Southern African Universities, 16-19 November 1988, Dehra Zeit, Ethiopia.

Eltom, M. A., *Geography at the University of Khartoum* (1975).

Howes, D. W., "Sudanese Geography: Recent Research at the University of Khartoum," *PG*, 41 (1989), 214-217.

• *Galal Eldin Eltayeb*

Sweden

Sweden has made and continues to make an impressive contribution to academic geography. The impact was delayed, but has gathered momentum out of all proportion to the number of its practicing geographers. The situation is inseparable from (1) the quantity and quality of research (and public support for it); (2) the contribution of geographers to official committees; (3) the volume of publications appearing in the English language, which greatly widens their circulation; and (4) the productive collaboration with foreign colleagues, especially those from the English-speaking world (Anne Buttimer, Peter Gould, Richard Morrill, and Allan Pred, for example), but not forgetting others such as the Estonian Edgar Kant.

Sources for the study of geography in Sweden strike deep; two are unique. The first is the archives and current output of the 350-year-old Lantmäteristyrelsen (The Land Survey Board), with its headquarters in Stockholm's Kungsträdgården and its technical plant in Gävle. The second source is the population statistics (Tabellverket), kept through ten generations by parish clergy and currently being given a new accessibility through the Demographic Data Base at Umeå University. Add to these the contributions of the Academy of Science (Vetenskapsakademien) (founded in 1737), of the natural scientists of the Enlightenment (such as Linnaeus and Celsius), of several generations of topographers, and of the hydrographical activity of Gustav af Klint, and the strength of the tradition is understandable. New elements were added through the Arctic explorations of A. E. Nordenskiöld and the central Asian expeditions of Sven Hedin.

The institutionalization of geography paralleled those undertakings. The Svenska Sällskapet för Antropologi och Geografi (Swedish Society for Anthropology and Geography) came into being in 1877; local societies such as those of Uppsala (1895) and Göteborg (1908) followed. A pioneering chair in political science and geography was created in Lund in 1895. In the four oldest departments of geography—Uppsala, Lund, Stockholm, and Göteborg—separate chairs of physical geography were established after World War II. Distinguished departments of geography also developed at the privately funded Schools of Economics (Handelshögskolar) of Stockholm and Göteborg. In addition, geography plays its part in the more recently founded Högskolar (university *filialer*) at Växsjö, Karlstad, Örebro and Linköping (now elevated to university status). It is noteworthy that the first Swedish woman to become a professor was the Uppsala geographer Gerd Enequist. Swedish geographers hold an annual meeting (Geografdagar) at which policy matters of national interest are debated. The reduction in status of geography in schools through its incorporation in environmental and social studies has tended to affect recruitment at the undergraduate level, although postgraduate numbers are maintained.

The longest running Swedish journal, *Ymer*, dates from 1881. *Geografiska Annaler* (founded 1919) was divided into a physical and a human series in 1965. *Svensk Geografisk Årsbok* has been published since 1925 in Lund, from where *Geografiska Notiser* is also issued. *Gerum*, the in-house publication of Umeå, is the most substantial of the new journals. All of the old-established departments of geography—Uppsala, Lund, Stockholm, and Göteborg—publish (or have published) a monograph series, consisting principally of doctoral dissertations. Since all doctoral theses are required to be published, the results of geographical research are likely to appear in the transactions of a variety of learned societies. The geographical texts that are published commercially are almost always in Swedish.

The first two Swedish geographers to make an international reputation (apart from Sven Hedin) were Rudolf Kjellén, with the new ideas embodied in his publications on *Geopolitik*, and Gerard De Geer, with his extended field studies in geochronology. To the next generation belong his son Sten De Geer (with innovative population mapping), Hans W:son Ahlmann (with macro-glaciological studies), William William-Olsson (with his forecasts of Stockholm's future population and pioneering economic maps of Europe), David Hannerberg (senior among the emergent school of historical geographers), and Sven Godlund (with transport studies). Valter Schytt stimulated a generation of glacial geomorphologists through his promotion of the Lapland field station at Tarfala. Eric Bylund is representative of investigators into the economic and social problems of northern Sweden, and Gunnar Törnquist of those who seek to unravel the contact systems that characterize the Swedish economy. Information systems are the special area of younger geographers such as Sture Öberg. Torsten Hägerstrand has become something of an international guru through his highly individual approach to the study of time and place. Recruitment of Swedish geographers into national and regional planning agencies has been long and strong, while Nordplan, with Gunnar Olsson as its chairman, is a distinctive Nordic post-graduate research center.

The collective energies of mid-twentieth-century geographers yielded the first *Atlas of Sweden* (1953–) which included an extended text and which was arguably the most impressive national atlas down to its time. Work on a second national atlas is proceeding under the chairmanship of Staffan Helmfrid, lately rector of Stockholm University.

Sweden operates much in a Nordic context. *The Yearbook of Nordic Statistics* (annual) is an invaluable source for up-to-date information. Useful essays on Swedish geography are found in Uuno Varjo and Wolf Tietze, *Norden, Man and Environment* (1987). A helpful little book is *Sverigefakta*, edited by Jan Jonason and Solveig Martensson (a geographer) and published annually in Lund.

Bibliography

Asheim, B. T., "A Critical Evaluation of Postwar Developments in Human Geography in Scandinavia," *PHG*, 11 (1987), 333–353.

Buttimer, A., *The Practice of Geography* (1983).

Hägerstrand, T., "Proclamations about Geography from the Pioneering Years in Sweden," *Geografiska Annaler*, 64B (1982), 119-125.

Hägerstrand, T., and A. Buttimer, eds., *Geographers of Norden* (1988).

Helmfrid, S., "Hundra år svensk geografi," *Ymer* (1976/77), 360-372.

• *W. R. Mead*

Swedish Society for Anthropology and Geography (Svenska Sällskapet för Antropologi och Geografi)

The Swedish Society for Anthropology and Geography was founded in 1877, and its publication *Ymer* was begun four years later. In recent years *Ymer* has appeared annually in the form of a theme volume devoted to a specific aspect of geography. The society is also responsible for *Geografiska Annaler* (Series A: Physical Geography, and Series B: Human Geography). The society's council includes representatives from all the other Swedish geographical societies as well as from cognate scientific fields, especially cartography and geophysics. It administers a number of research funds and, at intervals, awards three medals, which bear the names of Retzius, Wahlberg and Hedin, as well as a fourth, *Vega* (named after the ship in which A. E. Nordenskiöld navigated the North-East Passage). The annual meeting of the society is held in Stockholm in April. It is a social as well as business occasion and is called Vega Day.

• *W. R. Mead*

Switzerland

The development of geography in Switzerland took place in the universities of German Switzerland in the second half of the nineteenth century; French Switzerland followed a little later. At first physical and regional geography were in the forefront. This development lasted until 1960 and from then on was supplemented by planning and, since 1970, increasingly by regional research in Switzerland, Europe, and Third World countries. Simultaneously, practical application became stronger, especially by the ecological orientation of teaching and research.

Among the leading geographers can be counted the following: Eduard Brückner, Bern (1862-1927), the founder, with Albrecht Penck, of Alpine Pleistocene research; Jean Brunhes (1869-1930), human geography; Eduard Imhof (1895-1987), Zurich, the developer of Swiss relief cartography; Heinrich Gutersohn (1899-) and Ernst Winkler (1907-1987) from Zurich, who authoritatively promoted regional geography and the beginnings of regional planning in Switzerland; Hans Boesch (1911-1978), Zurich, an economic geographer who was for many years the general secretary of the IGU; Hans Carol (1915-1971), Zurich and Toronto, economic and theoretical geography; Bruno Messerli (1931-), Bern, ecology, high mountain research, and development aid; Dieter Steiner (1932-), Zurich, human ecology; Claude Raffestin (1936-), Geneva, political geography; Hartmut Leser (1939-), Basel, geo-ecology; Antoine Bailly (1944-), Geneva, economic geography; Jean-Bernard Racine (1940-), Lausanne, social geography; and Kurt Brassel (1943-), Zurich, digital cartography.

The most important geographical journals are *Geographica Helvetica* (Zurich, since 1946) and *Regio Basiliensis* (Basel, since 1959). The individual geographical institutes also publish their works in thematic institutional series.

At the present time geography is represented in nine universities: Basel (Natural Science Faculty, 2 professors); Bern (Natural Science Faculty, 5 professors); Fribourg (Natural Science Faculty, 3 professors); Geneva (Faculty of Economic and Social Sciences, 3 professors); Lausanne (Faculty of Letters, 3 professors); Neuchâtel (Faculty of Letters, 1 professor); St. Gallen (1 professor); Swiss Federal Institute of Technology, Zurich (Division of Natural Sciences, 2 professors); and the University of Zurich (Natural Philosophy Faculty, 6 professors). In addition to the universities there are professional associations: Applied Geography (Fachverein der Angewandten Geographie), the Swiss Geomorphological Society (Schweizerische Geomorphologische Gesellschaft), the Association of Swiss Geography Teachers (Verein Schweizerische Geographielehrer), and the Swiss Cartographic Society (Schweizerische Gesellschaft für Kartographie). Regional societies exist in Basel, Bern, Geneva, Neuchâtel, St. Gallen, and Zurich. All the aforementioned institutions are united in the Swiss Association of Geography (Association suisse de géographie).

After the war Swiss geography modernized, slowly at first but then more rapidly after 1970. Physical and cultural geography have turned to ecological and regional topics, and have included quantitative and socio-economic methods and remote sensing and digital cartography, thereby acquiring a new strength of expression for the discipline. Whereas the classroom was formerly almost the exclusive workplace of the geographer, today practical activity takes place in areas of environmental research, planning, and development aid. (*See also* Zurich Geographical and Ethnographic Society.

Bibliography

Brugger, E. A., G. Furrer, B. Messerli, and P. Messerli, *The Transformation of Swiss Mountain Regions* (1984).

"Geography in Switzerland/La géographie en Suisse," *Geographica Helvetica*, vol. 35, no. 5 (special issue) (1980).

Gutersohn, H., *Geographie der Schweiz* (5 vols., 1958–69).

Imhof, E., ed., *Atlas der Schweiz* (1965–78).

Schuler, M., M. Bopp, K. Brassel, and E. Brugger, *Strukturatlas der Schweiz* (1985).

• *Klaus Aerni and Walter Leimgruber (translated by Ursula Willenbrock Martin)*

Tamayo, Jorge Leonides (1912–1978)

Graduated in civil engineering, National Autonomous University of Mexico, 1936. National Autonomous University of Mexico, 1943–78. *Data para la Hidrología de la República Mexicana* (2 vols., 1946); *Geografía Física de Mexico* (1949); *Geografía Moderna de Mexico* (1953); *Geografía General de Mexico* (4 vols., 1962). M. Gutiérrez de MacGregor in *GBS*, 7 (1983), 125–128.

Tanzania. *See* East Africa.

Tarr, Ralph Stockman (1864–1912)

B.S., Harvard University, 1891. Cornell University, 1892–1912. *Elementary Physical Geography* (1895); *First Book of Physical Geography* (1897); *Physical Geography of New York State* (1902); *New Physical Geography* (1904). *WWWA*; W. Brice, "Ralph Stockman Tarr: Scientist, Writer, Teacher," pp. 215–235 in *Geologists and Ideas* (ed. by E. Drake and W. Jordan) (1985); obit. by A. P. Brigham in *AAAG*, 3 (1913), 93–98.

Taylor, Eva Germaine Rimington (1879–1966)

D.Sc., University of London, 1928. Birkbeck College London, 1921–44. *Tudor Geography 1485–1583* (1930); *Late Tudor and Early Stuart Geography 1583–1650* (1934); *Mathematical Practitioners of Tudor and Stuart England* (1954); *Mathematical Practitioners of Hanoverian England* (1966). *Who Was Who*; E. Campbell, "Geography at Birkbeck College, University of London, with Particular Reference to J. F. Unstead and E. G. R. Taylor," pp. 45–57 in *British Geography 1918–1945* (ed. by R. Steel) (1987); obits. by G. Crone in *GJ*, 132 (1966), 594–596; and anon. in *TIBG*, 45 (1968), 181–186.

Taylor, Thomas Griffith (1880–1963)

D.Sc., University of Sydney, 1916. University of Sydney, 1917–28; University of Chicago, 1929–35; University of Toronto, 1935–51. *Environment and Race* (1927); *Australia* (1940); *Canada* (1947); (ed.) *Geography in the Twentieth Century* (1951, 3rd ed. 1957). J. Powell in *GBS*, 3 (1979), 141–153; autobiography *Journeyman Taylor* (1958); M. Sanderson, "Griffith Taylor: A Geographer to Remember," *CG*, 26 (1982), 293–299;

M. Sanderson, *Griffith Taylor: Antarctic Scientist and Pioneer Geographer* (1988); obits. by M. Aurousseau in *Australian Geographer*, 9 (1964), 131-133; and J. Andrews in *Australian Geographical Studies*, 2 (1964), 1-9.

Teleki, Paul (1879-1941)

Doctorate, University of Budapest, 1903. University of Budapest, 1921-38; Prime Minister of Hungary, 1920-21, 1939-41. *Atlas zur Geschichte der Kartographie von Inseln Japonischen* (1909); *A földrajzi gondolat története* (1917); *The Evolution of Hungary and Its Place in European History* (1923); *A gazdasági élet földrajzi alapjai* (2 vols., 1936). G. Kish in *GBS*, 11 (1987), 139-143; L. Tilkovszky, *Pál Teleki* (1974); obits. by G. Kish in *GR*, 31 (1941), 514-515; and anon. in *PGM*, 87 (1941), 291-294.

Terán Alvarez, Manuel de (1904-1984)

Ph.D., Central University of Madrid, 1927. Central University of Madrid, 1928-30; 1941-75. *Imago Mundi* (2 vols., 1952); *La ciudad* (1967); *Pensamiento geográfico y espacio regional en España* (1982). A. Garcia Ballesteros in *GBS*, 11 (1987), 145-153; obits. by J. Vilà Valentí in *Revista de Geografía*, 18 (1984), 137-140; and M. Drain in *AG*, 94 (1985), 188-189.

Thornthwaite, Charles Warren (1899-1963)

Ph.D., University of California, Berkeley, 1930. University of Oklahoma, 1927-34; U. S. Soil Conservation Service, 1935-42; Seabrook Farms, 1946-63; Johns Hopkins University, 1947-54; Laboratory of Climatology, 1947-63; C. W. Thornthwaite Associates, 1952-63. "The Climates of North America According to a New Classification," *GR*, 21 (1931), 633-655; "The Climates of the Earth According to a New Classification," *GR*, 23 (1933), 433-440; *Atlas of Climatic Types in the United States, 1900-1939* (1941); "An Approach to a Rational Classification of Climate," *GR*, 39 (1949), 498-501. *60 Years of Berkeley Geography*; obits. by J. Leighly in *AAAG*, 54 (1964), 615-619; and F. Hare in *GR*, 53 (1963), 595.

Tobler, Waldo Rudolph (b. 1930)

Ph.D., University of Washington, 1961. University of Michigan, 1961-77; University of California, Santa Barbara, 1977 to the present. "Geographic Area and Map Projections," *GR*, 53 (1963), 59-78; "A Computer Movie Simulating Urban Growth in the Detroit Region," *EG*, 46 (1970), 234-240; "Smooth Pycnophylactic Interpretation for Geographical Regions," *Journal of the American Statistical Association*, 74 (1979), 519-536; "Push-Pull Migration Laws" (with G. Dorigo), *AAAG*, 73 (1983), 1-17. *WWA*.

Toponymy

Toponymy, the study of place-names, is, to date, the most advanced sector within the field of linguistic geography. But such a statement is more than a little misleading in that the systematic analysis of language by geographers remains, in general, woefully underdeveloped, while the pioneering work on place-names was carried on by other scholars, *e.g.*, Isaac Taylor in his durable 1864 study. The handful of geographers who have cultivated place-names have yet to produce a full-length monograph or atlas, a journal, or more than the occasional course. They are outnumbered by linguists and members of English and foreign-language faculties who find that mapping toponymic usage is essential in their research. Yet the published output by geographers does document the rich analytical potentialities of place-names for the broader concerns of the discipline. Much of the more interesting toponymic research by geographers and others appears in the American Name Society's quarterly journal *Names*.

The words attached to places bear priceless evidence concerning physical and human geographies of the past and later changes therein, about the ways people perceive and use their environs, the identification of culture areas, and the succession and movements of various ethnic groups and ideas, among other topics. Perhaps no one has done more to flesh out this sort of agenda than George Stewart, that polymath of a closet geographer, in books and articles encompassing both the United States and the world.

The availability of detailed, comprehensive data has been a persistent problem for the place-name specialist, and so too the origin and history of the names. In only a few countries, such as Great Britain, do we have definitive inventories of place-names for major and minor features. In the United States, adequate gazetteers are published for only a scattering of states; but the Geological Survey's monumental Geographic Names Information System project is helping to remedy the situation as its computer tapes become available. But even it will omit names of streets, individual buildings and business establishments, and the many vernacular place-terms that do not show up on maps, all of which can be richly rewarding to students of place and society. For most parts of the world, large-scale topographic maps and laborious field surveys remain the best sources of information.

Two approaches dominate the geographic literature on toponymy: the analysis of generic terms, *i.e.*, the words or affixes used to designate classes of objects; and the analysis of specific terms, the names for particular things. But the spatial clustering of either type can be exploited and mapped to plot culture areas or particular groups of people. Geographers have also found it worthwhile to examine special classes of names over time and space, and have done interesting work on religious terms, names borrowed from the classical world and foreign lands, nationalistic names, items associated with given ethnic communities, and the names applied to vernacular, or popular, regions. In general, it must be admitted that the future of geographic scholarship in the place-name arena looks much brighter than even its recent past. *See also* Language, Geography of.

BIBLIOGRAPHY

Sealock, R. B., M. M. Sealock, and M. S. Powell, *Bibliography of Place Name Literature, United States and Canada* (3rd ed., 1982).

Stewart, G. R., *Names on the Globe* (1975).

———, *Names on the Land: A Historical Account of Place-Naming in the United States* (4th ed., 1982).

Taylor, I., *Words and Places: Illustrations of History, Ethnology, and Geography* (4th ed., 1911).

Zelinsky, W. "Some Problems in the Distribution of Generic Terms in the Place-Names of the Northeastern United States," *AAAG*, 45 (1955), 319-349.

———, "Classical Town Names in the United States: The Historical Geography of an American Idea," *GR*, 57 (1967), 463-495.

• *Wilbur Zelinsky*

Toschi, Umberto (1897-1966)

Doctorate, University of Bologna, 1921. University of Catania, 1933-35; University of Bari, 1935-49; University of Venice, 1949-51; University of Bologna, 1951-66. *Geografia urbana* (1947); *Geografia economica* (1959); *Emilia-Romagna* (1961); *La città* (1966). P. Bonora in *GBS*, 11 (1987), 155-164; obits. by L. Candida in *RGI*, 66 (1966), 472-476; and G. Merlini in *BSGI*, 42 (1966), 229-326.

Transportation Geography

Transportation geography as a separate branch of the discipline is of recent origin in anything more than a descriptive sense. The itineraries used by the Romans on their thousands of miles of constructed roads, the descriptions of medieval pilgrim routes, the accounts of caravan tracks in North Africa and the Middle East, and the writings of topographers from the seventeenth century onward all are part of this descriptive transportation geography. The rise of oceanic navigation in the fourteenth and fifteenth centuries began the spread of this commercially known world until by the late eighteenth century it encompassed all the unfrozen surface of the sea. And by the last years of the nineteenth century even the deepest interiors of most of the continents were basically known, with ultimate geographical comprehension gained in the quarter century after the effective introduction of commercial aviation around 1910.

Such a description told us what was there without often giving much understanding of why this was true or how transportation related to other geographical concerns. In the 1940s and 1950s Edward Ullman examined the physical structure of railroads, Harold Mayer analyzed railroad terminal problems and the Great Lakes waterways, Edward Taaffe studied the emerging pattern of civil aviation, and J. H. Appleton analyzed the terrain base of railroads. Where previously there had been mostly a description of what was in existence, now there was a growing interest in why human mobility took the form it did. At first, classification largely sufficed, but by the mid-1970s deeper and more complex relationships were studied. Borrowing from other dynamic fields, notably engineering and economics, network analysis and service delivery took on a highly ordered and quantified expression. Seeking this level of association over extended periods, historical geography and transportation were related over a 500-year period in James Vance's *Capturing the Horizon*, effectively published in 1990. Today transportation geography is subdivided between passenger and freight, by time periods, by regional differentiation, and in many other ways. Major mediums of transport are distinguished, and broadly similar concerns such as social responsibility and economic cost are organizing themes in research and teaching.

Bibliography

Appleton, J. H., *The Geography of Communications in Great Britain* (1962).

Black, W. R., "Transportation Geography," pp. 316-332 in *Geography in America*, ed. by G. L. Gaile and C. J. Willmott (1989).

Taaffe, E. J., and H. L. Gauthier, *Geography of Transportation* (1973).

Ullman, E., *American Commodity Flow* (1957).

Vance, J. E., *Capturing the Horizon: The Historical Geography of Transportation* (1986, 1990).

• *James E. Vance Jr.*

Trewartha, Glenn Thomas (1896-1984)

Ph.D., University of Wisconsin, 1925. University of Wisconsin, 1922-66. *Elements of Geography* (with V. Finch) (1936); *Japan: A Physical, Cultural, and Regional Geography* (1945); *The Less Developed Realm: A Geography of Its Population* (1972); (ed.) *The More Developed Realm: A Geography of Its Population* (1978). Obit. by R. Hartshorne and J. Borchert in *AAAG*, 78 (1988), 728-735.

Tricart, Jean Léon François (b. 1920)

Doctorate, University of Paris, 1948. University of Paris, 1945-48; University of Strasbourg, 1948 to the present. *Traité de géomorphologie* (with A. Cailleux) (5 vols., 1962-68); *Précis de géomorphologie* (3 vols., 1968-81); *Ecogeography and Rural Management* (with C. Kiewiet de Jong) (1989). *WWF.*

Troll, Carl (1899-1975)

Dr.phil., University of Munich, 1921. University of Munich, 1922-30; University of Berlin, 1930-38; University of Bonn, 1938-65. "Die tropischen Andenländer," pp. 309-462 in *Handbuch der geographischen Wissenschaft Südamerika* (ed. by F. Klute) (1932); "Strukturboden, Solifluktion und Frostklimate der Erde," *Geologische Rundschau*, 34 (1944), 545-694; "Die tropischen Gebirge," *Bonner Geographische Abhandlungen*, 25 (1959); (ed.) *Geoecology of the High-Mountain Regions of Eurasia* (1972). P. Tilley in *GBS*, 8 (1984), 111-124; H. Beck, "Carl Troll . . . ," pp. 273-281 in *Grosse Geographen* (1982); obits. by U. Schweinfurth in *GZ*, 63 (1975), 170-176; and K. Lenz in *Die Erde*, 106 (1975), 225-227.

Tuan, Yi-Fu (b. 1930)

Ph.D., University of California, Berkeley, 1957. Indiana University, 1956-58; University of New Mexico, 1959-65; University of Toronto, 1966-68; University of Minnesota, 1968-83; University of Wisconsin, 1983 to the present. *Topophilia* (1974); *Space and Place* (1977); *Landscapes of Fear* (1979); *Morality and Imagination: Paradoxes of Progress* (1989). *GOF* (1972); *60 Years of Berkeley Geography; Conversations with Geographers* (ed. by C. Browning) (1982), pp. 114-127.

Tulippe, Omer (1896–1968)

Doctorate, University of Paris, 1934. University of Liège, 1933–66. *L'habitat rural en Seine-et-Oise* (1934), *Méthodologie de la géographie* (1947), *Initiation à la géographie humaine* (1949). J. Denis in *GBS*, 11 (1987), 165–172; *Mélanges de géographie . . . offerts à M. Omer Tulippe* (2 vols., 1967); obits. by G. Chabot in *AG*, 78 (1969), 473; and F. Dussart in *GJ*, 134 (1968), 622–623.

Tunisia. *See* North Africa.

Turkey

Geographic teaching in Turkey both at secondary and higher levels began with the return of Turkish students trained in France before World War I. During this period, the first books on geography were published: *Economic Geography* by H. S. Selen and *Ottoman Economic Geography* by F. Sabri. In 1915, when the Faculty of Letters in Istanbul was being reorganized, teachers were brought from Germany and a department of geography was established. At first, all the courses were taught by German geographers. Then Turkish professors trained in France and Austria began teaching. Thus, German and French influences have been important in the development of geography in Turkey for a long time.

After the "University Reform" that took place in 1933 in all Turkish universities, another geography department was established, in Ankara University in 1935. Before that time there was only a chair of physical geography and geology in Ankara. The departments of geography (for a time the term "Institute of Geography" was used, but the term "department" was reinstated in 1982) in Istanbul and Ankara gained power with the return of a new generation of scholars from their government-sponsored training in France and Germany. This group of scholars contributed in two ways to the development of the science of geography in Turkey: the French School reestablished its influence so that the human side of geography was not ignored, and the geological and geomorphological training became more geographic with the influence of the German School. By the end of the inter-war period, the departments of geography in Istanbul and Ankara had started to develop into their present state and were the only institutions where geography was taught at the university level. Although the situation has not greatly changed, there are now more geography departments (Erzurum University in Erzurum and Ege University in İzmir) and chairs of geography (in the universities of Samsun, Uludağ, Konya and Elaziğ) in Turkey. But the second group has been placed in the embryonic universities with only a few lecturers. Meanwhile, geography classes have also been offered in the faculties of economics and architecture (urban planning departments). Geography courses are included in social science classes with history and sociology in the first half of the secondary level (the first three years of the secondary level come after a five-year primary school); in the second half, courses are independent and include physical geography (first year), world geography (second year) and the geography of Turkey (last year before the university).

The publication of geographical research studies began in 1941 with the establishment of the National Geographic Council. This institution began to publish the *Turkish Geographical Review* (*Türk Coğrafya Dergisi*) in 1943. The journal ceased publication forty years later. A more important step was taken in 1951 when the Institute of Geography at the University of Istanbul began to publish two journals, one in Turkish and the other an international edition. The departments of geography in Ankara and İzmir now have similar periodicals. Besides the periodicals, research studies of Turkish geography are also published by the universities. Some of the research studies have also been published in languages other than Turkish (with Turkish translations) or in foreign periodicals. Studies by Turkish geographers fall mainly into two groups: studies of particular areas with a regional approach, and systematic studies of particular problems in the country, such as urbanization, industrialization, and settlement problems (especially squatter settlements).

Bibliography

Erinç, S., "Climatic Types and the Variations of Moisture Regions in Turkey," *GR*, 40 (1950), 224–235.

———, "Géoecologie de la région d'Istanbul," *Travaux de l'Institut de géographie de Reims*, no. 65–66 (1986), 7–16 (and other papers on Turkey by Turkish geographers in the same number).

Tümertekin, E., "The Development of Human Geography in Turkey," pp. 6–18 in *Turkey: Geographic and Social Perspectives*, ed. by P. Benedict, E. Tümertekin, and F. Mansur (1974).

———, *La distribution de la population en Istanbul* (Institute of Geography, University of Istanbul, 1979).

———, "Turkey," in *Essays on World Urbanization*, ed. by R. Jones (1975).

• *Erol Tümertekin*

Uganda. *See* East Africa.

Ullman, Edward Louis (1912-1976)

Ph.D., University of Chicago, 1942. Harvard University, 1946-51; University of Washington, 1951-76. *American Commodity Flow* (1957); *The Meramec Basin* (with others) (3 vols., 1962); *The Economic Base of American Cities* (with M. Dacey and H. Brodsky) (1969); *Geography as Spatial Interaction* (ed. by R. Boyce) (1980). R. Boyce in *GBS*, 9 (1985), 129-135; *GOF* (1972); *A Man for All Regions: The Contributions of Edward L. Ullman to Geography* (ed. by J. Eyre) (1977); obit. by C. Harris in *AAAG*, 67 (1977), 595-600.

Union of Soviet Socialist Republics

By any measure, the cumulative record of Russian geographical scholarship over the past two or three centuries compares well with that of any other country. However, in common with other branches of Russian science, it was little known in the rest of the world until recently and therefore not incorporated into the general history of geographical thought.

The progress and emphases of Russian geographical scholarship have been unusually responsive to the distinctive developments and needs of the country and its people. Notable geographical work in Russia dates from the time of Peter the Great, whose vigorous interest in cartography, expeditions and the discovery and appraisal of natural resources resulted in the first national atlas, edited by I. K. Kirilov. Many original studies appeared in the eighteenth and nineteenth centuries. They combined, in a distinctively Russian way, exploration and scientific appraisal with a reforming zeal for the improvement of peasant life, and they promoted regional development. The Imperial Russian Geographical Society, founded in 1845, was generally recognized as the most successful of the country's learned societies, and regional branches were set up in Siberia, Caucasia and Turkestan, as well as Russia proper, to promote local knowledge, publication, and lectures. P. P. Semënov Tian-Shansky (1827-1914), who directed this society for several of its most vigorous decades and who had been inspired by Ritter in Berlin, was at the same time exploring the Central Asian mountains, agitating for the emancipation of the serfs, and writing a comprehensive statistical survey of the Russian Empire.

Other major scholars in what may be called the Golden Age of Russian Geography from 1880 to World War I were A. I. Voeikov (1842-1916), V. V. Dokuchaev (1846-1903), and D. N. Anuchin (1843-1923). All of them were both scientists and humanists, and their work grew out of their experience with the distinctive natural landscapes, rural societies, and national problems of their vast country. Dokuchaev's theory of soil formation and natural zonation and Voiekov's theory of heat and water balance had considerable influence on world science. Voeikov's monumental *Klimaty zemnogo shara* (Climates of the World, 1884) established his reputation as a pioneer of climatology. The major organizing figure in university geography as such was Anuchin, who built it up (with anthropology) at Moscow University from 1884 to 1923. During this period Prince Peter Kropotkin (1842-1921), who had early in his life been prominent in Russian geography, and who kept up this scientific interest throughout his life, was in exile in Britain.

Thus by the time of the Bolshevik Revolution there was a vigorous body of geographical tradition in Russia, combining scientific and humanist approaches, with a national integration of human, environmental, and regional studies. It was continued during the 1920s, informing, and fitting in reasonably comfortably with, the needs of the young Soviet state. However, this tradition was severely disrupted in the Stalinist period, and submerged by the mobilization of large numbers of specialists who were directed toward practical tasks. Human-oriented studies tended to be politically vulnerable, and by the 1950s geography had been transformed into a predominantly physical science. It had also become highly fragmented, since official doctrine discouraged those integrated studies of man and his environment that had been central to the subject in Russia up to 1930. However, in spite of the difficulties, some of the spirit of the earlier heritage was kept alive, largely through the "Landscape School" of L. S. Berg (1876-1950) and the "Regional School" of N. N. Baransky (1881-1963).

The Krushchev period saw the release of a ferment of vigorous disputation about the nature and direction of Soviet geography. The most effective advocate of a more integrated and humanized geography, inspired by the broken heritage and by international comparisons, as well as practical considerations, was V. A. Anuchin (1913-1984), whose book *Theoretical Problems of Geography* appeared in 1960. Baransky was an influential supporter of his viewpoint, which was, however, strongly opposed by the powerful Institute of Geography of the Academy of Sciences, notably by I. P. Gerasimov (1905-1985), its director for three decades.

In the mid-sixties, following this debate, geography was officially "restructured" around a set of integrated studies of specific problems such as multipurpose regional projects, environmental protection, and resource evaluation. However, although good work along those lines was done by some scholars, actual restructuring of geography was slow, as elsewhere in what is now officially referred to as "the years of stagnation." The structure and scholarly output of such key institutions as the Academy of Sciences' Institute of Geography and Moscow University are still disproportionately weighted towards a rather narrow physical geography. However, the potential and resilience of Soviet geography is indicated in publications by excellent individual scholars, and the advent of new leadership in Soviet geography. Furthermore, the changes in the Soviet government since 1985 promise to release the energies, ideas, and aspirations of a bright young generation of Soviet geographers—heirs to a great tradition.

Bibliography

Anuchin, V. A., *Teoreticheskie Osnovy Geografii* (1960) (English edition ed. by G. Demko and R. Fuchs, *Theoretical Problems of Geography*, 1977).

Baransky, N., *Izbrannye Trudy* (2 vols., 1980).

Gerasimov, I. P., ed., *A Short History of Geographical Science in the Soviet Union* (1976).

Hooson, D. J. M., "The Development of Geography in Pre-Soviet Russia," *AAAG*, 58 (1968), 250-272.

———, "The Soviet Union," pp. 79-106 in *Geography since the Second World War*, ed. by R. J. Johnston and P. Claval (1984).

Lavrov, S. B., ed., *Soviet Geography Today: Aspects of Theory* (1981).

• *David J. M. Hooson*

USSR Academy of Sciences, Institute of Geography (Akademiia Nauk SSSR, Institut Geografii)

The Institute of Geography of the Academy of Sciences of the USSR (Institut Geografii

Akademii Nauk SSSR [IG AN]) is the most important center of geographical research in the Soviet Union. It also established the pattern typical in socialist countries of a geographical research institute within an Academy of Sciences and with a large, full-time professional research staff. The Geographical Section of the Commission for the Study of Natural Productive Forces of the USSR was founded in 1918 in Petrograd. It became the Geomorphological Institute in 1930, the Institute of Physical Geography in 1934, and the Institute of Geography in 1936. In 1934 it moved from Leningrad to Moscow. A. A. Grigor'yev, the Arctic specialist, was its director during its formative years and I. P. Gerasimov, the soil geographer, its director during the period of greatest expansion in scale and scope after World War II. The institute has a broad spectrum of research but is strongest on general physical geography, resource utilization, landscape studies, geomorphology, paleogeography, soil geography, hydrology, glaciology, climatology, biogeography, and ecosystems. In recent years it has increasingly developed programs in economic and social geography, urban geography, global problems, and the geobiosphere. The current (1989) director is Vladimir M. Kotlyakov, a glaciologist. Its *Izvestiya Akademii Nauk, Seriya Geograficheskaya*, published since 1951, appears in six issues a year. The institute is located at Staromonetnyi pereulok 29, Moscow 109017, USSR.

• *Chauncy D. Harris*

USSR Geographical Society (Geograficheskoe Obshchestvo SSSR)

The Geographical Society of the USSR (Geograficheskoe Obshchestvo SSSR) was founded in 1845 in St. Petersburg as the Russian Geographical Society. From 1850 it was called the Imperial Russian Geographical Society. After the revolution it was renamed successively the Russian Geographical Society in 1917, State Russian Geographical Society in 1926, State Geographical Society in 1931, and All-Union Geographical Society (VGO) or alternatively Geographical Society of the USSR (GO SSSR) in 1938, when the society became part of the Academy of Sciences of the USSR. In the nineteenth century the society sponsored many expeditions for geographical exploration of the Russian Empire and other parts of the world. Leading explorers were among its members and officers. Among its presidents have been F. P. Litke, polar explorer; P. P. Semenov-Tyan-Shanskiy, explorer of the Tyan-Shan in Central Asia, director of the first Russian census in 1897, and author of gazetteers and regional geographies; N. I. Vavilov, the plant explorer; L. S. Berg, the physical geographer and historian of geographical exploration and of the society; and S. V. Kalesnik, physical geographer. The current (1989) president is A. F. Treshnikov, polar scientist. The society has a membership of about 40,000; about 300 regional and local branches throughout the Soviet Union; about 600 scientific specialist sections, commissions or other units; a library and archives of more than 420,000 volumes and 100,000 documents; and an extensive lecture series. It holds an all-union geographical congress every five years. The most important of its publications, its *Izvestiya*, published since 1865, appears in six issues a year. The headquarters of the society are at pereulok Grivtsova 10, Leningrad 190000, USSR.

Bibliography

Berg, L. S., *Vsesoyuznoye Geograficheskoye Obshchestvo za Sto Let* (1946).

Kalesnik, S. V., ed., *Geograficheskoye Obshchestvo za 125 let* (1970).

———, "The Geographical Society of the USSR: Review of Activity," pp. 398–401 in *Soviet Geography: Accomplishments and Tasks*, ed. by I. P. Gerasimov *et al.* (English translation ed. by C. D. Harris) (1962).

• *Chauncy D. Harris*

United Kingdom

If institutional beginnings are to be taken as a guide, the modern development of British geography can be traced to the establishment of the Royal Geographical Society of London (RGS) in 1830. But this represents an arbitrary point of departure, for the society itself was the outgrowth of various earlier travelling clubs that in turn mirrored Britain's long-standing engagement with geographical exploration dating back at least to the scientific circumnavigations of men like James Cook in the eighteenth century. Moreover, important theoretical pronouncements on the nature and scope of geography and technical advances in the art of cartography had already been made during the period of the Scientific Revolution. In many ways these traditions—theoretical, exploratory, and cartographic—characterized the ethos of the RGS, as is illustrated by the involvement of several notable British natural scientists like Roderick Murchison, Charles Darwin, A. R. Wallace, and H. W. Bates, and by the society's sponsorship of expeditions in Australia, Africa, and elsewhere. Later, through the advocacy of John Scott Keltie and Douglas Freshfield, the RGS involved itself in a campaign for the improvement of geographical education in schools and for the establishment of the subject as a university discipline.

To equate the evolution of nineteenth-century British geography with the history of the RGS, however, would be to ignore the vibrancy of independent geographical traditions in the United Kingdom, and especially in Scotland. Edinburgh's heritage of mapmaking (where the Bartholomew and Johnston companies were based), its tradition of encyclopedia production, and its scientific achievements in geology (James Hutton, Hugh Miller, John Fleming, John Playfair, and Archibald and James Geikie were all Scottish), oceanography (the *Challenger* offices were located in Edinburgh), and natural history more generally made it a particularly congenial environment for the launching of the Royal Scottish Geographical Society in 1884. Indeed, it was within this scientific network that British geographers like H. R. Mill, H. N. Dickson, and A. J. Herbertson were first initiated into the study of geographical distributions.

The professionalization of British geography in the decades around the turn of the century owed much to the activities of Halford John Mackinder, through whom much of the RGS's higher educational aspirations were channelled. Mackinder's celebrated 1887 lecture on "The Scope and Methods of Geography" was a manifesto for the scholarly integrity and scientific coherence of the discipline, erected on environmental determinist foundations and legitimated by its practical relevance to political and commercial interests. Later, Mackinder's own career as an educationalist, a parliamentarian and as British High Commissioner to southern Russia demonstrated his own—and indeed geography's—twin commitment to scholarship *and* utility.

For all his logistic advocacy for geography, Mackinder's naturalistic and environmentalist conception of the subject was not universally appealing, particularly among those first-generation professional geographers who had come under the spell of the Scottish polymath and political radical Patrick Geddes. Geddes's work in natural science—biology and botany in particular—had brought him to an appreciation of Darwinian evolution, albeit tempered by the reinterpretative gloss of neo-Lamarckism. This, together with his enthusiasm for urban planning, regional survey, and practical socialism, provided a combination of motifs conducive to an interest in geography. And so he laid out his thoughts on the essentially synthetic nature of the subject in a number of places.

Of the geographers who came within the orbit of the Geddesian circle special mention must be made of A. J. Herbertson and H. J. Fleure. In Herbertson's case it was the centrality of the regional concept, especially of the great natural regions of the world, that he imbibed from Geddes, but a posthumously published paper revealed his growing sensitivity to regional consciousness and regional psychology, themes rooted in the Lamarckian emphasis on will and habit in evolutionary history. As for Fleure, his commitment to neo-Lamarckian evolutionary idealism came through in his opposition to the narrowness of Mackinder's necessitarianism, his emphasis on *human* regions, his historical interest in early civilization and cultural contacts, and his anthropological research on the racial characteristics of England and Wales. Many of these themes were to be perpetuated in the work of his student Estyn Evans, whose luminous writings on the personality of Ireland retained Fleure's enthusiasm for the Vidalian tradition of French regional geography.

The regional focus of these practitioners was to be enshrined, in large measure, as geography's intellectual *raison d'être* in the early decades of the twentieth century, and accordingly a range of regional classification schemes emerged, chief among which was that developed by J. F. Unstead in the early 1930s. At that time too Dudley Stamp's orchestration of the land utilization survey of Great Britain represented the public utility of the regional paradigm. Meanwhile the establishment of geomorphology as a sub-specialism of geography was being achieved through J. A. Steers's coastal studies and the magisterial reconstruction of the denudation chronology of Southeast England by S. W. Wooldridge and D. L. Linton. Historical research was also achieving prominence: Eva Taylor, for example, had begun to make a major contribution to the history of science by her works on the history of early geography and mathematical practice, while Clifford Darby was involved in reconstructing a Domesday geography of Anglo-Saxon England and recounting the story of the draining of the medieval fen. Besides these, geographers in the immediate aftermath of World War II concerned themselves with the problems of local government reorganization and urban and rural planning. And since 1935 the establishment of the Institute of British Geographers provided an additional forum and outlet for much of the new geographical discourse.

By the late 1950s, however, much of this previously uncontested regional writing began to be challenged, due to the import from the United States of a more "scientific" methodology grounded in statistical methods and bolstered by the language of positivist philosophy. British geography's quantitative baptism occurred largely through the contributions of Richard Chorley and Peter Haggett, who called for the articulation of a set of theoretical models relevant to the interrogation of spatially referenced data. Chorley himself turned to general systems theory as a means of reintegrating the subject, while Haggett introduced various theorems from earlier location theorists like Thünen and Lösch in the endeavor to transform geography into a robust spatial science. The philosophical legitimation of this quantitative turn in British geography did not come, however, until David Harvey's monumental *Explanation in Geography* was published in 1969, by which time the voices of critics were clearly to be heard. Indeed, throughout this whole period, dubbed the "Quantitative Revolution" in geographical vernacular, there were many who did not share the vision. And so those who turned towards a more humanistic geography by emphasizing the centrality of subjective values could look back to the earlier writings of William Kirk on the behavioral environment or to the apologia for the role of imagination provided by David Lowenthal and Hugh Prince. Besides this there were those who retained an uncompromising commitment to issues of welfare and social relevance.

Concurrent with these developments was the espousal of various forms of historical materialism, heralded much earlier in the anarchistic writings of Peter Kropotkin. Now, a new generation of radical geographers subjected questions of philosophical principle, industrial location, spatial divisions of labor, residential segregation, policy formulation and implementation, and so on to Marxist or neo-Marxist critique.

The excesses of the subjectivists with their encounter of literary texts, and the socioeconomic reductionism of the more fundamentalist structuralists, has led some to look for a more flexible articulation of the relationship between human agency and social structure. Accordingly, some among the present generation of geographers have turned to the structuration theory of the Cambridge sociologist Anthony Giddens for theoretical sustenance as they argue for the "spatializing" of social theory. Of course their revisionist calls have not been universally received and much work on geographic information systems and quantitative geomorphology proceeds largely unaffected, fired by the need to find practical answers to the crucial environmental problems of global change facing the modern world. And yet some recent physical geographers have also taken up the task of seeking for an appropriate philosophy of science within which to locate their own endeavors.

All in all there is much diversity in contemporary British geography, but this merely reinforces the fact that the evolution of British geography has always been the history of a contested tradition. *See also* British Library Map Library; Geographical Association; Institute of British Geographers; Royal Geographical Society; Royal Scottish Geographical Society.

Vallaux, Camille (1870–1945)

D.-ès-L., University of Paris, 1907. Naval College in Brest, 1901–13; Lycée Buffon and Ecole des hautes études sociales (Paris), 1913–19; Janson de Sailly Lycée and Ecole des hautes études commerciales, 1919–32. *La Basse-Bretagne* (1907); *Géographie sociale: la mer* (1908), *Géographie sociale: le sol et l'état* (1911); *Les sciences géographiques* (1925). F. Carré in *GBS*, 2 (1978), 119–126.

Veen, Hendrik Nicolaas ter (1883–1949)

Doctorate, University of Amsterdam, 1925. University of Amsterdam, 1927–49. *De Haarlemmermeer als kolonisatiegebied* (1925), *Van anthropogeografie tot sociografie* (1927), *Grenzen* (1947), *Japan, bakermat van het Aziatische imperialisme* (1949). A. C. de Vooys, *De ontwikkeling der sociale geografie in Nederland* (1950); W. H. Vermooten, *Sociografie en sociale geografie in Nederland na Steinmetz* (1968); W. F. Heinemeyer, *De sociale geografie in de rij van de sociale wetenschappen* (1968); W. J. van den Bremen, "De Nederlandse geografie van 1850 tot 1950," pp. 160–174 in *Algemene Sociale Geografie: Ontwikkelingslijnen en Standpunten*, ed. by A. G. J. Dietvorst, J. A. van Ginkel, *et al.* (1984).

Venezuela

Professional geographical studies began in 1956 with the creation of the geography section of the Faculty of Humanities and Education of the Central University of Venezuela at Caracas. This was reorganized in 1958 as the School of Geography, its graduates obtaining the degree of Licentiate in Geography after five years of study. The current program is structured in units of basic practical and theoretical courses, workshops on applied Venezuelan regional research, and projects with fellowships and special degree work. Another school of geography was set up in 1963 at Mérida, attached to the Faculty of Forest Sciences of the University of the Andes. Its program is organized according to a curriculum that takes into consideration basic matter and courses in physical geography, cartography, and human geography, as well as seminars on Venezuelan regional studies. Down to the present, these two schools have turned out more than 900 geographers, of whom the majority belong to the professional organization known as the Colegio de Géografos de Venezuela, established in 1969 and with active centers at Caracas, Mérida, Barquisimeto, and Ciudad Guayana.

Geographers work chiefly in the public sector: in government ministries, the Petróleos de Venezuela Company, autonomous institutes, municipal offices of urban planning, the National Agrarian Institute, the National Institute of Sanitation Works, regional development corporations, the National Parks Institute, and in universities and institutes of higher education. In the private sector they engage in cartographic enterprises, planning projects, and regional urban studies. The training of geography teachers for middle education has been handled since the 1930s by the National Pedagogical Institute of Caracas and other cities, which at present operate as institutions of higher education under the Libertador Experimental Pedagogical University.

University research started to develop in Caracas from 1954, with the foundation of the Institute of Geography of the Faculty of Humanities and Education of the Central University of Venezuela. After maintaining itself to 1958, this was reopened in 1972 as the Institute of Geography and Regional Development. It has paid particular attention to research in geopolitical problems and epistemology, having published in 1987 *Venezuela y su espacio fronterizo: El problema del Esequibo*, as well as studies on the agrarian and urban geography of the central regions. At present it is stressing applied research topics in climatology and geomorphology.

In 1959 there was established at Mérida the Institute of Geography and Conservation of Natural Resources of the University of the Andes. This has pursued lines of research on the river basins of the Chama, Santo Domingo, Uribante, Motatán, and the Ticoporo area, and basic research on El Vigía, to the south of Lake Maracaibo, as well as on Andean agrarian systems. It is at present designing geothematic research centers, giving emphasis to geodemographic dynamics and methods of territorial occupation, urban agglomerations and settlement systems, regional geo-economies and socio-spatial changes, geomorphology and natural hazards, hydro-agroclimatology, and mountain biogeography.

Integration of geography teaching and research has given rise to graduate courses. At the Central University of Venezuela a Master's degree in geographical theory and methodology was introduced in 1976, to be followed subsequently by Master's degrees in spatial analysis and climatology. At the University of the Andes a Master's program in land use analysis was started in 1978, and converted in 1983 into one of territorial organization. Both these institutions periodically offer short updating courses.

Outstanding among the official agencies that carry on geographical work is the Office of National Cartography, a branch of the Ministry of the Environment and Renewable Natural Resources, which has published the *Atlas de Venezuela* as well as plans on particular subjects. Of prime importance is the Bureau of Geography and Cartography of the Armed Forces, which has functioned since 1953 under various titles, and which devotes special attention to studies on the federal frontier districts of Apure, Amazonas, Bolívar and Delta Amacuro.

The country's output in geography has been large and diverse. From the end of the nineteenth century foreigners residing in the country made various contributions in physical geography, among them Adolfo Ernst (1832-1899), and Henri Pittier on plant geography, with his monumental *Manual de las Plantas Usuales de Venezuela* (1926). These were followed in the second half of this century by the contributions of the Venezuelan Francisco Tamayo. Of note also are the researches of Wilhelm Sievers (1860-1921), with his publications on the physiographic provinces; those of Alfredo Jahn (1867-1921), including his *Esbozo de las formaciones geológicas de Venezuela*; the various geophysical contributions of Eduardo Röhl (1891-1959); and those of Ralph Alexander Liddle, *e.g.*, his *The Geology of Venezuela and Trinidad* (1928). At the same time, general and regional topics in rural Venezuela and its export foodstuffs down to the 1920s were treated in such comprehensive works as those of Wilhelm Sievers, *Venezuela* (1888, 1921); Leonard W. Dalton, *Venezuela* (1912); and Pierre Denis, *Le Venezuela* (1927). Subsequently, the changes in geographic, economic and social structures due to the mono-export of oil were early analyzed in such works as E. Lieuwen, *Petroleo en Venezuela* (1964) and Guillermo Zuloaga, *Geografía Petrolera de Venezuela* (1969). In the same period studies also began in climatic systematization, such as that of Alfonso Freile, *Regiones Climáticas de Venezuela* (1969).

Among contemporary geographical treatments, there stand out the *Geografía de Venezuela*, edited by Pedro Vila, of which two volumes have appeared: I. *El territorio nacional y su ambiente físico* (1960), and II. *El Paisaje Natural y el Paisaje Humanizado* (1965); the textbook of Antonio Luis Cárdenas, *Geografía Física de Venezuela* (1969); the monographs on all the Venezuelan states, edited by Marco Aurelio Vila and completed in 1967; and the *Geografía de Venezuela Nueva* collection edited by Pedro Cunill in ten volumes, 1981-89, of multiple authorship. At the same time numerous works of some length for the general reader have circulated, *e.g.*, Leví Marrero, *Venezuela y sus recursos* (1964); that edited by Isbelia Segnini entitled *Conocer Venezuela* (1986); and Pedro Cunill's *Venezuela: El medio y la historia* and *Venezuela: El espejismo petrolero* (1988).

Among geographical periodicals of long standing there is the *Revista Geográfica Venezolana*, started in 1960, twenty-three volumes to date, which is a journal of the Institute of Geography and Conservation of Natural Resources in the Faculty of Forest Sciences of the University of the Andes at Mérida. At Caracas, in the Humanities Faculty of the Central University of Venezuela, two periodicals are published: *Síntesis Geográfica*, the journal of the School of Geography, started in 1977, fifteen volumes to date; and *Terra: Pensamiento Geográfico*, the review of the Institute of Geography and Regional Development and Graduate Degrees in Geography, from 1977, with eight numbers to date. There is also the occasional publication at Caracas since 1974 of the *Boletín del Centro de Investigaciones Geodidácticas*.

BIBLIOGRAPHY

Avilan Rovira, J., and H. Eder, *Sistemas y Regiones Agrícolas de Venezuela* (1986).

Carpio Castillo, R., *Geopolítica de Venezuela* (1981).

Cunill Grau, P., *Geografía del Poblamiento Venezolano en el Siglo XIX* (3 volumes, 1987).

Sequera de Segnini, I., *Pensamiento Geográfico* (1979).

Vila, P., *Geografía de Venezuela*, Vol. 1, *El territorio nacional y su ambiente físico* (1960).

Vila, P., R. Carpio, F. Brito Figueroa, and A. L. Cárdenas, *Geografía de Venezuela. El paisaje natural y el paisaje humanizado* (1965).

• *Pedro Cunill Grau*
(translated by C. Julian Bishko)

Vidal de la Blache, Paul Marie Joseph (1845-1918)

D.-ès-L., University of Paris, 1872. University of Nancy, 1872-77; Ecole Normale Supérieure, 1877-98; University of Paris, 1898-1914 (retired in 1909 but continued to teach part time until 1914). *Etats et nations de l'Europe, autour de la France* (1889); *Tableau de la géographie de la France* (1903); *La France de l'Est* (1917); *Principes de géographie humaine* (1922). S. Baker in *GBS*, 12 (1988), 189-201; Hanno Beck in *Grosse Geographen* (1982), pp. 180-190; P. Pinchemel in *Geographisches Taschenbuch 1970/72* (1970), 266-279 (reprinted in *LGF*, 9-23); H. Andrews, "L'oeuvre de Paul Vidal de la Blache: notes bibliographiques," *CG*, 28 (1984), 1-18; H. Andrews, "The Early Life of Paul Vidal de la Blache and the Makings of Modern Geography," *TIBG*, 11 (1986), 174-182; obit. by L. Gallois in *AG*, 27 (1918), 161-173.

Vilà Valentí, Juan (b. 1925)

Doctorate, University of Madrid, 1956. University of Murcia, 1958-65; University of Barcelona, 1965 to the present. *La Península ibérica* (1968), *El estudio teórico de la Geografía* (1983), *El conocimiento geográfico de España* (1989).

Voeikov (Woeikof), Aleksandr Ivanovich (1842-1916)

Doctorate, University of Göttingen, 1865. University of St. Petersburg, 1884-1916. *Klimaty zemnogo shara* (1884); "De l'influence de l'homme sur la terre," *AG*, 10 (1901), 97-114, 193-215; "La géographie de l'alimentation humaine," *La Géographie*, 20 (1909), 225-240, 281-296; *Le Turkestan russe* (1914). A. Fedosseyev in *GBS*, 2 (1978), 135-141; A. Fedoseev in *DSB*, 14 (1976), 52-54; G. Rikhter in *GSE*, 5 (1974), 546-547; obits. by W. Köppen in *PGM*, 62 (1916), 422-423; and anon. in *AG*, 25 (1916), 150-151.

Vooys, Adriaan Cornelis de (b. 1907)

Doctorate, University of Utrecht, 1932. University of Utrecht, 1949-73. *De trek van de plattelandsbevolking in Nederland* (1932); *De ontwikkeling der sociale geografie in Nederland* (1950); "Western Thessaly in Transition," *Tijdschrift van het Koninklijk Nederlands Aardrijkskundig Genootschap*, 76 (1959), 31-54; "Die Pendelwanderung: Typologie und Analyse," pp. 99-107 in *Zum Standort der Sozialgeographie, Wolfgang Hartke zum 60. Geburtstag* (1968). M. W. Heslinga, "Over de Vooys en de geografie," pp. xi-lii in *Een Sociaal-Geografisch Spectrum: Opstellen aangeboden aan Prof. Dr. A.C. de Vooys 1949-1973* (1974); H. van Ginkel, "Mirror of a Changing Society—The Development of Residential Geography at Utrecht," *TESG*, 74 (1983), 358-366; H. van Ginkel, "Nederlandse geografie na 1950," pp. 255-283 in *Algemene Sociale Geografie: Ontwikkelingslijnen en Standpunten*, ed. by A. G. J. Dietvorst, J. A. van Ginkel *et al.* (1984); W. M. Karreman and M. de Smidt, eds., *Redevoeringen en kleine geschriften van professor A.C. de Vooys* (1987).

Vries Reilingh, Hans Dirk de (b. 1908)

Doctorate, University of Amsterdam, 1945. University of Amsterdam, 1950-71. *De Volkshogeschool, een sociografische studie* (1945); "Soziographie," pp. 522-536 in *Handbuch der empirischen Sozialforschung*, vol. 1 (1961), ed. by R. König; "The Tension between Form and Function in the Inner City of Amsterdam," pp. 309-323 in *Urban Core and Inner City: Proceedings of the International Study Week Amsterdam* (September 1966); "Gedanken über die Konsistenz in der Sozialgeographie," pp. 109-118 in *Zum Standort der Sozialgeographie: Wolfgang Hartke zum 60. Geburtstag* (1968). W. F. Heinemeijer, "Relingh en de sociale geografie," *Geografisch Tijdschrift*, 5 (1971), 291-303; C. van Paassen, *Het begin van 75 jaar sociale geografie in Nederland* (1982); H. van Ginkel, "Nederlandse geografie na 1950," pp. 255-283 in *Algemene Sociale Geografie: Ontwikkelingslijnen en Standpunten*, ed. by A. G. J. Dietvorst, J. A. van Ginkel, *et al.* (1984).

Vuuren, Louis van (1873-1950)

Graduated from the Royal Military Academy, Breda. Encyclopaedic Bureau, Department of Internal Affairs, Netherlands East Indies, 1910-21; University of Amsterdam, 1923-27; University of Utrecht, 1927-45. "Practische geografie," pp. 346-353 in *Geografische bijzonderheden* (Feestschrift R. Schuiling), ed. by C. L. van Balen (1924); *De Merapi, bijdrage tot de sociaal-geografische kennis van dit gebied* (1932); *Rapport betreffende een onderzoek naar de sociaal-economische structuur van een gebied in de provincie Utrecht* (1938); "Warum Sozialgeographie?," *Zeitschrift der Gesellschaft für Erdkunde zu Berlin*, 76 (1941), 269-279. M. W. Heslinga, "Between German and French Geography: In search of the origins of the Utrecht School," *TESG*, 74 (1983), 317-334; W. J. van den Bremen, "De Nederlandse geografie van 1850 tot 1950," pp. 160-174 in *Algemene Sociale Geografie: Ontwikkelingslijnen en Standpunten*, ed. by A. G. J. Dietvorst, J. A. van Ginkel *et al.* (1984); obit. by A. C. de Vooys in *Tijdschrift van het KNAG*, 68 (1951), 365-368.

Wagner, Hermann (1840-1929)

Doctorate, University of Göttingen, 1864. University of Königsberg, 1876-80; University of Göttingen, 1880-1920. *Die Bevölkerung der Erde* (with E. Behm) (1866); *Lehrbuch der Geographie* (of H. Guthe, 4th edition [1879] and subsequent editions edited by H. Wagner, 10th ed. 1920). W. Behrmann, "Hermann Wagner als Akademischer Lehrer," *Die Erde*, 6 (1954), 362-368; obits. by L. Mecking in *GZ*, 35 (1929), 585-596; and W. Meinardus in *PGM*, 75 (1929), 225-229.

Wagner, Philip Laurence (b. 1921)

Ph.D., University of California, Berkeley, 1953. University of Chicago, 1954-60; University of California, Davis, 1960-67; Simon Fraser University, 1967-88. *Nicoya, a Cultural Geography* (1958); *The Human Use of the Earth* (1960); *Readings in Cultural Geography* (with M. Mikesell) (1962); *Environments and Peoples* (1972). *GOF* (1973); *60 Years of Berkeley Geography.*

Waibel, Leo Heinrich (1888-1951)

Doctorate, University of Heidelberg, 1911. University of Kiel, 1922-29; University of Bonn, 1929-37. *Urwald, Veld, Wüste* (1921); *Probleme der Landwirtschaftsgeographie* (1933); *Die Rohstoffgebiete des tropischen Afrika* (1937); *Die europäische Kolonisation Südbrasiliens* (1955). G. Pfeifer in *GBS*, 6 (1982), 139-147; Carl Troll, "Leo Waibel . . . ," pp. 223-230 in *Bonner Gelehrte* (1970); G. Pfeifer and G. Kohlhepp (eds.), *Leo Waibel als Forscher und Planer in Brasilien* (1984); obits. by J. Broek in *GR*, 42 (1952), 287-292; and H. Schmitthenner in *PGM*, 97 (1953), 161-169.

Ward, Robert DeCourcy (1867-1931)

M.A., Harvard University, 1893. Harvard University, 1895-1931. *Climate, Considered Especially in Relation to Man* (1908); *The Climates of the United States* (1925). W. Koelsch in *GBS*, 7 (1983), 145-150; obits. by C. Brooks in *AAAG*, 22 (1932), 33-43; and *GR*, 22 (1932), 161.

Warkentin, John Henry (b. 1928)

Ph.D., University of Toronto, 1961. University of Manitoba, 1959-63; York University, 1963 to the present. (ed.) *The Western Interior of Canada* (1964); (ed.) *Canada: A Geographical Interpretation* (1968); *Historical Atlas of Manitoba* (with R. Ruggles) (1970); *Canada before Confederation* (with R. C. Harris) (1974). *CWW.*

Warntz, William (1922–1988)

Ph.D., University of Pennsylvania, 1955. University of Pennsylvania, 1949–56; American Geographical Society, 1956–66; Harvard University, 1966–71; University of Western Ontario, 1971–88. *Toward a Geography of Price* (1959); *Geography Now and Then* (1964); *Macrogeography and Income Fronts* (1965); (ed.) *Breakthroughs in Geography* (with P. Wolff) (1971). *WWA*; *CWW*; *GOF* (1973); W. J. Coffey, ed., *Geographical Systems and Systems of Geography: Essays in Honour of William Warntz* (1988).

Watson, James Wreford (b. 1915)

Ph.D., University of Toronto, 1945. McMaster University, 1938–40, 1945–49; Department of Mines and Technical Surveys (Ottawa), 1949–54; Carleton University, 1950–54; University of Edinburgh, 1954–82. (ed.) *Geographical Essays in Honour of Alan G. Ogilvie* (with R. Miller) (1959); *North America, Its Countries and Regions* (1963, rev. 1967); (ed.) *The American Environment* (with T. O'Riordan) (1977); *Social Geography of the United States* (1979). *WW*; *CWW*; *GOF* (1982); G. Robinson, "James Wreford Watson: An Appreciation," pp. 374–384 in *A Social Geography of Canada: Essays in Honour of J. Wreford Watson* (ed. by G. Robinson) (1988).

Wellington, John Harold (1892–1981)

B.A., University of Cambridge, 1921. University of the Witwatersrand, 1921–57. *Southern Africa: A Geographical Study* (2 vols., 1955); plus articles on the physical and human geography of South Africa. S. Baker in *GBS*, 8 (1984), 135–140; obits. by S. Jackson in *South African Geographical Journal*, 63 (1981), 1–2; and anon. in *GJ*, 147 (1981), 275–276.

West, Robert Cooper (b. 1913)

Ph.D., University of California, Berkeley, 1946. Louisiana State University, 1948–82. *Cultural Geography of the Modern Tarascan Area* (1948); *The Pacific Lowlands of Colombia* (1957); (ed.) *Handbook of Middle American Indians* (vol. 1, 1964); *Middle America* (with J. Augelli) (1966). *GOF* (1973); *60 Years of Berkeley Geography*; W. Davidson and J. Parsons, "Robert C. West, Geographer," pp. 1–8 in *Historical Geography of Latin America: Papers in Honor of Robert C. West* (1980).

Wheatley, Paul (b. 1921)

D.Litt., University of London, 1958. University of Malaya, Singapore, 1951–58; University of California, Berkeley, 1958–66; University College London, 1966–71; University of Chicago, 1971 to the present. *The Golden Khersonese* (1961); *The Pivot of the Four Quarters* (1971); *From Court to Capital* (with Thomas See) (1978); *Nagara and Commandery* (1983). *WWA*.

White, Gilbert Fowler (b. 1911)

Ph.D., University of Chicago, 1942. Haverford College (President), 1946–55; University of Chicago, 1956–69; University of Colorado, 1970–78. *Human Adjustment to Floods* (1942); *Strategies of American Water Management* (1969); *Drawers of Water: Domestic Water Use in East Africa* (with D. Bradley and A. White) (1972); *The Environment as Hazard* (with I. Burton and R. Kates) (1977). *WWA*; *GOF* (1972, 1984); R. Kates and I. Burton (eds.), *Geography, Resources and Environment*: Vol. 1, *Selected Writings of Gilbert F. White*, Vol. 2, *Themes from the Work of Gilbert F. White* (1987).

Wilhelmy, Herbert (b. 1910)

Dr.phil., University of Leipzig, 1932; (habil.) University of Kiel, 1936. Technical University of Stuttgart, 1954–58; University of Tübingen, 1958–78. *Südamerika im Spiegel seiner Städte* (1952); *Klimamorphologie der Massengesteine* (1958); *Geographische Forschungen in Südamerika* (1980); *Die Städte Südamerikas* (2 vols., 1984–85). *Wer Ist Wer?*; *Beiträge zur Geographie der Tropen und Subtropen: Festschrift für Herbert Wilhelmy* (1970).

William-Olsson, William (b. 1902)

Fil.Dr., University of Stockholm, 1937. University of Stockholm, 1937-41; Stockholm School of Economics, 1941-69. *Huvuddragen av Stockholms geografiska utveckling 1850-1930* (1937); *Stockholms framtida utveckling* (1941); *Stockholm, Structure and Development* (1960); *An Economic Map of Europe West of the Soviet Union* (1974). *Vem är Det*; W. William-Olsson, "My Responsibility and My Joy," pp. 153-166 in *The Practice of Geography* (ed. by A. Buttimer) (1983).

Wise, Michael John (b. 1918)

Ph.D., University of Birmingham, 1951. University of Birmingham, 1946-51; London School of Economics, 1951-85; President of IGU, 1976-80. (ed.) *Birmingham and Its Regional Setting* (1950); (ed. consultant) *An Atlas of Earth Resources* (1979); (ed.) Ordnance Survey *Atlas of Great Britain* (with R. Butlin) (1982); (ed. consultant) *The Great Geographical Atlas* (1982). *WW*; *GOF* (1982).

Wolman, Markley Gordon (b. 1924)

Ph.D., Harvard University, 1953. United States Geological Survey, 1951-58; Johns Hopkins University, 1958 to the present. *Fluvial Processes in Geomorphology* (with L. Leopold and J. Miller) (1964); "Magnitude and Frequency of Forces in Geomorphic Processes" (with J. Miller), *Journal of Geology*, 68 (1960), 54-74; "Cycle of Sedimentation and Erosion in Urban River Channels," *Geografiska Annaler*, 49A (1967), 385-395; "The Nation's Rivers," *Science*, 174 (1971), 905-918. *WWA*.

Wolpert, Julian (b. 1932)

Ph.D., University of Wisconsin, 1963. University of Pennsylvania, 1963-73; Princeton University, 1973 to the present. "The Decision Process in Spatial Context," *AAAG*, 54 (1964), 537-558; "Behavioral Aspects of the Decision to Migrate," Regional Science Association, *Papers*, 15 (1965), 159-169; "Urban Neighborhoods as a National Resource" (with J. Seley), *Geographical Analysis*, 18 (1986), 81-93; "The Geography of Generosity," *AAAG*, 78 (1988), 665-679. *WWA*.

Wonders, William Clare (b. 1924)

Ph.D., University of Toronto, 1951. University of Toronto, 1948-53; University of Alberta, 1953-87. "Our Northward Course," *CG*, 6 (1962), 96-105; "Marginal Settlement," *SGM*, 91 (1975), 12-24; "Mot Kanadas Nordväst: Pioneer Settlement by Scandinavians in Central Alberta," *Geografiska Annaler*, 65B (1983), 129-152; "The Canadian North—Its Nature and Prospects," *JG*, 83 (1984), 226-233. *CWW*.

Wooldridge, Sidney William (1900-1963)

D.Sc., University of London, 1927. Birkbeck College London, 1944-47; King's College London, 1922-44, 1947-63. *The Physical Basis of Geography* (with R. Morgan) (1937); *Structure, Surface and Drainage in South-East England* (with D. Linton) (1939); *The Spirit and Purpose of Geography* (with W. East) (1951). W. Balchin in *GBS*, 8 (1984), 141-149; obits. by D. Linton in *GJ*, 129 (1963), 382; and M. Wise in *Geography*, 48 (1963), 329.

Wright, John Kirtland (1891-1969)

Ph.D., Harvard University, 1922. American Geographical Society, 1920-56. *Aids to Geographical Research* (1922, rev. ed. with E. Platt 1947); *The Geographical Lore of the Time of the Crusades* (1925); *Geography in the Making: The American Geographical Society, 1851-1951* (1952); *Human Nature in Geography* (1966). J. Wright, "Introduction," pp. 1-10 in *Human Nature in Geography*; D. Lowenthal, "Introduction," pp. 3-9 in *Geographies of the Mind: Essays in Historical Geosophy in Honor of John Kirtland Wright* (ed. by D. Lowenthal and M. Bowden) (1976) (Wright bibliography on pp. 225-256); obits. by D. Lowenthal in *GR*, 59 (1969), 598-604; and M. Bowden in *AAAG*, 60 (1970), 394-403.

Yamasaki, Naomasa (1870–1929)

D.Sc., Tokyo University, 1913. Tokyo University, 1908–29. *Dai Nippon Chishi* (with D. Sato *et al.*) (10 vols., 1903–15); *Waga Nanyo* (1916); *Seiyo mata Nanyo* (1926); *Yamasaki Naomasa Ronbunshu* (ed. by T. Karo) (2 vols., 1930). U. Tsujita in *GBS*, 1 (1977), 113–117.

Yugoslavia

The development of geography on the territory of present-day Yugoslavia has proceeded in concert with that country's political evolution. At the end of the eighteenth century Yugoslavia's present area was divided between the Ottoman and Habsburg empires. Serbia gained *de facto* independence from the Ottomans in the early nineteenth century, and, after World War I, became part of an independent Kingdom of Serbs, Croats, and Slovenes when it was joined to the south Slavic lands of the Habsburgs. The SCS kingdom was renamed Yugoslavia in 1928. Following fragmentation by Axis invaders in World War II, Yugoslavia reappeared on the map of Europe in 1945 as a socialist republic based on federal principles.

Higher education in geography began at the end of the nineteenth century in Croatia and Serbia. Petar Matković (1830–1898), a student of Carl Ritter at the University of Berlin, became instructor in the newly founded department of geography in Zagreb University in 1883. Much of his research work was in historical geography, but he also wrote noteworthy descriptions of terrain.

Jovan Cvijić (1865–1927) had earned a doctorate in physical geography in Vienna in 1893 under Suess and Penck. He returned to his native Serbia and became a professor of physical geography and ethnography in the Velika Škola (Great School) in Belgrade, where a faculty of history and geography was established in 1896. The Velika Škola became Belgrade University in 1905. It had separate physical geography and anthropogeography sections. Twice, in 1907 and 1919, Cvijić was named rector of the university.

Cvijić was extremely active in research and teaching. Soon after accepting his post in Belgrade he began publication of *Pregled geografske literature o Balkanskom poluostrvu* (Review of Geographic Literature of the Balkan Peninsula). Although his interests remained primarily in the physical aspects of geography, he did significant work in the area of human geography, studying settlements and ethnic characteristics of the different peoples of the Balkan peninsula. In 1915, escaping the Austrian occupation of Serbia, he made his way to Paris, where he received a post in the Sorbonne. In 1918 he published there his book *La péninsule balkanique*, which gained him international fame. He returned to Belgrade after the war, and his influence was felt long after his death in 1927, particularly in the continuing development of Serbian geography.

Slovenia did not have its own higher educational institution until the University of Ljubljana was founded in 1919. Artur Gavazzi (1861–1944) was its first geographer, being brought to the new university from the University of Zagreb, where he had been named a professor in 1914. However, since the mid-nineteenth century, a number of academic geographers, mainly from Vienna, had studied Slovenia. Perhaps the most outstanding early native Slovenian geographer was Simon Rudar, who used contemporary European geographic methodology in his works, including *Poknezena grofija Goriška in Gradiščanska* (The Imperial Counties of Goriska and Gradiscansko) in 1890.

Macedonia also lacked advanced work in geography until 1922, when a department of geography and a geographic institute were founded in Skopje under the sponsorship of Belgrade University. Petar Jovanović (1893–1957), who had earned his doctorate in Belgrade, was named its first professor and director.

Geographical societies became popular in the twentieth century. The earliest was the Serbian Geographical Society. It was founded in 1910 and began publishing its *Glasnik* (Review) two years later. In 1922 the Slovenian Geographical Society was established, and its journal *Geografski Vestnik* (Geographic Herald) appeared in 1925.

Artur Gavazzi returned to the University of Zagreb in 1925 as professor of a new department of physical geography. He founded a geographic institute and established the Croatian Geographical Society. In 1929 the society began publishing the journal *Hrvatski Geografski Glasnik* (Croatian Geographical Review).

Academic geography suffered in Yugoslavia during World War II. A notable exception was production of the two-volume *Zemljopis Hrvatske* (Geography of Croatia) by geographers in the short-lived Independent State of Croatia.

Yugoslav geography expanded in the postwar years under the new federal regime. Particularly strong leadership was exerted by Josip Roglić in Zagreb University and Anton Melik in Ljubljana University, and later by their successors, respectively, Ivan Crkvenčić and Vladimir Klemenčić. Institutes of geography were established in the Serbian and Slovenian Academies of Sciences. A geography department was established in the new University of Sarajevo in 1950. Other developments included a geographic museum founded in Ljubljana in 1946 and a Geographical Society of the Peoples Republic of Macedonia organized in 1949. A revitalized Croatian Geographical Society began functioning in Zagreb in 1947, and in 1949 the society resumed publishing its *Geografski Glasnik*. In 1957 the Geographical Society of Bosnia and Hercegovina began publishing the *Geografski Pregled* (Geographical Review).

Other new departments of geography began operating in 1961 in Novi Sad and in 1963 in Priština, capitals respectively of the Vojvodina and Kosovo autonomous regions of the Serbian republic. In 1982 a joint department of geography and history was established as a branch of the University of Titograd in the Montenegrin town of Nikšić. In recent years positions for geographers have been opened in Slovenia in the University of Maribor and in Croatia in the University of Split and its branches in Dubrovnik and Zadar, the University of Rijeka and its branch in Opatija, and the University of Osijek.

Bibliography

Enyedi, G. and A. Kertesz, "South-east Europe," pp. 64–78 in *Geography since the Second World War*, ed. by R. J. Johnston and P. Claval (1984).

"Geografija," *Enciklopedija Jugoslavije*, 3 (1958), 438–446.

Ilešić, S., "Géographie humaine dans les pays de la Yougoslavie à la fin du 19e et au commencement du 20e siècle," pp. 67–78 in *La naissance de la géographie humaine: Friedrich Ratzel et Paul Vidal de la Blache* (Geographical Research Institute of the Hungarian Academy of Sciences, 1974).

Jovanović, P. S., "Razvoj geografske nastave u najvišim našim školama," *Glasnik Srpskog geografskog društva*, 32 (1952), 277–289.

Kratkaya Geograficheskaya Entsiklopediya (5 vols., Moscow, 1960–66).

Vasović, M., "Jovan Cvijić, 1865-1927," *GBS*, 4 (1980), 25–32.

• *Thomas M. Poulsen*

Zelinsky, Wilbur (b. 1921)

Ph.D., University of California, Berkeley, 1953. University of Georgia, 1948–52; University of Wisconsin, 1952–54; Chesapeake and Ohio Railway Company, 1954–59; Southern Illinois University, 1959–63; Pennsylvania State University, 1963–87. *A Prologue to Population Geography* (1966); (ed.) *Geography and a Crowding World* (with L. Kosinski and R. M. Prothero) (1970); *The Cultural Geography of the United States* (1973); *Nation into State* (1988). *GOF* (1971, 1984); *60 Years of Berkeley Geography.*

Zimm, Alfred (b. 1926)

Ph.D., Humboldt University, 1955; (habil.) Humboldt University, 1960. Humboldt University, 1950 to the present. *Die Entwicklung des Industriestandortes Berlin* (1959); *Industriegeographie der Sowjetunion* (1963); *Ökonomische Geographie der Sowjetunion* (with J. U. Gerloff) (1978); (ed.) *Berlin und sein Umland* (1980).

Zoogeography. *See* Biogeography.

Zurich Geographical and Ethnographic Society (Geographisch-Ethnographische Gesellschaft Zürich)

The Ethnographic Society of Zurich, Switzerland, was founded in 1888, and the Geographical Society in 1897. The two were combined in 1899. Earlier Swiss geographical societies had been established in Geneva (1858), Bern (1873), and Neuchâtel (1885). One of the original tasks of the society was the development of the ethnographic collections. Since 1913 these collections have been looked after by the University of Zurich. Since its founding, the society has been concerned with providing popular lectures on foreign lands and peoples, in addition to the promotion of scientific or academic geography. These concerns are being looked after by the eight professors of geography in the University of Zurich and the Swiss Federal Institute of Technology (ETH). Holders of academic chairs have served as presidents of the society—*e.g.*, Conrad Keller, Hans Wehrli, Eduard Imhof, Otto Flückiger, Heinrich Gutersohn, Hans Boesch, and Emil Egli. The society's journal is especially noteworthy. It began as the *Jahresbericht* (1899–1917), continued under the name of *Mitteilungen* (to 1945), and since 1946 has been published under the title of *Geographica Helvetica. GH* is a collaborative effort of the Geographical and Ethnographic Society of Zurich and the Swiss Geographical Society, with the active support of the other geographical societies in both the German-speaking and French-speaking portions of Switzerland. It draws its contributors and readership from all areas and is recognized as Switzerland's leading geographical journal.

Bibliography

Jud, P., "100 Jahre Geographisch-Ethnographische Gesellschaft Zürich," *Geographica Helvetica*, 44 (1989), 113–151.

• *Klaus Aerni*
(translated by Ursula Willenbrock Martin)

Index of Personal Names